Russia and Europe

Russia–Europe political relations have always been problematic, and one of the main reasons for this is the two sides' different perspectives on even the very basic notions and concepts of political life. With a worldwide recession, the problems as well as opportunities in Russia–Europe relations are magnified. While most works on Russia–Europe, Russia–America and Russia–West relations focus on current policies and explain them from a standard set of explanatory variables, this book penetrates deeper into the structural and ideational differences that tend to bring about misperceptions, miscalculations, misinterpretations and misdeeds in this two-directional relationship. It applies a very broad conceptual framework to analyse differences that are as relevant for Europe and the EU as it is to Russia's immediate neighbours and, while doing so, identifies the key factors that will dominate Russia–EU ties in the next decade.

Kjell Engelbrekt and **Bertil Nygren** are Associate Professors of Political Science at the Swedish National Defence College and at the Department of Political Science, Stockholm University, Sweden.

Routledge contemporary Russia and Eastern Europe series

Russia and Europe

Building bridges, digging trenches

**Edited by
Kjell Engelbrekt
and Bertil Nygren**

Routledge
Taylor & Francis Group

LONDON AND NEW YORK

First published 2010
by Routledge
2 Park Square, Milton Park, Abingdon, Oxon, OX14 4RN

Simultaneously published in the USA and Canada
by Routledge
270 Madison Avenue, New York, NY 10016

Routledge is an imprint of the Taylor & Francis Group, an informa business

© 2010 Kjell Engelbrekt and Bertil Nygren selection and editorial matter; the contributors for their contributions

Typeset in Times New Roman by Glyph International Ltd.
Printed and bound in Great Britain by CPI Antony Rowe, Chippenham

British Library Cataloguing in Publication Data
A catalogue record for this book is available from the British Library

Library of Congress Cataloging in Publication Data
Engelbrekt, Kjell.
Russia and Europe : reaching agreements, digging trenches / Kjell Engelbrekt and Bertil Nygren.
p. cm. – (Routledge contemporary Russia and Eastern Europe series ; 21)
Includes bibliographical references and index.
1. Russia (Federation)–Foreign relations–Europe.
2. Europe–Foreign relations–Russia (Federation) 3. Political culture–Russia (Federation) 4. Political culture–Europe.
5. Russia (Federation)–Politics and government–1991- 6. Europe–Politics and government–1989- I. Nygren, Bertil. II. Title.
JZ1616.A57E8 2010
327.4704–dc22

2009040356

ISBN 13: 978–0–415–56105–1 (hbk)
ISBN 13: 978–0–415–85464–8 (ebk)

ISBN 10: 978–0–415–56105–1 (hbk)
ISBN 10: 978–0–203–85464–0 (ebk)

Contents

Contributors

Russell Bova is Professor of Political Science and International Studies at Dickinson College, Carlisle, PA, USA.

Petro Burkovsky is Chief analyst at the National Institute for Strategic Studies Kyiv.

Kjell Engelbrekt is Associate Professor at the Department of Political Science, Stockholm University, and the Swedish National Defense College.

Isabelle Facon is Senior Research Fellow at Fondation pour la Recherche Stratégique, Paris.

Yuri E. Fedorov is Associate Fellow, Russia and Eurasia Program, The Royal Institute of International Affairs (Chatham House), London.

Patricia Fogarty is an anthropologist at Emory University, Atlanta, Georgia, USA.

Boris Frumkin is Associate Professor of Economics at the Department of International Relations of Moscow State Institute of International Relations and at the European Studies Institute of MGIMO and Head of Department at the Institute of Economics Russian Academy of Science, Moscow.

Olexiy Haran is Professor of Comparative Politics at the University of Kyiv-Mohyla Academy (UKMA) and Founding Director of the UKMA School for Policy Analysis, Kyiv.

Bertil Nygren is Associate Professor of Political Science at the Department of Political Science, Stockholm University, and the Swedish National Defence College, Stockholm.

Sergei Prozorov is a research fellow at the Helsinki Collegium for Advanced Studies, Helsinki.

Tatiana Romanova is Associate Professor at the School of International Relations of St. Petersburg State University, St. Petersburg.

Angela Stent is Professor of Government and Director of the Center for Eurasian, Russian and East European Studies in the Georgetown University School of Foreign Service, Washington DC.

Ilian Vassilev is former Ambassador of Bulgaria to the Russian Federation (2000–2006) and consultant, Sofia.

Preface

This edited volume has developed out of the interests of the two editors, Kjell Engelbrekt and Bertil Nygren, both of whom in recent years were preoccupied with what can be described as two sides of the same coin: analysing to what extent Russia and the policies of its government can be rendered compatible with European interests, desires and values. An ancillary but related issue is whether and how Europe – and the European Union (EU) in particular – may be in a position to engage Russian authorities and other actors in working toward that end, while avoiding serious confrontation in the regions that borders both. After 1991, the EU's general approach was to slowly drag Russia into its normative sphere of democracy and free market principles, which given time would strengthen incentives for Russia to act 'European' or 'Western'. Some two decades after the fall of the Berlin Wall and the ensuing disintegration of the USSR, however, the word 'failure' is being painted on the Western side of a new kind of barrier between the two political entities. There are powerful autocratic tendencies in Russia and elsewhere in the Commonwealth of Independent States (CIS), hostility in Russia against foreign investments in key sectors of the Russian economy and, in mid-2008, there was the Kremlin's show of military force in Georgia. Today, the sentiment is that hundreds of years of separation between Russia and Europe manifest themselves in the way the two regard themselves and each other, and in terms of the goals and means perceived as necessary to promote their respective interests and values.

Put briefly, this volume is about these reinvigorated divisions and how they translate into stark policy variations, how great power relations mitigate and magnify tensions stemming from the former, and the way in which 'in-between' countries like Serbia, Ukraine and Moldova try to handle these differences when simultaneously pulled in two opposing directions. The full argument that serves as the overall thematic of this book is developed in Chapter 1. This chapter is followed by three sections that explore distinctive subsets of that overall thematic.

In the first section, three chapters address various aspects of the ideational tensions between Europe and Russia. In a chapter called 'Russia and Europe after the Cold War: Cultural Convergence or Civilizational Clash?', Russell Bova deals explicitly with the trend that Russian norms, values and institutions seem to be drifting further away from those of Europe at large and of the EU in particular, in real life as well as in political rhetoric. The question of Russia's cultural relationship to Europe is an old one, frequently debated by Russians and Europeans alike. For some, Russia is a European nation whose belated movement toward cultural convergence with the rest of Europe at the beginning of the twentieth century was temporarily interrupted by the Bolshevik Revolution. From this perspective, the end of communism should have led to the resumption of that process of integration of Russia into the larger European culture. For others, Russia and Europe represent two distinct cultures – indeed, two distinct civilizations. From this perspective, Bolshevism was a less temporary interruption of Russia's path to Europe but a reassertion of the essential chasm between Russian and European cultures that the end of communism is unlikely to bridge. The chapter examines these two competing perspectives on the cultural relationship of Russia and Europe in light of trends and developments in Russia since the fall of communism.

Also, Bertil Nygren analyses the ideational friction between Europe and Russia, but particularly in connection with the evolution of electoral democracy in the CIS area and the EU attempts at encouraging democratization. In 'Perceptions of Democracy and Democratic Institution-Building – Electoral Democracy the Russian Way', Nygren looks at growing normative differences between the Western and Russian understandings of democracy in the second presidential term of Vladimir Putin. The notion of 'sovereign democracy' became a catchword for the Russian (and other CIS) ruling elites in describing an alternative interpretation to Western- and EU-induced interpretations of the notions of electoral democracy. Against the backdrop of the 1993 Russian Constitution and norms promoted by the Organization for Security and Co-operation in Europe (OSCE) and the Council of Europe, the path along which Russia has changed its political system with respect to parliamentary and presidential elections since 1993 is traced. These changes have affected nomination procedures, registration procedures, media coverage and election campaigning. Nygren provides a detailed account of the road from a Yeltsin-led application of democratic rule to a neo-authoritarian understanding of electoral democracy according to which mockery is made of political parties as bearers of the will of the people. Nomination and registration of parties and candidates are constrained by election laws, and media and government resources are unequally distributed among parties and candidates.

Tatiana Romanova, in a chapter called 'The Theory and Practice of Reciprocity in EU–Russia Relations' then tackles an important normative notion and applies it in the context of Russia–EU relations. The principle of reciprocity is fundamental to international relations today, and it has increasingly gained currency in debates on EU–Russia relations. However, the two parties understand this principle very differently. These divergent interpretations lead to fundamental difficulties in EU–Russia relations, particularly at a time when the parties have been striving to reinvigorate relations by designing a new legal basis. The very fact that the same word (reciprocity) is used to describe different processes and ideas means that most of the agreements are essentially impossible to implement. The differences are first presented in several empirical cases and then analysed through a theoretical prism.

After contrasting EU and Russian interpretations of the principle of reciprocity, the chapter ends by offering some practical recommendations for future EU–Russia relations.

In Sergei Prozorov's chapter 'Beyond the Paradigm of Integration in EU–Russia Relations: Sovereignty and the Politics of Resentment', a relatively long historical perspective is adopted, acknowledging that EU–Russia relations in the aftermath of the Russia–Georgia war in August 2008 reached their lowest point since the end of the Cold War. The disillusionment in Russia's 'Europeanization' and the retreat of ambitious visions of EU–Russia cooperation, Prozorov argues, require a reassessment of the overall theoretical framework of analysing EU–Russia relations that remained dominated by the integration paradigm. This chapter goes on to analyse the ongoing crisis in EU–Russia relations in the aftermath of the Russia–Georgia war in terms of a psychoanalytically inspired discourse theory that focuses on the changing status of the other in the symbolic order that defines one's identity. It is concluded that recent events starkly manifest Russia's abandonment of the integrationist paradigm in foreign policy, which in turn is conditioned by the destitution of the 'idea of Europe' in the symbolic order of contemporary Russian politics. In rounding off the analysis, the implications for European policymaking towards Russia are addressed.

In the second section of this book, four chapters examine issues that arise between Russia and the most important European powers, in particular, the Union's 'Big Three'. Two chapters are predominantly thematically oriented, focusing on geo-political and geo-economic matters of high relevance. They are followed by another two chapters devoted to Russia–Germany and Russia–France relations, respectively. In 'The Return of History: Hard Security Issues in the Russia–Europe Relationship', Yuri Fedorov shows that hard security issues are becoming more important elements in Russia's relations with Europe. Fedorov analyses the present and future points of friction between Russia and the major Western countries, covering conventional arms matters, strategic positioning and diplomatic and military posturing by both sides. The first part of the chapter problematizes the validity of Russian concerns related to the conventional balance in Europe, US ballistic missile defence (BMD) in Poland and the Czech Republic, and Georgian 'aggression against South Ossetia'. The second part focuses on the intellectual and emotional mainstream of the Russian elites and the vested interests that seem to cause the Russian bellicosity. It is argued that Moscow's foreign policy in general and its foreign policy towards Europe in particular results from a blend of megalomania, paranoia and the yearning of a cohort of military command, masters of the military–industrial complex and chiefs of security services for confrontation with the West, seeking to justify a higher defence budget and repression of domestic opposition. The third part of the chapter analyses future trajectories of the 'hard security' relationship between Russia and Europe with special focus on the consequences of probable Russian 'countermeasures' to US BMDs in Central Europe.

From thereon, Boris Frumkin takes us into a set of geo-economic problems of high relevance to Europe and Russia. In a chapter called 'Russia–EU

Relations: The Economic Dimension', Frumkin examines Putin's second presidential term and notes that the economic strengthening of Russia has caused serious changes in EU–Russia economic cooperation. The main directions and effects of the growing role of the state in the economy and Russia's external economic relations, especially its impact on the dynamics and structure of the EU–Russia trade, are analysed. Particular attention is paid to the investment expansion of Russian business and the recent sovereign funds at the EU, problems and prospects of the Russia–EU partnership in energy, agriculture and innovation sectors. Russia's increasingly intense relations with the EU are considered on three levels: (1) relations with member-states including the Central and East European states and sub-regional initiatives such as the Northern Dimension, Eastern Partnership; (2) relations with the EU as part of a European macro-region, stressing the problems and prospects of the new Strategic Partnership Agreement and the formation of a Common Economic Space; (3) relations with the EU as a decisive partner in confronting globalization with regard to financial crises, world energy, food-related issues and adaptation to climate changes. The possible effects of the Medvedev–Putin governing tandem on Russia's internal and external economic policy are also considered.

In a chapter called 'Germany–Russia Relations, 1992–2009', Angela Stent begins by summarizing how interaction between the two countries in many ways defined the twentieth century, through military and political confrontation but also through peaceful cooperation and economic exchange. Germany also played a major role in the collapse of the Soviet Union. In the 1990s, Stent tells us that four bilateral issues dominated the relationship, namely troop withdrawals, ethnic Germans, economic ties and German support for Russia's domestic transformation. After a somewhat confused political situation in Russia during the late 1990s, Vladimir Putin assumed the reins of power, making Germany a pivotal element of his foreign policy strategy. As a German-speaker and former resident of (East) Germany, Putin engaged German Chancellor Gerhard Schröder and ensured that Berlin started viewing itself as Russia's main advocate in the EU. Angela Merkel, another former resident of East Germany (and a Russian-speaker), has taken a somewhat more balanced view than her predecessor but acknowledges that the economic dimension of Russia's ties to Germany guarantees a close relationship.

Turning to the relations between Russia and another of the EU 'Big Three', Isabelle Facon provides a number of insights in a chapter called 'Russia and the European Great Powers: France'. Compared with Germany, Facon notes that France is similarly preoccupied with sustaining an extensive, high-level diplomatic dialogue with Moscow, but that it is based a much 'thinner' economic relationship with the Russian Federation. Some plans to remedy the latter exist, partly through the agreed 25 per cent participation of the Total energy giant in the exploitation of the Shtokman oil fields. With regard to diplomacy, France's insistence on robust ties to Russia was apparent through President Sarkozy's personal engagement in resolving the Georgian crisis in 2008, and in Paris' scepticism of the 'Eastern Partnership' project promoted by most other EU members states (although it is unclear whether the scepticism stems from Moscow's opposition or an assessment that the project itself lacks merit). According to Facon, however, the France–Russia bilateral relationship is increasingly entangled in the broader framework of EU–France relations as well as affected by President Sarkozy's strong desire to bolster ties to Washington.

In the third section of the volume, three contributions illustrate the stark dilemmas that the countries of 'in-between-Europe' face in a part of Europe exposed to two powerful economic and political centres, Russia and Europe. In 'European Energy Policy Meets Russian Bilateralism: The Case of Southeastern Europe', Kjell Engelbrekt and Ilian Vassilev explore Russia's policy of consolidating its position as the main energy provider to EU countries through a series of attractive deals with Bulgaria, Serbia and Hungary. As European decision-makers debate energy security and the repercussions of enhanced reliance on Russian fossil fuels in the future, Engelbrekt and Vassilev note that Southeastern Europe is integral to any long-term solution of the continent's energy challenges. For the EU, it makes a lot of sense to encourage Central Asian, Middle East and Gulf producers to provide gas and oil directly to the European market, in part to balance the growing significance of supplies delivered from or via Russia. The authors question the prospects for the EU putting its patchy energy approach into practice, taking into account the complex technical challenges and the fact that individual countries pursue nationally defined objectives that undermine a coherent approach at the Union level. Indeed, Russia and its largest energy corporations recently worked hard to develop delivery systems in Southeastern Europe and secure long-term bilateral deals with various governments, and even gained majority stakes in key energy assets in Bulgaria and Serbia. On the other hand, Russia's assertive 'gasplomacy' also alienated some European partners, and cash-strapped Russian energy companies may not be able to follow through on their ambitious commitments.

In a chapter called 'Ukraine's Emerging Democracy and the Russia Factor', Petro Burkovsky and Olexiy Haran describe the particularly stark choices between adjusting to Moscow or to Brussels that are faced in Ukraine. Haran and Burkovsky demonstrate how Ukrainian independence helped transform the status of the previously rather provincial elite, with independence emerging as a dominant value. However, the past practices of communist elites as well as mass behaviour and institutions resulted in several setbacks for Ukraine's political, economic and foreign reforms. The 'blackmail state' and 'virtual politics', in coexistence with a multiparty system, a diversified economy and the relative freedom of the press, often confused Western observers about the nature of the Ukrainian political regime and its place in the European space. The 2004 'Orange revolution' seemed to be a breakthrough, but post-revolutionary processes rendered the picture more ambiguous. In this chapter, Burkovsky and Haran discuss the main trends of Ukrainian politics, exploring institution building, government performance, shifts in public opinion and the impact of decisions made by Russia and the EU on the political and public agenda in Ukraine.

In the final contribution to this section of the book, Patricia Fogarty applies an anthropological perspective to society and state in 'Riding Three Horses:

Moldova's Enduring Identity as a Strategy for Survival'. Moldova has had great difficulties in finding a niche as an independent state in the post-Soviet space. Nearly all ethnic groups in Moldova can claim affiliation with another state, e.g., Romanians, Ukrainians, Bulgarians and Russians. Being simply 'Moldovan', a citizen of the poor state of Moldova, holds few benefits. However, internationally funded development agencies have been working in Moldova since the early 1990s and have influenced discourses on citizenship and state–society relations in the direction of standards common to Western liberal democracies. Such discourses compete for legitimacy with ideologies standard during the Soviet period. Fogarty examines how Moldovans negotiate and navigate these discourses of development and citizenship on a daily basis, drawing on a case study of the Moldova Social Investment Fund (a multilateral development agency). It transpires that Moldovans – through development work and the ongoing redefinition of state–society relations – simultaneously (re)produce and challenge the concept of citizenship and of the shared culture that brings Moldovans together.

In the final chapter 'Conclusions and Outlook', the editors try to draw some preliminary conclusions on the basis of the individual chapters, eliciting insights into key aspects of the Europe–Russia relationship. Using the three parameters, as outlined in Chapter One, as the backbone of that assessment, the editors discuss how to proceed and what to expect of the prevailing differences, what their meaning is to the strategies of the EU and to Russia, and what to expect from Europe–Russia ties in the coming ten to fifteen years.

Acknowledgements

The Department of Security and Strategic Studies at the Swedish National Defence College (SNDC) has for several years received an annual grant from the Swedish Defence Ministry to conduct security studies relevant to Swedish defence planning. This volume is the fourth published by Routledge in a series that began with Hallenberg and Karlsson (eds): *Changing Transatlantic relations: Do the US, the EU and Russia Form A New Strategic Triangle?* in 2006, followed by Engelbrekt and Hallenberg (eds): *The European Union and Strategy: An Emerging Actor* in 2008, and by Wagnsson, Sperling and Hallenberg (eds): *European Security Governance. The European Union in a Westphalian World* in 2009. We greatly appreciate the financial and other support provided by the SNDC and our department to make this volume possible. We are similarly thankful to Professor Jan Hallenberg for endorsement of this particular book project and to the eleven authors for their valuable contributions. The editors also wish to thank Lisa Larsson, their research assistant, for her work on the book conference in November 2008 and for compiling the references. Special mention should be made of our expedient and thorough language editor, John Åkerman, who has saved more than one author from embarrassment. Finally, we are immensely grateful to our wives, Antonina Bakardjieva Engelbrekt and Olga Vlasenko, for their continuous support and understanding in the more intense phases of the production process.

Stockholm July 2009
Kjell Engelbrekt and Bertil Nygren

1 A reassertive Russia and an expanded European Union

Kjell Engelbrekt and Bertil Nygren

At the end of the first decade of the twenty-first century, Russia–Europe relations came under a cloud. Disruptions of Russian deliveries of gas to Ukraine and several member states of the European Union (EU) in 2006 and 2009, along with the Georgian war of August–September 2008, even prompted some observers to speak of a looming new Cold War. Others pointed to indications of a 'rearmament race' on the continent, with a new US-sponsored defence system against missiles launched by rogue states at the forefront, as well as to numerous signs of neo-authoritarianism in the way in which the Kremlin wields power over Russian society.

Meanwhile, economic cooperation between Russia and Europe continues to evolve, with interdependence deepening in a variety of areas. Because of the increasingly structural character of institutional and business relations, today's conflicts do not necessarily spill over into fields of established cooperation. However, that does not mean that the twists and turns of Russia–Europe ties do not have important repercussions for both sides, or that the current level of engagement is smooth and non-conflictual. In particular, the EU's aspirations to consolidate the European marketplace and to enhance cooperation in foreign, security and defence policy among its own member governments are objectives that require tighter coordination of policies with respect to Russia's activities in its 'near abroad' as well as on the global stage. For its part, Moscow seems increasingly ambivalent, sometimes even adamantly opposed, to either project.

The rise of Dimitry Medvedev to the post as President of the Russian Federation has not altered most of Moscow's fundamental views and the ways in which it goes about doing business with its European counterparts. With Vladimir Putin still running much of the day-to-day executive policymaking and quite possibly preparing for regaining the presidency in a coming election, anything else would be sensational. This is not to say that there are not very important 'unknowns' in the relationship between the two men, in their mutual or separate plans for the future, as well as in the political dynamic of a country where elections are regularly held and where the Kremlin no longer can control the entire press and the thickening flow of electronically transmitted information.

Nevertheless, it is clear that the worldwide economic recession sparked in 2008 poses a series of more profound challenges for Russia, its leadership and to Russia–Europe relations. Russian leaders will no longer be able to point to a steady growth in government revenues from which to draw legitimacy. Its neo-authoritarian modes of operation may have enhanced the government's capacity to contain discontent in the short term, but render some of its institutions increasingly brittle over the medium- to long-term, as the discrepancy between political rhetoric and reality possibly widens. What, if anything, can Europeans and the EU in particular do in this situation? Is it wise to try and further expand relations at present, or should one wait for a more opportune moment sometime in the future? Either way, what priorities should be set for the continued development of Russia–Europe ties?

This volume is based on the premise that relations between Russia and the EU inevitably involve tensions that influence the ways of doing business across the entire spectrum of multilateral and bilateral ties, yet have reached a point at which moderate political or economic fluctuations do not undermine the broader pattern of growing interdependence (Keohane and Nye 2001). From the vantage point of that premise, it seeks to disentangle the medium- and long-term trends from short-term occurrences and primarily to focus attention on the former. The ambition is to identify the key factors and types of relations that will dominate Russia–EU engagement in the next decade. As we see it, three parameters are poised to strongly influence that relationship.

The first parameter consists of the norms, values and institutions that Russia presently embodies both internally and externally, and which from time to time compete or clash with those of the EU. Most recently, there has been serious contention regarding the democratic process and respect for human rights in the countries situated west and south of Russia, but there are several layers of historical, cultural, diplomatic and other ideational aspects to this competition or 'friction of ideas'.

A second parameter is constituted by Russia's relationship to Brussels and, even more importantly, to European great powers such as Germany, France and Great Britain, each with a long historical lineage. To a lesser extent, it also pertains to Moscow's ties to Poland, Italy, as well as Spain, as significant players in the Union. In any case, here we are dealing with the so far limited potential for a cohesive Russia policy on the part of the EU as long as important and long-lasting interests of major European powers are at stake. The partial withdrawal of the United States from the European continent as a mitigating factor on intra-EU political relations should be noted in this context, along with economic considerations concerning trade, foreign investments and energy.

A third parameter concerns the relations between the EU and Russia by way of the states and areas geographically located between the latter two, regardless of formal EU membership on the part of those states. This parameter is related to some novel – but mainly to historically charged – problems regarding borders, minorities and lingering legacies of deep-seated distrust, though also an increasing need to regulate trade, migration and citizenship across such borders. Countries not encompassed by the EU's enlargement program are the most vulnerable, straddling the 'spheres' of Russia and the Union, respectively. Also here, of course, economic ties and energy issues constitute a substantial challenge because of existing or potential dependence on Russian energy supplies. In the rest of this introductory chapter, we will flesh out what we mean by the three parameters, in an attempt to set the scene for the contributing chapters and the questions that guide them.

The 'friction of ideas'

As just mentioned, the first parameter concerns the competing norms, values and institutions of Russia and of the EU, respectively. It is clear that, in recent years, there has been growing contention regarding the democratic process and respect for human rights in the Commonwealth of Independent States (CIS) countries, including Russia itself. For several of these countries, the ideational friction might not surprise us since the political and social cultures of Central Asian countries are deeply influenced by Islam and largely based on extended families and clans in terms of their social structure, rendering them distinctly different from Europe as well as the western portions of Russia. With respect to Europe and Russia, however, the political, social cultural differences are by far not as distinct. Indeed, the entire democratization and marketization effort after the demise of the USSR was built on the assumption that Russia was democratically, socially and economically 'transformable' to European standards precisely because of these similarities (Opalski 2001). Almost two decades later, domestic and outside observers are visibly less optimistic and perhaps somewhat wiser than before (Sakwa 2005; McAllister and White 2008; Wagnsson 2008), but still without an explanation as to why the distance between European and Russian norms and values seems to have widened.

Among the contentious policy issues of the last decade were: (1) the conflict with the United States over plans for national missile defence installations in the Czech Republic and in Poland, (2) the (second and then third) debate on the further enlargement of NATO (North Atlantic Treaty Organization), (3) the issue of Russian membership in the World Trade Organization (WTO), and (4) the continuing acrimony over human rights issues, in particular of how to deal with the situation in Chechnya. In addition, since 2004, there is the issue of electoral democracy in Russia and other CIS countries where Russian and some CIS leaders, notably in Central Asia and in Belarus, starkly contrast with European expectations on Russian democracy. Finally, in the recent past, gas deliveries to Europe, the Moscow's abandonment of the CFE (Conventional Armed Forces in Europe) Treaty and the Russia–Georgia war have surfaced on the agenda.

It is a commonplace to assert that some of the present conflict issues are linked to ideational friction that, in turn, stems from deep-seated cultural differences between Europe and Russia. This difference is alternatively formulated as concerning the 'European-ness' of Russia, how 'European' Russia actually is, to what extent Russia 'belongs' to Europe, whether Russia forms part of Europe or not, or, as Alexei Arbatov rhetorically asks, if Russia (geographically and culturally) constitutes the 'Eastern part of Europe or Western tip of Asia?' (Arbatov, 1998).

Such questions are virtually perennial and always hotly debated in Russia, although the answer to each of them is bound to be inconclusive. One reason for the inconclusiveness is that by asking whether Russia is one or the other, the question itself excludes the possibility that Russia is both European and Asian or neither – a particular and different (from both continents) entity. As a geographical fact, of course, Russia covers much of both European and Asian territory, so, in that respect, the answer is quite simply that it is both (Trofimenko 1999: 187).

However, most domestic commentators find the geographical dimension less intriguing to explore than notions of identity and national interests (Trofimenko 1999: 187, fn 3). Trofimenko refers to the 'theme of split orientation, of dualism in the national psyche of Russians', as a constant in Russian intellectual debates both before 1917 and after the demise of the Soviet Union, more precisely in the controversy between Westernizers and the Slavophiles. Trofimenko's conclusion is that there is an 'emerging consensus' that Russia 'has to follow its own unique ways and traditions because Russia is neither a purely Western nor purely Eastern nation, but blends in its nature a combination of both cultures, psyches and even genes' (Trofimenko 1999: 187–88, fn 3).

In recent years, under the Putin insignium, this notion of self-assertion seems to have become a virtual doctrine. It certainly is a leading idea in Russia's political life of today, embraced by the Unified Russia political party controlling the State Duma, the 'Nashi' movement with its near-fascist undertones, advocates of state ownership of 'strategic resources', and the 'Chekist nomenklatura' controlling a large amount of business ventures; in fact, many of these groups directly support the idea that there is a particular definition of Russian democracy, so-called 'sovereign democracy'. So, how do we find clues as to the reasons for this emerging cleavage between Russia and Europe, given that common values were widely promoted only a decade ago?

In 1994, Henry Kissinger noted that Russia 'never had an autonomous church; it missed the Reformation, the Enlightenment, the Age of Discovery, and modern market economics' (cited in Arbatov, 1998: 101, fn 5). He pointed to some undeniable facts, the consequences of which are nevertheless open to interpretation. Generally speaking, the Russia–Europe relationship has, ever since Russia's earliest days of statehood, been coloured by amity/enmity sentiments, as evidenced in documentation from the courts of Novgorod, Kiev and Moscow. Competition with the ambitions of other European powers intensified in the seventeenth century, as Russia strove to expand to the West, followed by confrontation in the south and southeast in the eighteenth and nineteenth centuries. Attempts to 'Europeanize' Russia by Tsar, Peter the Great, had by then increased Russia's imperial appetite, and Russia's fate had become inevitably linked to that of Europe.

Cherkasov has suggested that two varieties of the question exist, one posed in Russia and the other in Europe. The West traditionally does not accept Russia as an equal partner because of the different evolutionary paths of Christianity in the West and the East. This is the traditional argument that the dividing line between Russia and the rest of Europe came about as a result of two important developments. First, there was the schism in the Christian church between Rome and Constantinople, and later the effective closing-off of Russia from the rest of Europe in the mid-thirteenth century and for almost 400 years as a result of the Tartar–Mongol domination (Cherkasov 1998: 19–20, fn 5). When Russia returned to Europe, Europe itself had changed. The real re-engagement came only in the seventeenth century and was strongly reinforced by Peter the Great 100 years later. To later Russian tsars, the question whether Russia belonged to Europe was simply not put: 'Russia is a European power', as Catherine the Great stated in 1767, not only because of its participation in European affairs, but also because of the 'internal structure of the state' (the innovations introduced by Peter the Great).

The long period of Tartar domination in the medieval ages should be added to the observation that 'Russia did not experience the Renaissance and Humanism in the European sense of these words', Cherkasov writes (Cherkasov 1998: 20–21, fn 11). The Mongol rule 'put a deep stamp on the national character and on the entire Russian political culture' and, he continues, introduced 'the traits of despotism which destroyed the shoots of democracy grown in the period of Ancient Rus'. Eventually, this left Russia with many deficits: the lack of property and land rights for the majority of Russians (Cherkasov 1998: 25, fn 11). The Orthodox Church was a major actor in this development of the Russian state and culture and that explains the peculiar prevalence of social utopian ideas as well as its 'manifestation of messianism' inherited from Russia's Byzantine past, when Russia proclaimed itself the successor and leader of the entire Orthodox world after the fall of Byzantium in 1453. The outcome, according to Cherkasov, is an entrenched feeling of Holy Russia being repeatedly attacked by enemies out to destroy it, an idea reaffirmed by the twentieth-century communist regime (Cherkasov 1998: 26–27, fn 11).

However, what is then the present-day import of these historical pathways, and how are they perceived? Arbatov says Russians 'prefer collective labour, communal or state property and a more or less egalitarian distribution of wealth', in contrast to 'Western free market economy, clear-cut private property, and individualist self-interest'. And, instead of 'political pluralism, democracy and the division of state powers, Russia is associated with the overwhelming power of the state, relying on a mighty and huge army run by an authoritative and wise leader, guided not by laws but by conscience and enlightenment'. Further, 'in contrast to civil society and self-government, Russians are expected to live by communal wisdom and consensus, delegating the exercise of their will to higher authorities'. And, finally, 'instead of pragmatic adjustment to the inadequacies of the world and making the most of its opportunities, Russians are supposed to believe in the special mission of man and the people'. In short, 'Orthodox Christianity, Autocracy, Collectivism … is the traditional colloquial triad for the so called "Russian Idea" ' (Arbatov 1998: 102–3, fn 5).

The fact that the Church was subordinated to the state meant that state interests were prioritized, which in turn laid the foundation for state centralization. Furthermore, Russia's geopolitical location and lack of natural and defendable borders on the open plains forced it constantly to fight invaders from the south, east and west, for living space. The military might need to defend the territory, which in turn required state centralization, and defence therefore became 'the core and purpose of the administrative and economic organization of Russia'. In search of security, Russia expanded its influence, frequently imposing its will on neighbouring weaker societies (Arbatov 1998: 105–7, fn 5).

Both on the domestic and international political scene of today's Russia, democracy and free market ideology remain a foresworn ideal among the rulers. Putin and Medvedev promptly take to blowing up this dual-faced balloon on official occasions, especially when the arena is an international one. They undoubtedly wish to remain among equals, to be important members of the club of democratic great powers. At the same time, the restrictions to democracy and free market ideals have recently received a particularly Russian touch; 'sovereign democracy' as to the former and state intervention as to economic and public policy. In many economic sectors of a 'strategic nature', representatives of obscure state structures seem to have gained an advantage over private sector businessmen.

The Russian notion of 'sovereign democracy' was partly forged as a response to the colour revolutions, especially Ukraine's 'orange revolution', in late 2004. The most basic confrontation seemed to concern the actual voting, or the fact that the vote count was challenged both in the first and the second round of elections (since no one received more than 50 per cent). The demonstrations that followed in Ukrainian cities revealed a direct clash of official views. After Yanukovich, the Russian protégé, had won the second round, EU leaders could not hide their disappointment with the count, whereas Putin noted that 'the race was fierce – but open and fair – and the victory is convincing' (RFE/RL 23 November 2004). After several top-level European politicians had appealed to Putin to accept a renewed election, he did so, although reluctantly, complaining that Western interference was 'intolerable' (RFE/RL 7 December 2004).

It was a major concession and one that Putin later seemed to regret. Renewed elections brought Yushchenko to power and this defeat of 'managed democracy' style elections had major repercussions all over Eurasia. The parliamentary elections in Kyrgyztan in spring 2005 ended with demonstrations in the so-called 'tulip revolution', which in turn brought about new presidential elections, with the incumbent president fleeing to Moscow. Here too, Russia was deeply irritated over European election observers. The presidential elections in Belarus in the spring of 2006 finally brought the two opposing notions of democracy to the fore. Here, the West, both the United States and Europe, were accused of trying to overthrow Lukashenka.

From this point on, the official Russian attitude towards elections in Eurasia became that Europeans and Americans have no right to criticize the way in which elections in the CIS region are handled. The Russian Duma and presidential elections in 2007/8, similar to the election cycle in 2003/4, then affirmed its exclusive right to interpret elections that actually 'maintain the appearance of democracy but disdains its essence' (Kimmage 2005). What initially seemed to be successful Western norm diffusion into the 'managed democracies' of

Eurasia, precipitating a 'democratization wave', later backfired. In Russia, the epithet denoting an alternative to Western-style democracy in 2006 became known (although never expressly endorsed by Putin or Medvedev) as 'sovereign democracy' (Krastev 2007). The general and ideologically potent notion is simple if not trivial: sovereignty is a precondition for democracy, which makes sense in a state centred world. The notion is instrumental, reflects 'mobilization objectives' (Okara 2007), and suggests that Russia gets its own way (Amsterdam 2006).

The broader question is how we can expect Russia–Europe relations to evolve given a greater amount of friction over values, norms and institutions. How distinctive are the views of the present government and how malleable would they be under a new administration in which neither Putin, nor Medvedev takes part? In addition, how do these views, if still quite distinctive, translate into foreign policy doctrine? Is the EU's diplomatic approach, including its newly initiated 'Eastern Partnership' toward Russia's neighbours, compatible with a constructive engagement with Moscow as envisaged by many Union member states?

Russia, the EU and the Big Three

The Russian view of the EU has shifted over the years, from Soviet-era perceptions of the Union as an extension of the United States and NATO to a non-antagonistic but confused relationship in the 1990s. It began with the 1992 Partnership and Cooperation Agreement (PCA), through which the EU established new principles for dealing with Russia (assisting in its transformation towards a market economy and democracy). Given Russia's recent loss of its empire status, this minor 'carrot' was not particularly appetizing. Russia was essentially given the stark choice of either accepting or rejecting a ready-made formula to become a 'normal European state'. The further, disappointing trajectory of Russia–EU relations had thus been pre-programmed. The PCA was signed in December 1994 and entered into force in 1997, after a prolonged ratification process; but in most policy fields, there has been little concrete progress.

The EU Common Strategy for the first decade of the 2000s 'welcomes Russia's return to its rightful place in the European family in a spirit of friendship, cooperation, fair accommodation of interests and on the foundation of shared values enshrined in the common heritage of European civilization'. Despite the professed 'mutuality', there was a great number of irritatingly detailed instructions and advice on how to reform Russia. Russia's 'counterstrategy' towards the EU looked different, prioritizing political dialogue, trade and investment development, financial and transport cooperation and convergence of technical standards. Putin's own explanation of the differences between the strategies was that 'Russians emphasize geopolitics, great power interests, and the instrumental bases of cooperation'. The official Russian view was that cooperation should be established 'on an equal basis and with full respect for Russia's sovereignty' with no linkage to democratic reforms or common values (Wagnsson 2008: Chapter 6). Unmistakably, the main priority was to protect Russian key economic sectors.

In Putin's second presidential mandate, it became obvious that Russian foreign policy had changed and that the ideologically West-oriented (democracy and free market) stance had given in to a much more traditional Russian foreign policy, embracing power politics and geostrategic thinking. Thereby it gradually pushed aside Western-induced ideas on political and economic affairs, replacing them with neo-authoritarian understandings. This is an obvious backlash in any attempt to integrate Russia into the Western and European world. The most obvious mistake by EU and US leaders was their unwillingness – or sheer lack of ingenuity – when it comes to creating a common transatlantic approach toward Russia; instead, they left individual governments to dealing with Moscow on their own. Soon enough, Western countries were inclined to acknowledge and even reinforce Russia's own sense of alienation, in part self-inflicted and in part precipitated by post-1989 European actions (and non-actions).

Today, there are three areas of disagreement between Russia and the EU's 'Big Three' – Britain, Germany and France, areas that conveniently enough can be described as contingent on geopolitical, geo-economic and geo-cultural factors (Nygren 2008a). In the area of geopolitics, the most persistent issues have revolved around the possibility of a continued NATO expansion. Even if relations with NATO have changed for the better over the years, the first enlargement wave did cause a rift in relations. That divide was reinforced by the Kosovo war of 1999, and deepened even further as a result of the second wave of NATO enlargement. The NATO–Russia Council (NRC) established at the Reykjavik summit, in May 2002, provided Russia with new cooperation and consultation possibilities that did not exist in the predecessor instrument (the Permanent Joint Council – PJC) and today allows Russia to work on a near-equal basis. The war with Georgia prompted the suspension of that cooperation, but relations formally resumed in March 2009.

Obviously, Russia did not consider itself a threat to NATO neighbours and could consequently not see the need for an 'article five-NATO' along the lines of the Cold War. To the former Baltic Soviet Republics and the former Warsaw Pact members, conversely, 'article five' was the very reason for joining NATO, as it bolstered protection against Russia. Preventing further NATO expansion to include Ukraine and Georgia will likely be the highest priority of the Russian geopolitical agenda in the years to come. NATO will be considered a threat to Russia as long as two conditions are not met. The first is that Russia itself is granted a genuine say in the organization (or even membership), and the second is that NATO seriously downplays its article five orientation. Either condition is wholly unlikely to be met in the present international context.

A second, partly related, geopolitical issue where Russia and the West differ is the US nuclear missile defence plan, with its anti-missile radar base in the Czech Republic and anti-missile launching sites in Poland. True, European states do not actively support the US missile shield with the exception of two or three post-communist member countries. Although the extent to which the shield could be a threat to Russia is doubtful, to Russia, it suggests a powerful instrument in what is perceived as NATO's strategic positioning against Russia. Here it is

Russia's own, historically conditioned, zero-sum perspective that generates such an interpretation and so reinforces Russia's traditional perceptions of being under siege.

Another point of tension concerns the 'war on terror', in which Russia superficially remained on the side of the United States. The most serious differences pertain to the situation in Chechnya, where Russia consistently claims that it is fighting the same type of war as the United States in the Middle East and the eastern Subcontinent. Unlike Russia, Europe and the West primarily see the Chechnya problem through a human rights perspective. A similar situation applies to other 'anti-terrorism' issues in the CIS region (as, for example, in the bloody crackdown of Uzbek authorities on protesters in the city of Andijon, 12–13 May 2005). None of these three issues is likely to be met with any compromise proposals from any side and will therefore remain in the foreground for the foreseeable future.

After the Georgia war, the obvious difference in Russia–Europe ties with respect to how to treat neighbours came to the fore. The very fact that Russia used excessive military violence to punish Georgia for the South Ossetian war showed clearly that Russia's present leadership is unable to escape its old-fashioned great-power mode of reasoning, and that this position resonates with parts of the public. There are few other ways to explain why the political establishment in Moscow so wholeheartedly supported the local Russian military commanders even against the General Staff in Moscow. Putin, rather than Medvedev, was likely the driving force, since Moscow's actions fit the former's worldviews as well as his well-known animosity toward Georgia and President Mikhail Saakashvili (Bremner 2008).

Among the potentially conflict-laden geo-economic points of disagreement, energy is the major one. Generally speaking, Russia imports consumer and manufactured goods from Europe while exporting raw material and energy. Energy and fuels account for some 50 per cent of Russian exports to Europe, and forecasts predict that European energy imports from Russia could rise to some 70 per cent in 2030, with Germany and Italy as the major importers. Nevertheless, while Europe is Russia's main trading partner and receives 50 per cent of its exports, Russia accounts for only 5 per cent of the EU foreign trade, a little more than Norway. Whatever problems may emanate from this unequal situation, geo-economic issues play out most evidently in the energy sector, creating strong producer/consumer interdependence, as energy is the very foundation of Russian wealth. While the EU has attempted to regulate this interdependence since the early 1990s, for instance in the Energy Charter Treaty, Russia has refused to ratify the Charter.

In the area of 'geo-culture', the most pertinent issues are free elections, democratization and human rights. When the USSR crumbled, the few and small islands of democratic tradition in Russia that existed were weak, but Europe, nevertheless, hoped to encourage democratic developments. The EU and other European institutions have applied its 'normative power' based on its own notions of how to 'shape conceptions of "normal" in the areas of peace, liberty, democracy, rule of law and human rights (Manners 2002: 239; Bretherton and Vogler 2006: 42). Sjursen agrees that the EU largely conducts itself as 'a normative, civilizing or ethical power' (Sjursen 2006: 170). Subsequently, the EU has tried to extend the use of EU membership as a carrot to diffuse European norms among CIS states via the European Neighbourhood Policy (ENP). The latter has had mixed success, in part due to serious Russian resistance against the 'forcible democratization of the post-Soviet space' (RFE/RL 17 August 2005).

Even before Ukraine's 'orange revolution', the diplomatic tone between Russia and Europe had deteriorated. EU support for the 1999 NATO intervention in the Kosovo conflict was described as 'a betrayal of the ultimate European political ideal of state sovereignty' (Prozorov 2006: 5). The increased European criticism of Russia's second Chechen war made the situation worse, followed by the Yukos affair in 2003 and the 2003/4 Russian election cycle. The 'colour revolutions' in Georgia and Ukraine spurred tensions in the elections that followed in several CIS countries in 2005 and 2006. A struggle between a Eurasian 'managed democracy' and a European 'electoral democracy' emerged, in which 'normative trenches' were dug out and increasingly diplomatic justifications elaborated (Nygren 2008a). In this area, however, the 'Big Three' were often pulling their punches and preferred the EU collective to act or speak up. While the United Kingdom is host to a number of exile and human rights non-governmental organizations (NGOs), the government soon reverted to repair the damage done to relations with Russia by the spying and assassination incidents a few years back. France and Germany, meanwhile, turned to developing closer economic ties with Russian companies, became more restrained in their public criticism of Moscow than in the recent past.

In order to properly gauge the prospects of Russian ties to Europe and the EU 'Big Three', though, any analysis needs to be broken down into subsets of issues and bilateral relationships. How are Europe–Russia relations developing in terms of defence and security policy, with regard to disputes over missile defence, the possible renewal of the Conventional Forces in Europe (CFE) Treaty (1990), and military diplomacy? How are economic links evolving at a time of great financial stress, and which areas seem more and less promising in this respect? How stable is the ambitious Russia–Germany partnership, founded under the Yeltsin–Kohl era and carefully maintained under Putin, Schröder, Merkel, and Medvedev? Moreover, did the diplomatic offensive conducted during the 2008 French EU presidency make up for that country's previous neglect of relations with Moscow?

The eastward shift of 'in-between-Europe'

From an historical perspective, the novel dimension of Russia–Europe relations is the greatly enhanced role of countries and peoples of the former Soviet Union and Central and Eastern Europe straddling the area between Russia and Western Europe. The majority of these countries today find themselves beyond the immediate sphere of Russian influence by virtue of their NATO and EU memberships. Others remain more directly exposed to Moscow's interests and desires, even if the latter are muted and mitigated compared with the communist era. In that sense, the category of most vulnerable countries of 'in between-Europe' (or *Zwischeneuropa* in classical historiography) has shifted to the east. Nonetheless, both EU members and non-members are poised to profoundly affect the future of the broader relationship (Malfliet, Verpoest and Vinokurov 2007).

As indicated in the foregoing sections, Russia's struggle for pre-eminence in the Euro-Asian landmass has been a priority for the country's leaders ever since the collapse of the Soviet Union in the fall of 1991. The founding of the CIS in late 1991 stabilized the relationship between Moscow and the former Soviet periphery in the sense that both sides could extricate themselves from complex legal and organizational interdependencies, while, in the meantime, economic and political ties were reconstructed on a new basis. By December 1993, the CIS encompassed twelve out of the previously fifteen member states of the Soviet Union, as only the three newly independent Baltic states had exercised the exit option. Belarus and Kazakhstan have since then participated in most economic and security-related initiatives associated with the CIS and Russia, while Kyrgyzstan, Uzbekistan, Tajikistan, Azerbaijan, Armenia and Moldova have tended to be selective about CIS-related collaboration. Turkmenistan and Ukraine, the latter being a founding member of the CIS, never ratified the formal accord and can therefore be said to have a low profile in the organization.

The reinvigorated Russian economy in the 2002–7 period and Vladimir Putin's increasingly assertive foreign policy towards the CIS region and the EU gave Moscow the opportunity to consolidate its position and to reverse the earlier trend to fragmentation both inside and outside the Russian Federation (Nygren 2008c). Out of the eighty-three federal subjects of the Russian Federation, a fair number of republics, counties (oblasts) and districts (okrugs) established, in the 1990s, their own mechanisms for electing governors and so bolstered their autonomy toward Moscow. As President and Prime Minister, Putin worked to subvert such decentralization, they have in some instances – such as in Karachay–Cherkessia – even reinstated direct appointment of regional leaders. No doubt, within Russia proper, the Chechen war constituted the single biggest and most direct challenge to Moscow's supremacy. In 2007–8, after a long and bloody conflict, the situation in Chechnya calmed somewhat under the Russian-backed president Ramzan Kadyrov.

Many have testified that the 'colour revolutions' of 2003–5, challenging authoritarian leaders in the region, prompted the Kremlin's deep concern that it was losing influence over countries just outside the borders of the Russian federation. In Georgia, Ukraine and Kyrgyzstan, Moscow-friendly governments fell from power and were replaced by more nationally oriented political leaders. Throughout the military intervention in South Ossetia and Georgia proper during the fall of 2008, Prime Minister Putin and President Medvedev made the point that Georgian sovereignty was circumscribed when it comes to how Tbilisi conducts itself toward Russian political leaders, Russian-speaking minorities, as well as how it expands economic ties to the West. The signal was not lost on neighbouring governments.

Notably, in 2006 and 2009, the Russian leadership confronted Ukraine, the only weighty counterpart to Russian power in the 'near abroad', over gas deliveries and gas prices. Moscow's explicit charge was two-fold, and concerned gas deliveries to Ukraine and to Europe through the massive Druzhba pipeline: first of all, Kiev had refused to accept a raise in the price on imported Russian gas and had ignored demands to service the debt accumulated during the long period of renegotiation; second, Ukrainian companies had for domestic use siphoned off a substantial portion of the gas destined for EU countries, where it fetches a higher price. Looming in the background, however, were the extraordinarily thorny issues of stationing Russian warships in the port of Sevastopol and the status of Russian-speakers in Ukraine.

By comparison, the majority of Central and East European countries today find themselves at a significant 'political distance' from Moscow's immediate grasp. Following the demise of communist regimes and the subsequent disintegration of the Soviet Union, de facto sovereignty was reinforced substantially through NATO's eastern enlargement in the 1990s. It should be recalled that Moscow's initial resistance to expansion of the alliance to any of the Central and East European states was quite vehement, with foreign minister Yevgeny Primakov in the mid-1990s repeatedly warning that Russia would be thrown into political turmoil. Later, a debate on the difficulty of defending the integrity of the three Baltic countries against a revanchist Russia, at some point in the future, almost derailed the process of NATO enlargement. In order to placate hostile Russian sentiments, the so-called NATO–Russia agreement, providing the Kremlin's envoy with a unique status vis-à-vis the alliance, was signed (Asmus 2002: 142–74).

There is little doubt that Russian elites as well as the population at large over time grew to accept the independence of Central Europe and, to a lesser extent, that of the three Baltic countries. Sensitivity over the stationing of potent weapons systems near the Russian border remains, but NATO's presence in- and of itself is seemingly perceived as much less threatening than that in the past. In fact, the EU has increasingly replaced NATO as the primary object of Russian criticism directed against the West (Bordachev and Moshes 2004). Given that Russian leaders have a long history of serious (and well-founded) apprehension over the military aspirations of powerful neighbours, one explanation for a more critical attitude towards the EU is the latter's development of strategic leverage of a more traditional kind from 1998 onwards, in terms of a European Security and Defence Policy (ESDP).

A second explanation is the growing realization that former satellite states of the Soviet empire, as full-fledged member states of NATO but – above all – of the EU, now wield considerable influence over the Union's policies towards Russia. The May 2004 accession of eight Central and East European countries soon rendered negotiations on a revised Partnership and Cooperation Agreement (PCA) between the EU and Russia a complex exercise, and the 2007 January expansion to Romania and Bulgaria shifted power in the Union even further to the east. By mid-2007, the Russian ambassador to Brussels was openly complaining about some countries, especially Poland and the Baltic states, slowing up progress in important deliberations on energy security, climate change and human rights, painstakingly prepared by Germany as EU chair (Traynor and Harding 2007). Needless to say, the Georgian war of August–September 2008 did nothing to alleviate the situation. Sensitive about Russian claims to protect minorities in border areas, the governments of the Baltic States in particular helped push the entire Union into stronger condemnation of Moscow's actions in the Caucasus.

The strategic energy game has featured most prominently in controversies concerning gas pipelines under construction/prospection, mainly the North European Gas Pipeline (the underwater Baltic Sea pipeline from Petersburg to northern Germany to begin operations in 2010, also known as Nord Stream) and the South Stream gas pipeline in the Black Sea (from southern Russia to Bulgaria). Both are meant to diminish Russian dependence on transit routes for gas to Europe, especially avoiding Ukraine and Poland. European moves to mitigate a growing energy dependence on Russian supplies – so far largely unsuccessful – have been to promote the construction of the Nabucco pipeline to potentially tie in Central Asian (and later perhaps Iranian) gas fields via Turkey to Central and Eastern Europe. In most of these developments, Russia pays lip service to free market economic rules but in effect strives for state control of resources and transport routes, aiming to secure safe and stable revenues from the European energy market. To be sure, Russian natural resources remain the basis for Russia's successful re-entrance onto the global arena as a great political and economic state power (Nygren 2008c).

Nevertheless, there is also a set of deeper concerns mobilized by recent Russian actions. Alongside 'ideational friction' and the uninterrupted practice of great-power diplomacy separate from those of the EU, a number of specific historical legacies weigh heavily on Russia–Europe relations. This parameter embodies a wide array of issues, events and narratives related to conflicts, borders, minorities and properties that through various actors are brought to bear on Russia's ties to individual regions and institutions, but with implications for others. Some are historical in the proper sense that they mainly pertain to grievances over past (mis)deeds that cannot be undone and that many believe need to be forgotten, for the sake of a more prosperous and constructive future relationship. Other such issues, events and narratives are directly linked to the present and to pending decisions – on border arrangements (DeBardeleben 2005), the status of Russian-speaking minorities or communist-era monuments – that cannot be postponed indefinitely.

Understandably, the hegemony of the Russian component in the Soviet Union over the rest of the communist world was never conducive to a nuanced discussion about controversial subjects such as the government-induced famines in Ukraine of the 1930s, the Katyn massacre of Polish officers during World War II, mass deportations of Baltic citizens to Siberia or show trials against dissenting Central and East European political leaders in the late 1940s. The post-communist leaders of most countries have over the past fifteen to twenty years helped to 're-nationalize' historiography in public speeches, programs and education policies, almost infrequently at the expense of previous accounts more sympathetic to Russia as well as to Soviet communism. Most importantly, perhaps, the 'de-Russified' stories now dominate textbooks in schools and in institutions of higher learning.

At least two features of the history of 'in-between-Europe' have been down-played through these developments. One is the Russophile streak in pre-communist society and politics that exists in much of the regions' classical literature and other cultural artefacts. A particularly potent variant is traditional pan-Slavism, without which the political map of Europe would no doubt look very different today. The Croat priest Juraj Križanic (1618–83) and the Polish prince Adam Czartoryski (1770–1861) belong to the precursors of pan-Slavism (Kohn 1953), though most important were arguably the writings of Johann Gottfried Herder in fomenting cultural and political aspirations among the Slavic peoples over the long term (Chirot 1996).

Right from the emergence of pan-Slavism, however, the position of Russia toward the other, smaller Slavic peoples was a matter of controversy. In the mid-nineteenth century, Austro-Slavism was widespread throughout the Habsburg Empire, in contradistinction to Russian Slavophilism that emphasized the cultural contribution of eastern Slavic nations along the lines of Herder's romantic ideas, and Moscow as the Third Rome. According to Hans Kohn, a gulf opened between the two conceptions of pan-Slavism at the Austro-Slav congress in Prague in 1848, with the anarchist Bakunin emerging as the chief advocate of a Russian-led federation of free and equal Slavic nations (Kohn 1953: 90–95).

A second, neglected aspect of the history of 'in-between-Europe' is the powerful impulses that originated from within the core of the Soviet Empire, from the government of Mikhail Gorbachev and from the cultural intelligentsia, in the 1980s. The notion of glasnost launched by Gorbachev and his close associates was soon utilized effectively by Russian and other Soviet intellectuals to address the cracks in official historiography. The Chernobyl nuclear disaster was presumably the single most important event in convincing the administration to unleash the forces of critical media. When it did, an important part of glasnost coverage turned to controversial aspects of communist history. Domestic Russian phenomena like the death of members of the last Tsar and his family and Stalin's purges in the 1930s and 1940s became prominent, but also the exact circumstances surrounding the 1956 and 1968 military interventions in Hungary and Czechoslovakia.

Are the grievances of Central Europeans likely to linger in the medium-to long-term? And could the recent transformation of Central and Eastern Europe spell the end of 'in-between-Europe' as an historical phenomenon? At the very least, formal memberships in the EU, NATO, the OECD, the Council of Europe, along with the restructuring of the economies of Central and East European states, already serve to mitigate the influence of great powers like Russia, Germany, France and Britain over the region. The dependence on great power patronage, the weakness of domestic political and business structures, were the structural conditions under which CEE (Central East Europe) governments used to operate in the nineteenth and early twentieth centuries. None of these features exists in the beginning of the twenty-first century, thus paradoxically improving the prospects for greater integrity of individual states through far-reaching EU integration (Franzinetti 2008).

Certainly, there are plenty of challenges to the further consolidation of states, markets and civil society institutions in the CEE region. The worldwide economic recession that began in 2008–9 hit several CEE economies hard as they had relied heavily on foreign loans and credits to rebuild industries and establish new businesses. With the economic slowdown the relative dependence of smaller CEE countries on the larger economies in its vicinity may grow, and thereby somewhat limit the freedom to manoeuvre for the former. At the end of the day, however, there is nothing to suggest a reversal of the wider trend towards political and economic consolidation that characterized the past fifteen or so years.

For non-EU countries in the 'in-between' category, the basic challenge is how to mitigate or offset Russia's political and economic influence. Lacking important natural resources or other sources of substantial revenues, there are good reasons for aspiring to cultivate close political ties with the EU, NATO and core European states. However, from there on, the problems faced by various governments and communities are quite different depending on circumstances and characteristics of their 'in-betweeness'. A stark illustration is the comparison between Ukraine and Moldova, two countries that straddle the EU–Russia border. How can Ukraine meet the future, as it seems to harbour two competing visions for its citizens? Nevertheless, although Ukraine's choice is an arduous one, that of Moldova is existential. A more pertinent question may be: which vision of the future is at all viable?

Compared with this, EU member states are more comfortably situated in an overarching institutional architecture that provides a level of certainty for business relations and social life alike. The likelihood of them being torn apart, literally or metaphorically, is remote. However, the positive influence of the 'EU enlargement dividend' has been waning fast in conjunction with the sharp economic recession, and scepticism towards the project of European integration is likely to grow in the coming years. How will CEE governments act in this, less benign political and economic context? Will solidarity with other EU member states increase or subside? Will long-term strategic considerations, in a field such as energy security, stand up to short-term needs and opportunities?

Part 1

Norms, values, and institutions

2 Russia and Europe after the Cold War

Cultural convergence or civilizational clash?

Russell Bova

Some in the West are trying to 'exclude' the Soviet Union from Europe. Now and then, as if inadvertently, they equate 'Europe' with 'Western Europe'. Such ploys, however, cannot change the geographic and historical realities. Russia's trade, cultural and political links with other European nations and states have deep roots in history. We are Europeans.

Mikhail Gorbachev, 1987

We are a part of Western European culture. In fact, we derive our worth precisely from this. Wherever our people might happen to live – in the Far East or in the South – we are Europeans.

Vladimir Putin, 2000

The question of Russia's cultural relationship to Europe is an old one debated by Russians and Europeans alike (Bova 2003: 243–77). For some, Russia is a European nation whose belated movement towards cultural convergence with the rest of Europe at the beginning of the twentieth century was temporarily interrupted by the Bolshevik Revolution. From this perspective, the end of communism should have led to the resumption of that process of integration of Russia into the larger European culture. For others, Russia and Europe represent two distinct cultures – indeed, two distinct civilizations. From this perspective, Bolshevism was less a temporary interruption of Russia's path to Europe than a reassertion of the essential chasm between Russian and European cultures that the end of communism is unlikely to bridge.

This chapter will examine these two competing perspectives on the cultural relationship of Russia to Europe and examine them in the light of trends and developments in Russia since the fall of communism. Specifically the focus will be on political culture, and the paper will address three fundamental questions: (1) is there a European (or 'Western') political culture that transcends the variations among the national political cultures of European nations, (2) does European political culture remain different in important ways from Russian political culture, and (3) are European and Russian political cultures doomed to difference? This essay will answer 'yes' to the first two questions and 'no' to the third. Stated most generally, the essay argues against naïve expectations that the end of the Cold War meant immediate cultural convergence between Russia and the West, but it also weighs in against essentialist arguments that Russia is 'sui generis'.

European and Russian political culture

The concept of 'culture' refers to the 'goals, values, and pictures of the world' shared by a particular group (Shweder 2001: 211).[1] It follows that 'political culture' consists specifically of shared goals, values and pictures of the *political* world. Cultures and political cultures include shared normative predispositions concerning how the world *should* work. For example, political cultures can be distinguished by the relative priority assigned to freedom versus order or individual versus community rights. Political cultures are also defined by shared empirical understandings of how the world *does* work. For example, Francis Fukuyama's comparative study of the levels of trust in different societies is an examination based on the premise that there are cross-cultural differences in what one might expect in the behaviour of others (i.e. how will others behave and can one afford to be trustful?) (Fukuyama 1996).

To be meaningful as a concept in political science that can help explain variations in political behaviour, political culture is premised on assumptions of both commonality and difference. That is, it assumes that there are communities of people who share essential cultural predispositions that simultaneously bind them to one another while separating and distinguishing them from other communities in meaningful ways.

Within the boundaries of a single state, one can typically find various regional political cultures that differ in significant ways from one another. Thus, the political culture of the American south differs from that of New England just as the political culture of Sicily is not the same as that of Northern parts of Italy. Still, despite those subnational disparities one might still reasonably speak of an Italian or an American political culture insofar as shared language, shared historical experiences as a nation, and free and easy movement of people within the boundaries of the state also lead to the evolution of common 'goals, values and pictures of the world' that overlay regional variation.

There is empirical evidence to support the idea of 'national political cultures'. Using data from the *World Values Survey*, Ronald Inglehart and Christian Welzel note that the values differences among people of different religions, ethnicities, generations, occupations and educational status are much smaller among people living within a particular state than those among peoples of different states (Inglehart and Welzel 2005: 69). Thus, an elderly French farmer with only a basic education is likely to share many values in common with a young college-educated Parisian, perhaps more so than with a farmer of similar age and education who happens to live in Germany. In effect, political culture can be thought of as a set of Russian *matryoshka* dolls with local and regional cultures subsumed within one another, as well as within a larger national culture.

More controversial is the idea of supranational political cultures that extend beyond the boundaries of a single nation or a single state. Is the largest of the *matryoshka* dolls found at the national level or is there yet a larger layer of political culture – short of humanity as a whole – within which national cultures might be reasonably subsumed? Specifically, in the case of Europe, can one speak meaningfully of a European political culture that incorporates common orientations and values that transcend, for example, the cultural differences between Scandinavia and Mediterranean Europe or between the British Isles and Germany? Moreover, just as importantly, do the common European orientations and values that one finds clearly suffice to differentiate European political culture from that of the non-European world?

In the introduction to his massive study of Europe, historian Norman Davies noted that the idea of Europe is not unique to the period of formal European integration that began in the post-World War II era. Davies cites Rousseau whose comment that, '/t/here are no longer Frenchmen, Germans, and Spaniards, or even English, but only Europeans' could easily have been spoken in the celebratory climate of the Maastricht Treaty. Likewise, he notes the words of Edmund Burke who noted that, 'No European can be a complete exile in any part of Europe' (Davies 1996: 8). Davies himself echoes such sentiments in noting that Europeans 'forget all too often things that we have in common under the surface and tend to take these schematic political science views of whatever communist countries, or backward economies or whatever it is, and not see that in the past and in the consciousness of people, there's much more links than meet the eye' (Davies 2000).

If the idea of Europe as a cultural entity long preceded and even helps account for the emergence and specific boundaries of Europe as a set of institutions collectively known as the European Union, the creation of those all-European institutions no doubt reinforces the sense of cultural connectedness. Indeed, following the logic of 'congruence theory', as developed by political scientist Harry Eckstein, one would anticipate a reciprocal relationship between the structures and processes of governance (what Eckstein refers to as 'authority patterns') and the larger norms and patterns of authority found more broadly in society (what we might call 'political culture') (Eckstein 1998: 3–33). As Eckstein argues, 'incongruent authority patterns tend to change toward increased congruence' (Eckstein 1998: 23). In other words, where the structures, institutions and processes of governance are incongruent with the norms of the larger political culture, then one or the other (or both) will evolve until congruence is achieved. In short, just as Europeans can create the European Union, the European Union can create Europeans.

Given the existence of something called Europe as more than a piece of geography but as a meaningful cultural entity, it remains to ask: (1) what are the core, unifying elements of the European political culture and (2) how far do the boundaries of European political culture extend?

In his book, *From Plato to NATO*, David Gress argues that the West or 'Western Civilization' resulted from the synthesis of Greek and Roman civilizations with Christianity and Germanic culture (Gress 1998: 22). From those very same foundational sources, Samuel Huntington notes the derivation of some of the essential values and orientations of 'Western' political culture. They include, according to Huntington:

- *Separation of Church and State* – the idea of two separate spheres of influence, one spiritual and one secular.
- *Rule of Law* – the idea that governance is via laws to which all, including leaders, are subject.
- *Social Pluralism* – diverse groups, classes and interests exist and are accepted as legitimate.
- *Representative Bodies* – parliaments and other institutions to represent the interests of those diverse groups. The basis of modern democracy.
- *Individualism* – and a commitment to individual rights and liberties. The basis of the modern concept of human rights (Huntington 1996: 69–72).

As Huntington notes, one might, to a greater or lesser degree, find some of these attributes of Western political culture present in other, non-Western societies. However, what distinguishes 'the West from the rest', for Huntington, is the totality of their influence. For example, in much of the Islamic world the idea of separation of church and state is clearly a controversial idea, while some Asian critics of the West have questioned the Western emphasis on social pluralism and individualism at the expense of larger community interests.

In discussing the idea of the 'West' or 'Western civilization' rather than the narrower notion of 'European civilization', both Gress and Huntington extend the notion of that civilization beyond the geographical boundaries of Europe to include West European-settled countries like the United States. The more interesting question, in the context of this essay, is how far to the east Western culture extends. For Huntington, the answer to the question, 'Where does Europe end?' is simple. As he puts it, 'Europe ends where Western Christianity ends and Islam and Orthodoxy begin?' (Huntington 1996: 158). Citing William Wallace, Huntington suggests that the end of the Cold War simply returned the eastern boundary of the West to its natural cultural line starting, in the north, on the Russian Finnish border and separating Russia from the Baltic states before cutting through Belarus, Ukraine and Romania and finally cutting west through the former Yugoslavia leaving Serbia and most of Bosnia on one side and Croatia and Slovenia on the other (Huntington 1996: 159).

The former communist states of central Europe – Poland, Hungary, and Czechoslovakia – are on the Western side of that cultural divide. When the Berlin Wall came down, they were relatively quick to implement Western models of democratic politics and market economics and to seek membership in Western institutions such as NATO and the European Union. With a bit more hesitation, they were accepted into the Western fold. Milan Kundera perhaps put the cultural connection of Central Europeans to their Western neighbours most eloquently when he noted:

In fact, what does Europe mean to a Hungarian, a Czech or a Pole? For a thousand years, their nations have belonged to the part of Europe rooted in Roman Christianity. They have participated in every period of its history. For them, the word 'Europe' does not represent a phenomenon of geography but a spiritual notion synonymous with the word 'West'. The moment Hungary is no longer European – that is, no longer Western – it is driven from its own destiny, beyond its own history: it loses the essence of its identity (Kundera 1984: 33).

Kundera's words might apply elsewhere in the former Soviet bloc to Croatia, Slovenia and, perhaps, to the Baltic states. Nevertheless, the question is whether it applies to Russia itself. Kundera himself clearly does not think so. He notes that from the perspective of Central Europe, 'Russia is seen not as one more European power but as a singular civilization, an *other* civilization' (Kundera 1984: 34). Thus, Kundera clearly anticipates Huntington's cultural division between the culture of Western Europe and that of Russia and other 'Slavic/Orthodox' states. As such, both have a view of Russia's cultural relationship to Western Europe that is at odds with that suggested by the quotes from Mikhail Gorbachev and Vladimir Putin at the beginning of this chapter. For Gorbachev and Putin, the implication is that the cultural differences between Russia and countries of Western Europe are akin to the cultural differences between Sweden and Italy – real but within the context of an overriding European civilization.

If Gorbachev and Putin are right, then expectations of cultural convergence between Russia and the rest of Europe might still be expected even if the process has not been as simple, smooth or quick as some naively expected at the end of the Cold War. However, if Kundera and Huntington are right then such cultural convergence is unlikely, and 1990s efforts to export Western European political and economic models to Russia were doomed to fail.

The cultural relationship of Russia to Western Europe has been, of course, a central point of contention among Russian intellectuals for centuries as the nineteenth-century debate between Slavophiles and Westernizers has been reprised many times over. Indeed, some recent discussions of global cultural politics – and specifically discussion of Russia's location on the global cultural map – seem, knowingly or not, to be recycling centuries old arguments made by Russian thinkers of the past. For example, long before Samuel Huntington first articulated his controversial – and seemingly novel – clash of civilizations thesis with its posited distinction between Western and Slavic/Orthodox civilizations, Russian pan-Slavic thinker Nikolai Danilevsky had made a similar argument. In his 1869 work, *Russia and Europe*, Danilevsky argued that humanity was subdivided into several 'cultural historical types' including Romano-Germanic (what Huntington calls Western civilization) and Slavonic (what Huntington calls Slavic/Orthodox) (Grier 2003: 23–77).

In Danilevsky's pan-Slavism there was the assumption, adopted by Huntington, that there are indeed cultural groupings found at a level somewhere between the individual nation-state and humanity as a whole. Thus, for both Danilevsky and for Huntington the cultural distinctions between France and Sweden, for example, are fundamentally different from those between France and Russia. While the former are intra-civilizational variations, the latter presumably run deeper and are less subject to convergence and reconciliation.

So what are the attributes commonly associated with Russian political culture, and how different are they from those of Western Europe? In the eyes of many Western observers, Russian political culture is the antithesis of Western political culture in many essential respects. The 'goals, values, and pictures of the world' frequently associated with Russian political culture by many Western and Russian observers seem quite different from those of the 'West'. The conventional wisdom includes many of the following attributes of Russian political culture:

- *Personalized Authority*: Political scientist Stephen White notes a Russian predisposition to 'a succession of strong, autocratic monarchs' and a 'personalized attachment to political authority' (White 1979: 22, 30–33). Illustrative of this tendency are the words of a Russian soldier to the British ambassador in 1917: 'Oh yes, we must have a Republic, but we must have a good Tsar at its head' (White 1979: 33). For many Western observers, the Russian tradition of personalized political authority seems at odds with the Western concept of 'rule of law'.

- *Statism*: In his 1999 Millennium Manifesto, Vladimir Putin reflected an attachment to state power that he perceives as intrinsic to traditional Russian political culture and as a point of differentiation with Western political culture. He notes, 'Russia will not become a second edition of, say, the United States or Britain, where liberal values have deep historic traditions. Our state and its institutions and structures have always played an exceptionally important role in the life of the country and its people ...' (Sakwa 2008: 55).

- *Sobornost'*: The assumption of Western political culture is that different societal groups with different interests legitimately compete. That is at odds with the Russian concept of *sobornost'* with its implications of a collective will and an organic unity of social purpose. As Russian scholars Victor Sereyev and Nikolai Biryukov argued, 'The lesson of Russia's history must be learned: namely that a modern democratic society will not emerge unless the political mentality of *sobornost'* is overcome (Sergeyev and Biryukov 1993: 207–8).

- *Order*: A Russian fear of disorder and chaos, borne out of centuries of historical experiences in which Russia has seemed to exceed its plan for both, has been noted by many observers – Russian and non-Russian alike. Thus, Russian political culture is said to be characterized by a continual emphasis on the search for order (*poryadok*) even if it must come at the expense of freedom and liberty. This manifests as a fear of 'tsarlessness' (*bestsarstvie*) and in the desire for a 'firm hand'.

- *Unity of Church and State*: While the Enlightenment and Protestant Reformation led to a separation between church and state in the societies of Western Europe, a more symbiotic relationship has been maintained between the Orthodox Church and the Russian state. The continuation of this relationship was manifested in the 1997 'Law on Freedom of Conscience and Religious Association' whose preamble noted 'the special role of Orthodoxy in the history of Russia and in the establishment and development of its spirituality and culture'.

Comparing this list of orientations of Russian political culture with Huntington's list (noted earlier) of the central features of Western political culture reveals sharp disparities. Instead of rule of law, one finds personalized authority; instead of pluralism, one finds statism; instead of individualism, one finds sobornost'; in lieu of the emphasis on individual rights, one finds a hunger for order; instead of separation of church and state, one finds a notion of a more organic unity of church, state and the larger society.

It must be emphasized that not everyone would accept these characterizations of Russian political culture as valid. For some they are oversimplified cultural stereotypes that are at least exaggerated, if not completely inaccurate. Nevertheless, they are commonplace among observers both inside and outside of Russia. Moreover, if they are valid, even in part, they make the odds of political convergence of Russia with the West and the odds of successful Russian adoption of Western-style parliamentary democracy very slim. As one Western scholar has argued, 'Russian political culture is most often viewed not as an asset to establishing democracy but as something that needs to be overcome' (Petro 1995: 176).

The record of post-communist Russia

Almost two decades of post-communism in Russia provide considerable empirical evidence with which to examine the extent of convergence of Russian and European political cultures. That evidence comes in two forms: behavioural and attitudinal. In both cases, the pattern is the same. In the early years, following the collapse of the Soviet Union, a reduction in what the editors of this volume call the 'friction of ideas' between Russia and Europe seemed apparent in both Russian political behaviour and attitudes. However, over time, the distance between Russian and European political behaviour and attitudes expanded once again.

Russian political behaviour

The early years of Russia's post-communist era seemed to belie those cultural determinists who argued that Russian political culture was infertile soil for planting Western democratic political institutions. Beginning in the Gorbachev era, and accelerating following the Soviet collapse, Russian political institutions began increasingly to reflect and embrace the essential attributes of democratic governance: a) a universal right of political participation, b) meaningful political contestation and competition and c) civil rights and liberties including rights to free expression, free assembly and a free press.

Elections, beginning in December 1993, were truly competitive with multiple candidates and parties that spanned a wide ideological spectrum. The fact that the election results sometimes surprised and displeased the Yeltsin government (for example, the underperformance of Russia's Choice and the emergence of Zhirinovsky's Liberal-Democratic Party in Duma elections in 1993; the broad gains for the Zyuganov's Communist Party in Duma elections in 1995) provided evidence that Russian politics was more competitive and democratic than at any point in its history. Likewise, following the Soviet collapse, movement towards a free media that operated independent of direct government control accelerated and represented a significant advance from Gorbachev's policy of *glasnost*.

Russia's democratic progress was reflected in the improving scores it received from the US-based NGO (non-governmental organization), Freedom House. In its annual 'Freedom in the World' survey, Freedom House scores countries on both political rights and civil liberties (assigning a score of 1–7 for each with 1 indicating a high level of respect for political rights and civil liberties and 7 very low levels of respect). The combined score of 2–14 provides a rough approximation of the widely accepted procedural definition of democracy as a political system characterized by participation, competition and civil liberties. In the mid 1980s, the Soviet Union was receiving combined scores of 14 – the worst rating possible. A decade later – from 1992 through 1997 – post-communist Russia received a combined score each year of 7 – representing significant progress in democratizing the Russian political order (see Table 2.1).

Table 2.1 Trends in Russian Democracy: Freedom House Scores (possible range from 2 to 14; lower # = more democratic)

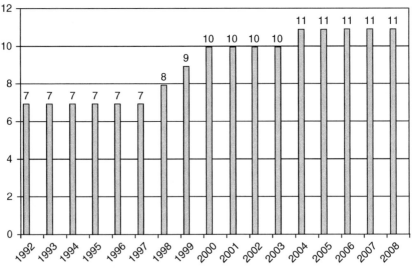

Source: Freedom House, *Freedom in the World*, http://www.freedomhouse.org.

While that 1992–97 period arguably represented the high point of post-communist Russian democracy, Russia's democratic behaviour during that period continued to suggest the persistence of a lingering and significant gap between Russian democracy and the democratic practices of its West European neighbours. Viewed most generously, Russian democracy in the 1990s remained limited to what scholars of democratization call 'electoral democracy'. An electoral democracy is one in which leaders come to power via reasonably competitive elections, but, once elected, they face few restraints on their power and authority.

In West European liberal democracies, in contrast, elections are not blank checks conveying unlimited power on the victors. On the contrary, leaders are constrained by: (a) rule of law, (b) competing political institutions and (c) an active and vibrant civil society. In post-communist Russia, in contrast, rule of law is precarious, institutions (e.g., the legislature, courts and political parties) that might compete with presidential power remain weak, and civil society is, in the words of one observer, 'anemic' (Simons 2008: 33).

By the end of the 1990s, and especially following the rise to power of Vladimir Putin, the democratic achievements of the early post-Soviet era, limited though they might have been, were increasingly threatened. As indicated in Table 2.1, the Freedom House scores for Russia began ticking upwards in 1998 and by 2004 reached the highest level (11) since the mid-1980s. In fact, the last years of Gorbachev's USSR received better scores for political rights and civil liberties from Freedom House – a 9 for 1990 and an 8 for 1991 – than any year in Putin's Russia since 2000.

The downgrading of Russia's democratic performance by Freedom House over the course of the past decade is a reflection of significant challenges to the democratic process in Russia during the Putin era. They include the following:

- The reassertion of state control over the Russian media. Especially important in this regard is the reassertion of control of Russian television where state control has not only been strengthened but also utilized to the advantage of Putin and his legislative supporters in recent elections.
- The crackdown on the oligarchs, including, perhaps most significantly, the arrest and imprisonment of Mikhail Khodorkovsky and the break-up of his Yukos empire.
- The dominance of United Russia in the Russian Duma and the apparent reduction in the competitiveness of Duma elections. Some of this is due to changes in laws that made it more difficult to register parties, that eliminated single member districts, and that raised the threshold for parties to earn seats as a result of the party list vote from five per cent to seven per cent.
- Putin's assumption of the post of Prime Minister as an end-run around the constitutional limit on serving a third term as Russian president. What seems especially problematic for Russian democracy about this is less his initial appointment as Prime Minister, but the fact that in transferring powers, including oversight of some of the key ministries of the Russian government,

from the president to the prime minister without formal constitutional amendment, it reinforces the personalization of power in Russia in a way that is contrary to democracy and rule of law (Ryabov 2008).

Taken individually, many of these Putin-era measures could be justified on their own terms and are not inherently antithetical to democracy. For example, many Western parliaments are elected entirely by party list votes. Likewise, a seven per cent threshold for being awarded Duma seats is not unreasonable. Indeed, the dizzying array of political parties, characteristic of Russian elections in the 1990s was arguably too much of a good thing that could sow voter confusion and produce fractured legislatures incapable of governing. As for the oligarchs, there is little doubt that in many cases their acquisition of assets during the 1900s era of privatization was less than fair and transparent or that their behaviour has been entirely consistent with Russian law.

That said, the challenge to Russian democracy posed by the Putin-era 'reforms' is that they have the cumulative effect of reining in many of the key competing, countervailing centres of power that are essential if democracy is to thrive. Most importantly, they serve not only to augment Putin's power once in office, but also to undermine the competitiveness of the electoral process itself. In effect, the shift in the pattern of Russian political behaviour that is reflected in the deteriorating Freedom House numbers represents a shift in Russia from a regime of 'electoral democracy' to one that can be labelled as 'electoral authoritarianism'. Whereas elections in the former still have some significance in giving voters a meaningful choice in determining who comes to power, in the latter electoral outcomes are largely predetermined (Schedler 2006). As Larry Diamond puts it, 'the distinction between electoral democracy and electoral authoritarianism turns crucially on the freedom, fairness, inclusiveness, and meaningfulness of elections' (Diamond 2002: 28).

This shift is readily apparent in the declining uncertainty, over time, of the outcome of Russian elections. Uncertainty, and the inability to predict outcomes with assurance, is central to the character of a democratic regime (Przeworski 1991: 12–14). Despite the limits of Yeltsin-era electoral democracy, at the presidential level the 1996 election was the only one where no candidate achieved the required majority vote in the first round and was thus the only one where a run-off vote was required. Indeed, in the months preceding the vote, there were moments when there was real doubt among Yeltsin supporters in both Russia and the West over his ability to prevail. By the time we get to the 2008 presidential election, uncertainty had all but been eliminated as Dmitry Medvedev had no significant competition and his victory was assured. Likewise in Duma elections, the last real surprise we saw was the 1993 strength of Zhirinovsky's Liberal Democratic Party, and the last real electoral victory over the 'party of power' was the 1995 showing of the Communists. Since then, Duma elections have become increasingly predictable and in keeping with the interests of the Russian president.

Russia's democratic backsliding under Putin is thus a qualitative change in regime type from the electoral democracy of the 1990s, and it raises doubts about the ability of the Russian political order to allow and foster a peaceful, electoral transition of power to a political opposition. As Huntington argued in the 1990s, the best indicator that democratic government has become consolidated in a democratizing country is the peaceful transfer of power from incumbents to the opposition following an election in which the incumbent loses. While Russia has experienced shifts in the balance of power within the legislative branch, in a system like Russia's, where the president is dominant, it is turnover in the executive branch that is most significant. On that score, Russia has yet to see its first turnover. Both the transition from Yeltsin to Putin and, subsequently, from Putin to Medvedev were transitions from an incumbent to an anointed successor. Indeed, to the extent that Putin has remained on the stage as a powerful Prime Minister, the transition from Putin to Medvedev is even less a turnover than it superficially appears.

Russian political attitudes

The evolution of Russian public attitudes toward Western-style democratic institutions has paralleled the record of Russian democratic behaviour in practice. Once again, the early indications were promising regarding the narrowing of the gap between Russian and European political culture. A fall 1991 survey conducted just a few months prior to the disintegration of the Soviet Union suggested that Russians were ready to embrace the idea of democracy. Despite the difficult economic situation in the country at the time, the challenge to political stability and order represented by the ongoing workers' strikes, the independence movements in the Soviet republics, and the aborted August 1991 coup, most Russians (51 per cent) chose 'democracy' over a 'leader with a strong hand' (39 per cent) as the better option for dealing with the country's problems (Pew Research Center 2006). The results of this and similar surveys of the time suggested that cultural stereotypes about a Russian penchant for strong, autocratic leadership needed some rethinking.

Fast-forward a decade and a half, and public opinion on that same question appears to have dramatically reversed. As illustrated in Table 2.2, the majority preference for democracy over a strong, non-democratic leader that was reported by the fall of 1991 survey had largely evaporated. By 2005, two-thirds of those polled indicated a preference for a strong leader with less than a third indicating a preference for democracy.

A 2007 study of 1600 Russians conducted by the Levada Center confirms this trend (Levada Center 2007). On the one hand, the survey found that the vast majority of respondents (94 per cent) believed that they had little or no influence over what happens in Russia. Yet, that sense of powerlessness did not seem to translate into a pent up demand for more democracy. On the contrary, while 12 per cent suggested that Western-style democracy is the ideal political system for Russia, almost three times that number (35 per cent) indicated that they preferred the Soviet-era system of the pre-1990s era.

Table 2.2 Best Kind of Government for Russia (Respondents were asked whether Russia should rely on a democratic form of government to solve the country's problems or on a leader with a strong hand.)

	Democratic	Strong leader	Don't know
1991	51%	39%	10%
2005	28%	66%	6%

Source: 2005 Pew Global Attitudes Survey, 'Russia's Weakened Democratic Embrace', January 2006, http://pewglobal.org/reports/display.php?ReportID=250.

Despite that attitude towards Western-style democracy, about half of the respondents suggested that there are many things that can be taken from Western culture. Yet, for most of those respondents, in doing so one would be borrowing from a culture that is alien to Russia, as almost three quarters (71 per cent) indicated that they did not think of themselves as European. Indeed, only about a third saw Europe as a partner with whom a long-term relationship could be developed and enhanced, while almost half (45 per cent) considered Europe as a potential threat to Russia.

The evolution of Russian attitudes towards democracy, in particular, and towards Europe, in general, is likely rooted in the track record of Russian political and economic performance in the 1990s. To the extent that democracy and an embrace of Europe were anticipated to represent the path to political stability and economic prosperity for Russia, the achievement of such results proved elusive. The economic pain that resulted from early 1990s economic shock therapy, the impact on Russians of the 1998 financial crisis, and the increasing evidence of political incompetence on the part of the Yeltsin government by the late 1990s, contributed to a discrediting of the democratic idea among the Russian public.

The democratic retreat reflected in the Freedom House scores (see Table 2.1) coincided, in contrast, with the post-1998 Russian economic recovery and the political stabilization of Russia in the Putin era. That is not to say that democratization was the cause of the 1990s-era economic pain or that the Putin-era democratic retrenchment stimulated economic recovery. The post-1998 devaluation of the Russian currency and the Putin-era spike in global oil prices arguably had a lot more to do with the recent, improved fortunes of the Russian economy. However, that Russian citizens would interpret correlation between the evolution of the political order and economic performance as a cause and effect relationship between the two is understandable and, indeed, predictable.

A significant part of the declining confidence of the Russian public in democratic institutions stems from the changing attitudes of the 'revolutionary generation' of Russians, i.e. those 18–34 years of age in 1991 when the USSR disintegrated (Pew Research Center 2006). In the Pew study cited earlier, that cohort was overrepresented among Russians who preferred democracy to a strong leader in the 1991 survey. At the time 58 per cent (compared with 51 per cent of the population as a whole) indicated more confidence in democracy than in a strong leader. By

Table 2.3 Evaluation of the present system of governing Russia

	1992–1998 (average of 7 surveys)	2000–2007 (average of 8 surveys)
Positive	30%	53%
Negative	52%	34%
Neutral	17%	13%

Source: 'New Russia Barometers II to XV' as reported in *Russia Votes*, http://www.russiavotes.org/national_issues/national_issues_politics.php?PHPSESSID=20da0aec67024ac9116dd6b08c76cd06#2nrb.

2005, that same cohort, now 32–48 years of age, saw its support for democracy over a strong leader fall to 29 per cent – about the same as the population as a whole in that year.

What all of this public opinion data suggests is that the recent anti-democratic tendencies in the practice of Russian politics are in accord with trends in the attitudes and orientations to politics of the majority of Russians. The Russian public seems less to be chafing under Putin's resurgent authoritarianism than largely supportive of, or at least tolerant toward, the anti-democratic trend. As indicated in Table 2.3, when asked to evaluate how well the Russian political system works, Russians became significantly more positive in the post-2000 era as democracy was in retreat. To use Eckstein's language once again, there appears throughout the post-Soviet period to be congruence between the authority patterns of Russian politics and the attitudes of the majority of the Russian public.

The fluidity, indeed the fickleness, of Russian political attitudes on such fundamental questions as the merits of democracy, Russia's cultural relationship to Europe, or the merits of the Russian political system is unsatisfying if one is trying to dig to the roots of contemporary Russian political culture. Asking Russians if they support democracy or if they think of themselves as Europeans, only provides snapshots of Russian public opinion at a given point in time. As noted, the picture developed is susceptible to change in response to specific events in Russia or in the Russia–Europe relationship and may be less a reflection of deeply rooted cultural values than of transitory political developments and the shifts in public opinion that accompany them. More revealing is an examination of attitudes and values unrelated in any direct way to the headlines of the day.

Such a perspective is provided by data assembled in the *World Values Survey*. Using that data, Inglehart and Welzel compare national cultures across two basic dimensions (Inglehart and Welzel 2005: Chapters 1 and 2). The first dimension compares countries according to their embrace of either traditional or secular–rational values. The former are values commonly associated with pre-industrialized societies and feature a heavy emphasis on religious sources of authority. The latter are the values that emerge as a result of modernization and industrialization and are largely related to the secularization of authority. The movement from traditional values to secular–rational values is essentially the replacement of religion and superstition with science and bureaucracy as the basis of behaviour and authority relations in a society. On this dimension, Russia and much of Western Europe rank very high on the secular–rational end of the scale. Indeed, on this dimension, there is more significant variation among the countries of the West than there is between Russia and the West. The United States in particular, along with other English-speaking countries, have less in common with the rest of Europe on this dimension than does Russia. Thus, in this respect, Russia is very much a European country.

The second dimension identified by Inglehart and Welzel is the dimension that measures embrace of either survival or self-expression values. While survival values prioritize concern with economic and physical security, self-expression values emphasize autonomy and freedom from centralized authority. According to Inglehart and Welzel, these self-expression values are associated with post-industrial societies where workers no longer, by and large, work in factories and where cognitive skills and creativity are highly valued and essential to economic success. The real difference between Russia and Europe, and indeed between Russia and the West, is along this second dimension. While the West European countries and West European-settled countries (i.e. the United States, Canada and Australia) all rank high on self-expression values, Russia is at the other end of the spectrum. Indeed of the 81 countries (representing 85 per cent of the world's population) included in the survey, only Moldova ranked lower on self-expression values than did Russia (Inglehart and Welzel 2005: 63). (At the other extreme, Sweden ranked highest on self-expression values of all the 81 countries included).

While, as noted, socioeconomic development explains much of the cultural variation found by Inglehart and Welzel, it is not the whole story. As they put it:

Do these cultural clusters simply reflect socioeconomic differences? For example, do the rich societies of Protestant Europe have similar values simply because they are rich? The answer is no. ... Whether a society has a Catholic or Protestant or Confucian or Orthodox or communist heritage makes an independent contribution to its position on the global cultural map (Inglehart and Welzel 2005: 66).

In fact, when Inglehart and Welzel overlay Samuel Huntington's civilizational division of the world over the two-dimensional plot of countries, civilizational clusters that are consistent with Huntington's expectations become apparent.

Table 2.4 summarizes some of the data on both the development of democratic institutions and of self-expression in Russia and the other post-communist states of the former USSR and Central and Eastern Europe. From that data, a number of conclusions are suggested. First, Russia's road to democratic government has been rocky and slow – not only by the standards of long-established West European democratic systems but also in comparison to the record and pace of democratic transformation established elsewhere in the post-communist world. Of 28 post-communist states, 21 do better than Russia in the Freedom House rankings. Of the

Table 2.4 Religion, democracy, and self-expression values in post-communist states*

Western Christian			Orthodox			Muslim		
Country	*FH Score*	*S-E Values*	*Country*	*FH Score*	*S-E Values*	*Country*	*FH Score*	*S-E Values*
Czech Rep	2	Moderate	Slovakia	2	Moderate	Albania	6	Low
Hungary	2	Low	Bulgaria	3	Low	Bosnia	7	Low
Poland	2	Low	Romania	4	Low	Kyrgyz Rep	9	N/a
Slovenia	2	Moderate	Ukraine	5	Very low	Kazakstam	11	N/a
Lithuania	2	Low	Serbia	5	Low	Azerbaijan	11	N/a
Estonia	2	Low	Macedonia	6	Low	Tajikistan	11	N/a
Latvia	3	Low	Montenegro	6	Low	Uzbekistan	14	N/a
Croatia	4	Moderate	Moldova	7	Very low	Turkmenistan	14	N/a
			Georgia	8	Low			
			Armenia	9	Low			
			Russia	11	Very low			
			Belarus	13	Low			
Average	2.4		*Average*	6.6		*Average*	10.4	

*Notes:
1: Freedom House scores range from 2 to 14. The lower the score, the more free and democratic.
2: Self-Expression or S-E values can be ranked as: very high, high, moderate, low, or very low.
3: All Western European countries received a score of 2 from Freedom House in 2008. All were categorized as either high or very high on self-expression (S-E) values.

Sources: Freedom House, *Freedom in the World 2008*, available online at http://www.freedomhouse.org. Self-expression categories developed by the author of this essay based on data found in Ronald Inglehart and Christian Welzel, *Modernization, Cultural Change, and Democracy*, Cambridge University Press, 2005, p. 63.

remaining six that score as poorly as or worse than Russia, five are relatively poor Islamic states.

Second, the gap between the political culture of Russia and of Western Europe runs deeper than disparities over the virtues of democracy. More fundamentally, there is a wide gap between Russia and Europe on the self-expression values that Inglehart and Welzel see as defining elements of West European political culture. Once again, Russia's distance from Europe in this regard is larger than it is for much of the rest of the post-communist world. Not including Azerbaijan and the central Asian states that were not included in the *World Values Survey*, Russia's score on self-expression values was lower than 19 of the other 21 countries in the post-communist world.

Third, though high self-expression values are conducive to democracy, the situation in the rest of the post-communist world suggests that democracy can emerge even in places where those self-expression values are undeveloped. Consolidated democratic regimes have emerged in Poland and Hungary, for example, despite self-expression scores that lag considerably behind those of their neighbours to the West. Nevertheless, in those cases, geography, cultural and religious affinity, and the promise and reality of integration into Western institutions like NATO and the EU can be compensatory factors. In Russia, in contrast, the political, geographic and cultural factors all seem, at least for the moment, to be stacked against democratic consolidation

Perspectives on democratic backsliding

The post-1998 democratic backtracking in Russia is subject to competing explanations. At the most general level, there are two possible ways in which to view the trends of the past decade. The pessimistic view (for advocates of Russian democratization and Europeanization) is that the early evidence of 'Europeanization' of Russian politics in the 1990s was ephemeral. From this perspective, the Russian embrace of Europe was born of a moment of historical weakness and chaos as Russia struggled to come to terms with the end of more than seventy years of communist rule. In due course, however, as Russia found its bearings and restored its sense of historical confidence, politics in Russia returned to its more 'natural' historical path. As Arnold Toynbee, borrowing a line from the Roman poet Horace, once said of efforts to change Russia: 'You may throw nature out with a pitchfork, but she will keep coming back' (Toynbee 1948: 164).

A second, more optimistic view is that it is the Putin-era democratic retrenchment that will prove to be the short-term blip on the long march of Russian history. From this perspective, Russia after 1991 was back on the path to Europe after the seventy-year interruption represented by communism. That path was inevitably to be bumpy and subject to setbacks. Indeed, in this view, what is most notable about Putin-era democratic backsliding is less that it occurred but more that it has been as limited as it has been. Though falling far short of the democratic gold standard of Western Europe, Russia remains today significantly more democratic than was the Soviet regime and significantly more democratic and open than the emerging Chinese superpower to its east. To the extent that Putin represents the Thermidorian reaction to the Gorbachev and Yeltsin revolutions in Russia it is, by historical standards, a relatively mild one especially when seen against the backdrop of the economic collapse and political disorder that characterized the 1990s. In looking at Russia today, optimistic democrats might suggest that Russia has taken 'two steps forward and one step backwards' on the road to democracy.

Predictions of the future are inherently filled with peril, and thus no firm predictions will be attempted here. Suffice it to note that, current trends notwithstanding, there is reason for the optimists noted previously to remain hopeful. Two points in particular need emphasis.

First, political cultures are not static, and history provides many examples of discontinuity over time. Western Europe itself is arguably the best example. Once characterized by monarchy and autocracy, it became, in time, the source of the modern practice of democratic government. Specific examples within Western Europe underline the point. In Germany, the Weimar Republic collapsed under the weight of economic and social upheaval to the Third Reich, only to be replaced by a democratic regime that outlived the Cold War division of Germany and eventually led to the unification of Germany as a democratic state. Similarly, Spain and Portugal, once deemed unlikely candidates for democracy because of cultural traditions, were instrumental in catalyzing by their example the global wave of democratization that began in the early 1970s.

The process of cultural change is often slow and rarely in a straight line. To the extent that one thinks of culture as the manifestation of attitudes and values developed over centuries, change is not only slow but also unsettling and even painful. In the Russian case, a political culture that evolved over centuries of autocratic rule, first under czars and then under commissars, cannot be transformed in an historical instant. In fact, viewed from this perspective, democratization of Russia – when measured against the Soviet experience rather than the contemporary West European standard – seems rapid indeed. Even today, despite the setbacks of the Putin era, democratic values remain in place among significant segments of the population.

Particularly among younger people a considerable commitment to the idea of democracy and freedom remains. When asked whether the primary responsibility of government is to maintain order or to respect individual freedom, 59 per cent of those over age 50 said to maintain order, but only 44 per cent of those 15–29 years of age preferred order over freedom. Similarly, when asked whether democracies are indecisive and have too much squabbling, 82 per cent over the age of 50 said yes, but only 63 per cent of those 15–29 years of age answered in the affirmative (World Values Survey 2009). This, even if the 'revolutionary generation' has become increasingly disillusioned, their children – none of whom experienced the communist era as adults – seem less nostalgic for the 'stability' of Brezhnev's Russia.

Similar disparities in attitude on such questions are found among people of different educational levels with expression of support for democratic values strongest, predictably, among the most highly educated. Indeed, on both of the aforementioned questions, the *World Values Survey* suggests that higher education correlated even more strongly with democratic values than did age. Perhaps just as important, Richard Rose notes that higher education correlates with patience towards the process of transition in post-communist states (Rose 2009: 41–42). As Rose argues, patience is crucial insofar as dismantling the legacy of communism would take time and would lead to pain along the way. While this is true, in Rose's view, of the post-communist world as a whole, it is arguably most an issue in Russia where communism existed for the longest period and where it was, unlike in most of Central and Eastern Europe, home grown.

A second reason for optimism is the fact that there are many different variations on the democratic theme. Democratic institutions and practices can be selected and tailored to match national political cultures and particular historical circumstances. The architecture of democratic systems can vary from parliamentary to presidential, from federal to unitary, from party list system to first-past-the-post single-member districts. Indeed, those broad categories only scratch the surface of the possibilities that exist in democratic design. There are, however, limits to this variation. To be considered a liberal democracy, one has to operate within certain unchallengeable parameters. Specifically there must be: (1) a universal right of participation in the democratic process without exclusion based on such things as ethnicity, religion, political affiliation and so forth, (2) a right of fair competition among candidates, political parties and interest groups as they work to influence policy and win elections, (3) a guarantee of those essential civil liberties – e.g., freedom of speech, press and assembly – necessary to participate and compete in the political process on an equal footing.

The line between the essential, uncompromisable attributes of a democratic system and the reasonable variations within systems that can be considered equally democratic is exactly what is at issue in the murky concept of 'sovereign democracy' that has gained considerable currency in Russian political discussions. In the view of Ivan Krastev, the intellectual foundation of 'sovereign democracy' is a 'fundamental mistrust towards the two concepts of the present democratic age – the idea of representation as the expression of the pluralist nature of the modern society and the idea of popular sovereignty that defines democracy as the rule of the popular will' (Krastev 2006). However, insofar as pluralism and popular sovereignty represent the essence of the democratic idea, 'sovereign democracy' cannot therefore be considered a legitimate variation on the democratic theme. On the contrary, as Krastev describes it, 'sovereign democracy' becomes just as meaningless as the concept of 'socialist democracy' was during the communist era.

However, if 'sovereign democracy' in Russia is simply, as another Russian analyst described it, the idea that 'the people should have the power to govern themselves, using democratic procedures and without being influenced from abroad' then the concept is little more than the recognition that all democracies are 'sovereign' in two related senses (Seregin 2006). First, in a world divided into 'sovereign states' democratic governments must be responsive first and foremost to their own people. Second, there is room for all sovereign nations to design their democratic institutions and procedures in ways that make sense in their particular national contexts. Viewed with this meaning in mind, French, British, German, American and Japanese democracies, to take just a few examples, are all 'sovereign democracies'.

Conclusion

After two decades of post-communism, it is evident that a large gap between the political cultures of Russia and Western Europe remains. After a brief flirtation with Western democratic institutions in the 1990s, considerable democratic backsliding has characterized Putin's Russia. Likewise, the enthusiasm for the democratic idea that seemed to capture the imagination of a majority of Russians in the late 1980s and early 1990s has clearly waned. On both counts, the evidence of a retreat from democracy in Russia in the Putin-era is hard to dispute. The vision and goal, once touted by Mikhail Gorbachev and others on both sides of the East–West divide, of a common European home stretching from the Atlantic to the Urals has not yet been achieved, as a significant gulf between the politics of Russia and Western Europe remains apparent.

At the same time, the twin projects of democratization and Europeanization of Russia have not been conclusively abandoned. The debate that continues today in Russia over the nature of its relationship to Europe is of a sort that one cannot imagine taking place in the Middle East or Asia. No one in China, India or the Arab world, to take but a few examples, would ask whether they were Europeans, let alone state as unequivocally as did Gorbachev and Putin in the quotations that opened this chapter that 'we are Europeans'. If Huntington is right that the clash of civilizations paradigm best explains the twenty-first-century world, it is a paradigm that can be fit more easily, if still not entirely convincingly, over Muslim–European relations than over Russia–Europe relations. Likewise, the debate over the meaning of 'sovereign democracy' betrays the continued power of the democratic idea for many Russians insofar as that debate is over the meaning of democracy rather than a complete rejection of the concept of democracy itself.

In Western Europe, the transition from monarchy to modern democracy took place not over the course of decades but over the course of centuries. Nor did all Western European countries democratize on the same timeline. For much of the twentieth century, the line between authoritarianism and democracy in Europe was not a simple North–South line dividing the Soviet Union and its allies from the West. On the contrary, for much of the century, the West European democracies were for the most part limited to just the northern and western parts of the continent to the exclusion of Germany and much of Southern Europe from the Iberian Peninsula to Greece.

Indeed, in historical terms, Russia's transformation over the past two decades seems comparatively rapid. The command economy is no more. Despite limitations, the access to information, freedom of movement and diversity of political discourse characteristic of Putin's Russia continue to represent, both quantitatively and qualitatively, significant advances in freedom from the Soviet era. Moreover, despite recentralization of political power over time, the one-party totalitarian state has not been reconstituted.

Thus, to those who see in the still incomplete convergence in political values and practices between Russia and Europe a confirmation of essentialist assumptions of an eternal Russian character that cannot readily accommodate Western democratic institutions, Richard Rose's counsel of patience ought to be kept in mind. Though Rose was addressing the need for patience on the part of the people of Russia and other post-communist societies, there is perhaps a lesson not only for the citizens of those countries but also for scholars of post-communism as well. Robert Dahl's conclusion, cited by Rose, that democratization is 'a slow process, measured in generations' is one to which scholars of Russia need to pay heed (Rose 2009: 36).

Note

1 Phrase originally attributed to German philosopher Johann Gottfried von Herder and later used by Isaiah Berlin to define culture.

3 Perceptions of democracy and democratic institution-building – electoral democracy the Russian way

Bertil Nygren

The starting point – what is electoral democracy?

When the Soviet Union broke up, a *Third wave* of democracy development spread to all former Soviet republics. Russia itself and especially its ruling elite at the time, including its first President, were very much attuned to this general development, among other things introducing a Constitution very similar to some *democratic prototypes*. In the early post-Cold War period, there were attempts by the most all-encompassing organization, the OSCE (Organization for Security and Co-operation in Europe), to take responsibility for the democratization process in Central and East Europe by setting the standards. And since all CEE (Central and East Europe) states were members, the strength of the developed arguments was evident. In the last half-decade, there have been two contrary general developments with respect to elections in the CIS (Commonwealth of Independent States) region. In Georgia, Ukraine and Kyrgyzstan, on the one hand, colour revolutions took place, all driven by fraudulent elections. These countries were semi-democratic or mildly authoritarian polities. The three colour revolutions led to counter-revolutions in other CIS states with authoritarian regimes. Later, the inclusion of most Central and East European countries into the so-called Western structures, and those remaining outside the more intimate Union embrace, were handed ENP (European Neighbourhood Policy) status, which emphasized not only market conditions but also a democracy dimension. Even before the Orange revolution in Ukraine in late 2004, leaders in Russia and several other CIS states rejected the notion of generally accepted democratic rules and norms. Instead, they claimed that democracy develops differently due to particularities in political traditions and culture.

In Russia, central state power was reinforced and the rather extreme decentralization tendency of the Yeltsin era was broken already in the first years of Putin's reign.[1] In his first presidential year, Putin took control of the constitutional mess that resulted from Yeltsin's wheeling and dealing with the 89 federation subjects and launched an offensive against regional leaders. The result was that in the end, they gave their approval to a change in election rules to the Federation Council. Within Putin's first term, effective close to one-party rule had been imposed, and all-important independent media had been brought under

Kremlin control. Later, other areas of democratic life were infected as well: a new law on NGOs (non-governmental organizations) was passed, information laws threatened freedom of speech, independent election monitors were forbidden, and judicial reforms halted (cf. Åslund 2007: 223–24). In short, in strangling in its infancy any attempts at 'colour revolutions' also in Russia, control was tightened over all factors that might breed effective political opposition in Russia too.[2]

Electoral democracy is not always easy to distinguish from more substantial democracy issues or from notions like liberal democracy. A definition of electoral democracy would stress that state government is somehow derived from the will of the people and that voters somehow decide who will govern them. The general problem with formal definitions is, of course, that even starkly formalized elections may be faked or that real political competition is simply lacking. In transitional countries, the introduction of free elections alone does not necessarily result in democratic procedures and rule. Something more, and much less tangible, is needed. Therefore, the approach taken here that election democracy is the most essential element of democratic development is not altogether uncontested even among analysts focusing on Russia.[3]

The question of whether or not there is a difference in European and Russian notions of democracy is an important one, as is the understanding of differences even when similar words, terms and concepts are used. Nevertheless, there are also differences with respect to whether or not we are concerned with a *presidential or parliamentary system*, and the extent to which there is a real division of power, i.e. a separation of the executive from the *legislative branch*. Generally, in a *parliamentary system*, the government is dependent on the *parliament*, while in a presidential system, the government is in the hands of the President. In transition countries generally, presidential systems have prevailed in Africa, Latin America as well as in the CIS countries, while the parliamentary system prevails in West European countries.

The procedures of an electoral system are important since they may engineer desired outcomes, the influential think tank IDEA posits. An electoral system correspond to the rules and mechanisms by which votes are translated into seats in the legislature or presidency. Electoral systems also 'influence other aspects of the political system, (e.g., development of the party system)'. There are three tasks in any electoral system: 'to translate votes in elections into seats in a legislature, the conduit to hold the elected accountable and to shape incentives among the contestants for power to 'frame their appeals to the electorate in distinct ways' (IDEA 2002: 22). On the other hand, one of the basic axioms, that of a division of power between the three tasks, has not been readily accepted in the CIS. This is best seen in the fact that dramatic constitutional conflicts took place on the matter and on the conflicts emanating from a division of power between the president and parliament (Belarus, Ukraine, Moldova, Russia, Kazakhstan, Kyrgyzstan, Georgia and Armenia). The general explanation is probably that a division of power was never accepted by Communist regimes, and even public understanding of the point with a division of power was lacking. Violent struggles between president and parliament took place in Russia in October 1993 and later in Armenia, Belarus,

Moldova, Ukraine, Kazakhstan and Kyrgyzstan. In fact, '/t/he less democratic a country is, the stronger its presidential powers'; presidential rule promotes authoritarian rule (Åslund 2007: 226–27).

Any attempt to compare democratic rule and electoral democracy of different countries needs a set standard for comparison. This could be done with the help of an 'ideal model' of electoral democracy. This is not the approach taken here. There are a few other potential starting points. One is to turn to the norms of one international organization where both 'Old Europe' and Russia and the other CIS countries are members, the OSCE, and especially its Office for Democratic Institutions and Human Rights (ODIHR).[4] Another is to turn to handbooks for judging elections and democracy development, for measuring democratic development, and the International IDEA – the Institute for Democracy and Electoral Assistance. A third possibility is to start from the Russian Constitution itself, the letter of which the Russian political leadership has not tried to change. In the section that follows, an attempt is made to find a starting point with all three inroads.

Democracy, generally, is an integral part of the OSCE member states in their commitments that 'pluralistic democracy based on the rule of law is the only system of government suitable to guarantee human rights effectively'. The OSCE is described as a 'community of values' (OSCE 2009a). These commitments are politically rather than legally binding, they are 'political promises' and in this sense more than declarations of good will and intentions, they are consensually based and represent binding promises according to the 'universality principle' (OSCE 2009b). The conditions for democratic rule set by the OSCE ODIHR is relevant to the election norms under scrutiny here because of its observation of elections to 'enhance meaningful participatory democracy' and 'support in building up democratic institutions' in order to 'strengthen the rule of law, democratic governance, and civil society' (OSCE 2005:xxii). The OSCE is also concerned with the media in this context and has a Representative on Freedom of the Media, the purpose of whom is to further 'free, independent, and pluralistic media as one of the basic elements of a functioning pluralistic democracy' (OSCE 2005:xxiii).

A useful standard for *electoral democracy* upon which to lean in judgments about specific countries is the use of election observation handbooks. In the ODIHR terminology, election observation is 'one of the most transparent and methodical ways to promote and encourage democracy and human rights', covering 'all the elements necessary for a democratic electoral process' (OSCE 2009c). The election observer's list of tasks include: 'a review of the election-related legislation, following which it monitors: the registration of candidates and voters; the campaign; the coverage provided by both publicly and privately owned media; the work of the election administration; the handling and resolution of complaints and appeals, including the functioning of the judiciary; and, finally, the instalment in office of elected officials'. Further, on election day, among other things, short-term observers monitor the opening of polling stations to check whether ballot boxes 'are empty and properly sealed' and if there are ballots.

Then, 'observers monitor how voters are processed, whether they are accurately listed in the voter register, and whether they are able to vote in secrecy and free from intimidation. The vote count is an important part of the election process and is also observed' (OSCE 2009d).

Another source for, or for assessing the standards of electoral democracy in Russia today is to use the guidelines of the Institute for Democracy and Electoral Assistance (IDEA 2002).[5] The legal framework for elections includes the constitution, civil and criminal codes, nationality and citizenship laws, laws related to the media and such items (IDEA 2002: 9). The Constitution and the Electoral laws are the most important parts of the legal framework for elections, both of which are directly controlled by the Legislature, the main difference between the two being the difficulty with which amendments can me made (IDEA 2002: 12).[6]

An electoral checklist suggests that all seats in at least one chamber of the legislature are distributed by regular elections, that there are electoral formula for converting votes into legislative seats, that seats do not benefit one party at the expense of others, the length of term and minority conflicts (IDEA 2002: 25). In practice, this might include not depriving nominees of the possibility to get elected, especially if 'disproportionately large deposits are required for nomination' (IDEA 2002: 34). With respect to voter registration, international standards require that the voter register is 'comprehensive, inclusive, accurate and up to date, and the process … fully transparent' (IDEA 2002: 45).[7]

To create and run *political parties* is an important feature of electoral democracy.[8] Therefore, one thing which is needed is 'a structure for the registration of political parties', details of registration procedures and the necessary dates (IDEA 2002: 50). Furthermore, if support for a nominee requires the collection of signatures, a reasonable timeframe is needed. One or several of the following criteria are possible: monetary deposits to avoid frivolous candidates and a minimum number of validated signatures (IDEA 2002: 51). With respect to election campaigning, there should be 'no unreasonable restrictions on the right to freedom of expression', 'equitable access to the media, especially the electronic media', 'equitable access to resources' for campaigning, 'no party or candidate is favoured, financially or otherwise through the availability or use of state resources', no one 'threatens or does violence to another party or candidate', and one has to 'cease active campaigning one or two days prior to polling day' (IDEA 2002: 55–56). With respect to *media access*, the state should ensure that 'all political parties and candidates have access to the media and are treated equitably by media owned or controlled by the state' (IDEA 2002: 61, 62). With respect to *government interference*, the necessary guarantees include that 'political parties and candidates are given the necessary legal guarantees to enable them to compete with each other on a basis of equitable treatment before the law and by the authorities' and that 'no legal or administrative obstacle stands in the way of access to the media on a non-discriminatory basis for all political groupings and individuals wishing to participate in the electoral process' (IDEA 2002: 62). With respect to *campaign financing*, the state should ensure that 'all political parties and candidates are equitably treated by legal provisions governing campaign finances and expenditures'. The main forms of public funding consist of free broadcasting time, state payments and facilities, the use of government facilities, state grants, tax reliefs and credits. Of particular importance to our empirical case is the notion that the legal framework 'should ensure that state resources are not used or misused for campaign purposes by the party in power' (IDEA 2002: 65–67).

A third and very different approach to judge upon the status of the Russian electoral democracy is to start from the *Russian Constitution* itself. Section 1, Chapter 1 ('The Fundamentals of the Constitutional System'), Article 1, states that 'Russia shall be a democratic federal rule-of-law state with the republican form of government'. Article 2 states that 'man, his rights and freedoms shall be the supreme value. It shall be a duty of the state to recognize, respect and protect the rights and liberties of man and citizen'. Article 3, among other things, states that 'the referendum and free elections shall be the supreme direct manifestation of the power of the people' (Russian Constitution 1993). With respect to separation of powers, the same section and chapter, Article 10, states that '/s/tate power in the Russian Federation shall be exercised on the basis of the separation of the legislative, executive and judiciary branches. The bodies of legislative, executive and judiciary powers shall be independent'. Article 13 notes that '/n/o ideology may be instituted as a state-sponsored or mandatory ideology', that '/p/olitical plurality and the multi-party system shall be recognized' (Russian Constitution 1993). There is reason to return to these general paragraphs later.

In Chapter 2 of Section 1, the rights and liberties are dealt with. It is the President that is 'the guarantor of the Constitution of the Russian Federation, and of human and civil rights and freedoms' (Section 1, Chapter 4, Article 80). The importance of the Constitution with respect to *presidential elections* is low. It states only that the procedure 'shall be determined by federal law' (Section 1, Chapter 4, Article 82). Despite this, presidential powers are overwhelmingly many and strong. The second pillar of constitutional power is the *Federal Assembly*, dealt with in Section 1, Chapter 5: the Federation Council and the State Duma. The Federation Council should have two deputies from each of the 89 subjects and the State Duma 450 deputies. The State Duma shall be elected for a term of four years, while the procedures for forming the Federation Council and the procedure for electing deputies to the State Duma shall be established by federal law (Article 96).

So, what should we look for in passing judgements on the development of Russian electoral democracy? My choice rests on several criteria. The first thing to look for is the general voting behaviour of the electorate and the general context in which elections take place. A second criterion is the general nomination procedures and the restrictions and opportunities that have developed over time both in presidential and Duma elections. A third is the campaigning itself, how financial and administrative resources for election work have been distributed and especially access to important Russian mass media. A fourth is to evaluate the extent to which today's Russian political system corresponds to the general spirit and soul of Russia's own constitution, especially with respect to the proclaimed division of powers, since such an approach would be formally accepted also in the Kremlin.

Changes in voting participation and in election rules and election procedures

Changes over time in the voting environment

Voting participation varies greatly also in 'old' European democracies. In comparing parliamentary and presidential elections, voting participation in Russia is of some interest in the light of the old Soviet habits of achieving a 99 per cent voter participation. It reveals the extent to which Russians really react to the very possibility of taking part in elections which they can abstain from. The number of votes cast in parliamentary and presidential elections in Russia is not in itself strange, in fact they seem fairly normal – the fairly stable figures for parliamentary and presidential elections around 62–64 per cent and 69–70 per cent, respectively, point in the direction of a trained and adapted voter cadre.[9]

The most challenging issue to pinpoint lies not in election details as much as in the general political context of the Putin reign – Putin has indeed changed not only some rules of the game but also indeed the very context of the game. While he has made a point in not changing the Constitution, several election laws and information and media-related laws have been changed, albeit in a lawful manner. The result of this is that the very democratic framework is very different today as compared with ten years ago. The three last election cycles (1999/2000, 2003/2004 and 2007/2008) have all resulted in large majorities for the 'party of power' to the Duma and the Kremlin presidential candidate. Although there have been some cases of election fraud, the election results would not have been substantially different had the elections been altogether 'free' and even 'fair'. The Russian political landscape has changed to the extent that it is today best characterized as 'authoritarian' rather than 'democratic'. Under the slogan of 'strengthening the vertical', Putin has not only centralized and rebuilt a hierarchical political administrative system, he has virtually taken control of the Russian society at large; it is difficult to think of any important issue, national or regional or even local, political or economic, social or cultural, in which the not always visible Kremlin hand is involved one way or the other. The rule is certainly not totalitarian, but it is to a large extent controlled by a new 'nomenklatura system' in which former or current employees of power ministries and agencies entrusted by the Putin administration are acting on behalf of the state or elite groups.

Duma party and candidate election rules

There have been changes over the years with respect to the number of parties that eventually ended up in the *State Duma*. The most obvious change is quantitative: the number of parties in the Duma has diminished from the first Duma elections in 1993 – with its eight parties ending up in the Duma as a result of elections on the party list – to four parties in 1995, six in 1999, four again in the 2003 and 2007 elections (only from the party lists). Taken together, i.e. the number of representatives allied in parties and blocs in the actual Duma after the single-mandate districts are included, the change is even more obvious: from eight in 1993 (plus a large number of independents – 141), ten in 1995 (plus a somewhat lower number of independents – 77), seven in 1999, to four in 2003 and 2007. These changes, from a fairly rich fauna of parties in the earlier elections to a fairly limited number in later elections, are at least in part a result of changes in election rules, which, in turn, have been adopted precisely in order to limit the number of parties in the Duma as a result of election results.

The first of these measures for diminishing the number of parties in the Duma is obviously the level set for the *barrier*, i.e. the percentage of votes needed in order to get any seats at all in the Duma (from party list elections, up to 2007 for 225 of the 450 seats), established by laws taken before each upcoming Duma election. In the first 1993 Duma elections, the barrier was set to five per cent, which was kept up to the 2007 Duma elections when it was raised to seven per cent. Furthermore, the possibility of 'voting against all' that existed in 1995 was later abandoned.[10]

Another way to diminish the number of parties in the elections and in the Duma was introduced by the electoral laws in the 1999 Duma elections where several amendments to limit the number of parties were made. First, the Central Election Committee should verify 20 per cent of the necessary 200,000 signatures to register a party. Second, the CEC (Central Election Commission) should disqualify parties with more than 15 per cent invalid signatures. Third, the right of the CEC to disqualify candidates due to incorrect income information also reduced the number of contestants. Fourth, from the 2003 Duma elections, the parties and blocs had to gather 200,000 signatures in support of their party lists or pay a fee of 37 million roubles ($1.2 million at the time). Fifth, in the 2007 Duma elections, only nationally based parties were allowed to compete, which squeezed out regional parties and parties with small numbers in some regions. Sixth, in the 2007 elections, the single mandate districts were simply abandoned, which reinforced the difficulties for party registration. Parties thus had to submit 200,000 signatures from the entire country (not only from particular regions) supporting the list of candidates, or deposit 60 million roubles (appr. $2.3 million at the time), to be returned only if the party received at least four per cent of the votes. Needless to say, the four parties already in the Duma were exempted from these rules (after all, they were the ones who introduced the amended election laws).[11] In addition, in the 2003 elections, the CEC should disqualify candidates who did not submit a correct income and property statement, or had a criminal record, or foreign citizenship, or if one of the three candidates on the list withdrew from it. The election law also abolished the validation requirement for a 25 per cent participation in Duma elections and 50 per cent in presidential elections.

In Russia, Duma elections are important also because of their role for Presidential elections, especially if there are only few parties in the Duma elections. Furthermore, the Duma election of 1999 offered a better indication of the upcoming presidential election than the Duma election of 1995 because of the 'Putin factor' and the connection to the Unity party; the success of Unity in 1999 also foresaw the success of Putin in 2000.[12] In 2004 and 2008, as in 2000, the main battles in the presidential election had already been won with the preceding Duma elections; the tandem locomotives of a *party of power* and an acting president have been a success story. In 2004, the enormous (and unexpected) success of its follower party of power Unified Russia was certainly due in large part to Putin's open support of it, and therefore its success also predicted Putin's own success three months later. Similarly, Putin's support of the party ascertained Medvedev's victory in 2008.

Presidential election campaign rules

Presidential election rules are important, because they set the stage for the strategies of candidates as well as voters. There are detailed rules of campaigning, especially with respect to paid media time, the publication of opinion polls and the financing of the campaigns. The rules for presidential elections have also changed over the years. In 1996, one million signatures were necessary for being nominated. In 2000, only 500,000 signatures were required (because the elections had been advanced after Yeltsin early retirement). In 2004 and 2008, two million nominating signatures were necessary for nomination. Signatures were not needed for candidates from a Duma party – Duma parties have a general right to nominate a presidential candidate.[13]

The new election laws adopted in December 1999 are fairly close to international standards and sufficient to guarantee free and fair elections: the state should pay most of the campaign costs, the candidates may have an election staff of 600 (salaried by the Election Commission), candidates may offer small contributions to their campaigns and may accept smaller contributions from those who nominate them (from private sources and from judicial actors, but only such that have been in existence for at least a year prior to the elections). Candidates may not accept contributions from foreign citizens or companies, international organizations, state agencies, charities or religious organizations, and there are limits to the amount of money a candidate may spend in the campaign – 26 million roubles in the first round and 34 million roubles in total for the two rounds. Candidates who are in government office may not use government agencies, government employees or other government resources in their campaigns (Federal Law 1999).

The use of government resources in Presidential elections

The strong political power of the Russian president is obviously a great advantage in presidential elections. Both Yeltsin and Putin enjoyed the presidential privilege of travelling (the country and abroad) and of taking part in official life. They were highly visible, which in itself favoured their election efforts. They also had ample opportunity to choose their appearances and thereby get the attention of the media. In addition, both state-owned and privately owned supporting media helped to put the acting presidents in the very living rooms of most Russians almost every day. This not only neutralized the lack of a nationwide election organization of the sort available to the Communist Party, but also ultimately proved more important than party organization.

The advantage of already being at the presidential helm in elections was thus obviously to Yeltsin's and Putin's advantage in their campaigns in 1996, 2000 and 2004 (although for different reasons).[14] The new election laws in the 2000 presidential election were designed precisely to eliminate the most obvious forms of misuse of the presidential position (such as access to government funds and the presidential administration), and the media law debate in the autumn of 2003 at least shows that the issue was still alive. In practice, however, control of the media is more important than the money spent on campaigning (in media and elsewhere). The role of the exiting president in elections is also generally great, a very prominent feature in the Russian presidential elections in 2000, in which Yeltsin's choice of Putin as his heir gave Putin an enormous advantage, and in 2008 when Medvedev was virtually lifted into the high office by Putin.[15]

Other means of increasing support were also available to the president in office. Both Yeltsin (in 1996) and Putin (in 2004) needed some success in office. To Yeltsin, this was a difficult task since his government policies were highly unpopular. To Putin, government policies had not yet become a burden in 2000, and in 2004, the government reshuffle was aimed precisely at showing where he was heading without denying the relative success of his first presidential period. Both Yeltsin and Putin also used their decree power to enforce laws that increased their popularity, and both were able to make decisions related to the respective Chechnya wars – decisions that supported their campaigns (White 2000: 97). This influential role of the Russian acting president shows features similar to those in other presidential systems and is in itself not surprising or suspicious in a democracy.

The role of the media

The role of the media is regulated in the *election laws* before each election. They, too, have changed since 1993, particularly because of the obvious abuse of the media in the 1995/96 election cycle. The fact that Russian media became almost entirely politicised in the mid-nineties was a result of the media interest of the oligarchs. In the 1996 campaign, the oligarchs together with the state-owned media were united against the threat of communism and Yeltsin in all practical terms enjoyed a monopoly on the nationwide television broadcasts.

The role of the media was therefore much more regulated in the election laws in 1999. Media outlets are required to offer all registered candidates equal opportunities, in principle, to run their campaigns. Nevertheless, at the same time, radio and television channels as well as newspapers have the right to sell advertising time and space, although principally on equal conditions.[16] Furthermore, state owned media should offer candidates free advertising, divided equally between them.[17] Commercial media may decide how much time and space should be offered to the campaigns of different candidates, under the general condition that all candidates should be offered equal opportunities. One of the more general rules that have remained is that election campaigns may not be run in the media prior to the registration of a candidate. What has not been part of the media rules or norms in the election laws is the way in which presidential candidates have actually been depicted and especially in the news during campaign periods. In the 2000 campaign, the oligarchs sided with different candidates and 'smear campaigns' were mounted in several directions; there was no total monopoly. The Kremlin's grip on the media increased further in 2004 and 2008, to the obvious advantage of the sitting or exiting president; this time, it was hard to find media in opposition to the sitting president and the national television broadcasts had learned the lessons of Putin's fight against the 'politically interested' oligarchs Berezovsky, Guzinsky and Khodorkovsky.

More than any other election campaign tool, the *television media* strongly contributed to the success of Yeltsin in 1996 and of Putin in 2000 and 2004 and of Medvedev in 2008. In Russia, television is by far the most important opinion medium, and the official and unofficial presidential campaigns have basically been run in the three nationwide television channels ORT (Channel 1), RTR (Channel 2) and NTV, the first two of which reach approximately 98 per cent of Russian households.[18] Both Yeltsin and Putin made their way into the headlines of television news on very broad and general messages, on travelling extensively and on promising a better future, or simply by news reporting on the president's activities.

The strategies of Putin's media supporters in 2004 were almost the same as in 1999/2000 (although the smearing was much less frequent): the visibility of Putin and his government on national television was extremely high, while that of other candidates was very low – Putin simply dominated television news-casts (Yablokova 2004a; Yablokova 2004b).[19] The two presidential candidates Khakamada and Kharitonov even filed complaints with the CEC accusing federal media of violating election laws by giving too much attention to a meeting with Putin and the CEC acknowledged that 'certain state TV channels slightly overdid it'. The state controlled TV channels explained that they 'cannot stop covering the activities of the presidency, while Vladimir Putin has not stopped (fulfilling) his responsibilities as president' (RFE/RL 17 February 2004).[20] This conciliatory stand on behalf of national air media can only be explained by the great influence of the President and his administration in conjunction with the *siloviki* structure that has developed during Putin's first presidential period. In the 2008 presidential elections, the same strategy was used by Medvedev – to be seen rather than taking part in discussions.

With respect to media treatment of the favourite candidate, we may draw some conclusions. In the 2000 elections, the media pictured Putin's rivals (especially Primakov in the early stage of the 2000 campaign and later also Ziuganov) as historic relics from the Soviet era. In the 2004 election, the media made a point of showing Putin among the people, working for the people, as opposed to his contestants (who were working, assumedly, only to their own benefit). This pattern was also prevalent in the 2008 elections, where Medvedev was frequently flashed together with the outgoing president Putin.

This strategy of voter neglect would not have been successful if there had been really independent media capable of bringing critical issues to the agenda.

The role of the media as an independent force, after all, is the most important thing in election times, and neither the free and fair emanation of standpoints of the candidates, nor the equal opportunity of candidates in this respect was the hallmark of the current Russian political scene.[21] In 2004, the freedom of the press situation had significantly worsened (in comparison with 2000), and Russian national air media had begun to resemble Soviet-era media, at least in the sense that a tacit ban seemed to exist on criticism of the political leadership (RFE/RL 14 January 2004).[22] In 2008, the situation was even worse and no really independent national television channel was left, none that was independent from the Kremlin or from state-controlled companies.

While *paid TV advertising* was important in the 1996 election campaign, in the 2000, 2004 and 2008 campaigns, it was not. Not to run a campaign in itself was, naturally, a campaign strategy and Putin's refusal to participate in simple debates (instead of running the country) enhanced his image of a serious and responsible servant of the people, both in 2000 and 2004. Medvedev followed suit and used the same strategy. Support from Russian national television was decisive for the victories in all four presidential elections. In fact, in all of them (in 1996, 2000, 2004 and 2008), as well as in the preceding State Duma elections in 1995, 1999, 2003 and 2007, the state-owned broadcast television openly supported the sitting president and the parties of power. The support consisted mostly of extensive news coverage of the President, while at the same time neglecting or at times smearing the opponents of the Kremlin.

To the extent that there have been actual campaigning attempts (like in 1996), they have basically been run in the three nationwide television channels, the state-owned ORT (Channel One), the 51 per cent state owned RTR Rossiia (in 1999, 49 per cent owned by Berezovsky who fled Russia in 2001) and the Gazprom-owned NTV (which, up to 2001, belonged to Vladimir Guzinsky who also had to flee the country). The ORT and RTR sided with Yeltsin, Putin and Medvedev in 1996, 2000, 2004 and 2008, while NTV supported Yeltsin in 1996 (but Fatherland–All Russia in the 1999 Duma elections), and, hesitantly, Putin in 2000. When Berezovsky and Guzinsky had been chased out of Russia in the early Putin assault on the oligarchs, control also of the NTV was transferred to Gazprom and support for Putin and Medvedev became evident in 2004 and 2008. In addition, the largest cable TV channel Ren-TV is today owned by a private company, Severostal, which is close to the Kremlin. National television has thus been under Kremlin control ever since 2003.

There were many formal breaches of these media related laws, not only in 1996 but also in the elections of 1999/2000 and the following election cycles in 2003/4 and 2007/8, where rules were supposed to be different. The state-owned media have largely supported the existing Kremlin circle, which has been able to use nationwide and government-owned media to support its candidates. Generally, the ability of important media to act independently of owner interests has greatly diminished, both state owned and privately owned (powerful interest groups) and such media cannot be considered 'independent' – they are, rather, an integral part of politics than objective recorders of politics.

The role of the media, especially the broadcast media is extremely important in understanding the issue of 'fairness', since it has a decisive role in attracting votes in Russia. The unveiled partisanship of the national television media in the four presidential elections of 1996, 2000, 2004 and 2008 has clearly reduced confidence in the Russian media. Furthermore, the fact that Russian media, especially television, has been a major object in power struggles among interest groups in Russia suggests that Russian media are extremely important to contestants for the highest power positions in Russia.

Electoral democracy in Russia – conclusions

A short comparative analysis of presidential elections

There have been four presidential elections in post-Soviet Russia. In the first, in 1996, a very unpopular president was elected (Yeltsin), in the second and third, a very popular one was elected (Putin), and in the fourth, the popularity was transferred to the chosen heir by the exiting president. The first president was well known to Russians in 1996 when he was elected, the second basically unknown when he entered the scene in 1999/2000, later to become extremely popular, the third was known as a contender for power only for some very short time before he was elected. With respect to voter participation, there is not much to complain about – the figures have been fairly normal by European standards.

Before going into details, let us look at more general similarities and dissimilarities of the presidential elections. There are some *obvious similarities* between three of them. Firstly, in two of the elections (1996 and 2004), the acting president was running for office. In the second and fourth, a 'crony' had been secured by the exiting president (2000 and 2008). A second more implicit factor was the role of the two Chechnya wars in the first three elections, in shaping popular opinion about the acting president (negatively in the case of Yeltsin and positively in the case of Putin, especially in 2000). A third similarity was the escalating use of parties in the preceding State Duma elections to support the acting president in the upcoming presidential elections already from 1996. This connection between the Duma and presidential elections has today become firmly cemented and is similar to the French situation.

There are also several obvious *dissimilarities*. In the 1996 campaign, there were two rounds of elections while only one was needed in the 2000, 2004 and 2008 elections. Secondly, while there was an obvious favourite in the 2000, 2004 and 2008 elections, the 1996 election results were very uncertain. Thirdly, the formal election campaigns were very different; in the 1996 election, there actually *was* a campaign among several of the contestants, while in the 2000, 2004 and 2008 elections, it was difficult even to detect one. Fourthly, both in 1996 and 2000, the Communist Party leader Ziuganov was the main contestant to the acting president, while there was no obvious 'opposite' in 2004 and 2008. Fifthly, in 1996 (and to some small degree also in 2000), an ideological struggle between two general ideological platforms (on continued reform policies and on communism) was explicit. The 2004 and 2008 elections differed in this respect too, since the Communist Party was no longer able to recruit new voters.

The more worrying feature of this general development of election cycles over time is that they reveal something very fundamental. For example, while the 1996 presidential campaign was a hair-raising thrill all through to the very end, neither the 2000 nor the 2004 or 2008 official campaign showed much of a campaign at all. The official campaign in 1996 did play a decisive role for the election result, but in 2000, 2004 and 2008 the very little campaigning there was did not seem to play any role at all. This indicates that electoral democracy in Russia has developed into a direction not at all familiar to European or Western election standards. What is wrong with Russian elections, then? Is Russia not an electoral democracy today?

General conclusions on electoral democracy in Russia – from hesitant democracy via managed democracy to authoritarian democracy?

The most important conclusions with respect to election democracy in Russia may very well be, first, that the principle of democratic elections to *choose* Duma representatives and the president has been firmly established, and second, that whatever there is of election campaigning does mirror the issues which seem important to the larger parts of the Russian electorate, largely because they are consensual issues. Third, the elections show the more general moods in society in pointing to what features are held important – the closing of the 'communism' chapter in Russian history in the 1996 and 2000 elections, and economic and political stability in 2004 and 2008. Fourth, from the 2000 election a definite generation shift with respect both to the electorate and to the style of leadership is evident. Fifth, while Russia is changing, the question is *how much*, and whether it is all for the best: while the institution of elections seems to be well established, the way in which they are being used (or misused) certainly should be of concern to any holder of (European) democratic values. The established *managed, controlled, captured, oligarchic* or even *authoritarian* democracy may well turn out to be a lasting stage of the Russian democratic development: after all – elections were also held during Soviet times.[23] Such notions of a managed, controlled or authoritarian democracy are today very strong in descriptions of the Russian political system.[24] Even a cursory look at Russian elections today indicates that the Russian electoral system has gone astray. Some go quite far in their judgements: Lukin suggests that 'Russia's second experiment with free elections is now over' (Lukin 2004b) (the first was the short interval in the summer of 1917). I will return to the general development of democracy in the following section but first summarize some of the conclusions of the preceding pages.

The *general development of the nomination procedures* and the restrictions and opportunities that have developed over time both in presidential and Duma elections tell us part of the story of the ailing Russian electoral democracy. The most obvious trait in the development of electoral democracy over the years is the fact that political parties have become much fewer, both in the Duma and

in society at large. This is not simply a result of raising the threshold for Duma elections, although this tool, too, aims precisely to eliminate small parties.[25] More importantly, the difficulties of creating and running parties, and to be offered a real chance of ever reaching the State Duma, have its origin in the very Duma itself, in the regulations of elections through election laws adopted by parties already established in the Duma.

One of the obvious ways in which the Putin administration has made it more difficult for political parties to function in Russia is to restrict the *registration of parties*.[26] Before 1999, 100 registered members were all that was needed (Freedom House 1998: 423). In 2001, a new law raised that number of signatures to 10,000. In addition, parties should be found in all of the federation and have at least 100 members in each of the 89 regions of the country (Federal Law 2001). In 2006, the membership requirement was raised to 50,000 members. Obviously, this is a serious limit which shuts out small and/or regional parties. Prior to the December 2007 Duma elections, the barrier was raised to seven per cent (all in due order, i.e. by the Duma in which Unified Russia enjoyed total domination). To form election blocs among small parties was also forbidden at this time (Federal Law 2005). Furthermore, the voter option of 'against all' disappeared from the ballot, an option which obviously had been used previously to express discontent (Freedom House 2008).

Since Putin's coming to power, the Kremlin has reformed elections to consolidate presidential control over the State Duma with the help of the *party of power*, Unified Russia. The creation of a mass party to support the president (not in itself an invention by Putin) took a decisive turn after he had been elected president in 2000. The party opposing Putin and initially (just after the December 1999 Duma elections) supporting his contestants – Fatherland – All Russia – was merged with the Unified Russia party (Hale 2004). After the December 2007 Duma elections, Unified Russia fully controls the Duma. There is today a definite risk that this total dominance and its strong connections to the Russian presidency as such could ultimately lead to a collapse into authoritarianism (Dawisha and Deets 2006: 691). In addition, the youth organization tied to Unified Russia also attracts new voters at the same time as it makes it more difficult for other youth organizations to be effectively formed.[27] Another phenomenon in the young Russian democratic development was the invention of '*bogus parties*'. From the beginning (i.e. the early 1990s), a great many political parties as well as individual candidates turned up, in some cases seemingly from nowhere, to confuse the public. Some of these had extremely few members, lacked policies and were funded by non-transparent sources, some of which emanated from the old nomenklatura remaining from the Gorbachev period 'who quickly changed their coats, but stayed in place – the supposedly enfeebled KGB – and the new entrepreneur class' (Council for the National Interest 2006). In addition, funding itself from political parties and election campaigns is an interesting field of study, where the Kremlin administration has had a say on almost every aspect of elections.[28] They use government money 'to help particular candidates, parties, and pressure groups' (Council for the National Interest 2006).

In this context, the development of access to important Russian mass media is of particular importance. As seen in the foregoing section, this particular Russian development is the most troublesome. The fact that important media were controlled by a few oligarchs in the mid 1990s was one of the basic reasons why Putin set out to crush their influence soon after being elected. The fight against the oligarchs began in 2000 with Berezovsky and Guzinsky and continued with Khodorkovsky in the summer of 2003. Today, the media is much less of an opposition force than ever since Soviet days. All the important mass media, especially national television, is fully controlled by the Kremlin. Furthermore, in later years, there have also been several laws that limit the freedom of the media: the Law on Terrorism, the Law on Information, the Law on Information Technology and the Law on Mass Media (Konnander 2008: 51).

When it comes to actually controlling elections in Russia, the Russian *Central Election Commission*, which is responsible for the campaign procedures and that the elections are run according to the election laws, can also disqualify individual candidates and declare election results invalid. Its powers are thus large. The CEC does not control the system for reporting election results, however: this lies with the Kremlin administration (Konnander 2008: 65). The independence of the CEC was questioned in the elections of 2007 when Putin appointed a friend of his as its chairman, but, in general, the CEC does not need to be involved in election fraud.

With respect to the *general development of democracy*, judgements could be harsh: Russians 'never had a real chance to grasp, and hold on to, their new democracy long enough before the process of stealing their freedoms took off, and by now has snuffed out all hope, all optimism, on the part of ordinary Russians, and revived and magnified that deep corrosive cynicism which was there, just under the surface, during the time of Communism' (Council for the National Interest 2006). Russia was deprived of the true meaning of democratic life. On the outside, everything might seem fine. Already in 2000, Putin announced 'the dictatorship of the law' as a reaction to the lawlessness of the 1990s. Putin and Medvedev are legalists and they advocate adherence to the law. However, the problem is that, apart from the general one of there being a lack of separation of powers, laws are themselves vague and their application is often arbitrary. This makes the regime unpredictable and rather un-constitutional (van Bladel 2008: 36). It is, in fact, hard to see the difference between the real and the fake because of this (Shevtsova 2008: 37, cited in van Bladel 2008: 36).[29] According to Lukin, there are three explanations for today's lack of democracy in Russia. The first is that free elections have been seen as a means, not a goal in its own right. The second has to do with the lack of actual separation of powers, which in Russia leads to a situation where 'elections quickly descend into farce' (since election commissions and courts are controlled by the executive). Thirdly, the electorate has changed its mood with respect to elections and the 'disillusionment with the policies of the democrats and the obvious rigging of elections' has made the electorate 'sceptical and cynical' (Lukin 2004b). Lilia Shevtsova also strikes a pessimistic tone when she says that 2004 saw the 'end of Russia's liberal democratic experience', or '/the/end of period of spontaneous development in Russia' (Shevtsova 2004: 8). The lack of a clear division of power between the executive, lawmakers and the judiciary is, trivial as it may seem, the greatest deficit of Russian democracy today. All this runs contrary to Putin's own description of the situation.[30]

If we try to evaluate the extent to which today's Russian political system corresponds to the general soul and spirit of Russia's own constitution, especially with respect to the proclaimed division of powers, we do that because we understand that the Russian electoral democracy is a derivate of its general democracy development. First of all, there was nothing wrong about the democracy aspirations which once (1993) were very clear in Russia.[31] Several researchers note that today, the initial euphoria of democratization has been overtaken by 'a more sobering view' and that many Central and East European political regimes 'will remain in the grey zone between liberal democracy and outright authoritarianism' (Carothers 2002). Today, the dividing line is between 'a minimal electoral democracy and an electoral autocracy' defined by the 'quality of their electoral processes' (Hartlyn *et al.* 2008: 74). In fact, *authoritarianism* is a common catchword for today's Russian political system. Ekiert (*et al.*) refers to the Russian political system of today as 'competitive authoritarianism', and the quest for democracy may very well end up in authoritarianism (Ekiert *et al.* 2008: 11,13). Åslund also makes similar judgements.[32] Many equally strange words are used to characterize today's Russian political system: 'pseudo', 'imitation', 'illiberal', 'guided', 'managed', or 'democratura' or 'soft orientation regime'. All these epithets are used to hide the fact that Russia today is an 'authoritarian regime' (van Bladel 2008: 34). Putin's regime is a 'counter-democracy'. Since authoritarian regimes differ from totalitarian regimes, Russia also lacks a totalistic ideology and tight monitoring of interpersonal contacts and cannot be said to be a totalitarian state (van Bladel 2008: 34, 40, 41).

In order to explain what an authoritarian system actually is, one could construct a continuum where competitive democracy is on one side and a totalitarian regime on the other, and where an authoritarian regime is a 'hybrid system' somewhere in between. Putin's regime today compares with the ideal authoritarian model developed by Linz (2000), where three features of an authoritarian regime are particularly interesting: the degree of pluralism, the formation of coalitions and the prevailing mentality. An authoritarian regime is one 'with limited, not responsible, political pluralism, without elaborated and guided ideology, but with distinctive mentalities, without extensive or intensive political mobilization, ... and in which a leader ... exercises power within formally ill-defined limits but actually quite predicable ones' (Linz, cited in van Bladel 2008: 34). '/T/he most distinctive feature of an authoritarian regime is the existence of limited pluralism'. While in a democracy, there is almost unlimited pluralism and in a totalitarian monism or one party, in an authoritarian regime there are 'a limited number of independent groups or institutions active in society and politics' (van Bladel 2008: 35).

The future of electoral democracy in Russia – the concept of Sovereign Democracy

Any predictions regarding future developments of electoral democracy in Russia would wisely be based on the notion of *sovereign democracy* promoted by the Kremlin administration since 2006, especially since it is also suggested as a key to understanding tensions in Russian relations with the European Union. This notion is a fundamental element of the Putin era, which 'embodies Russia's ideological ambition to provide an alternative Europe' (Krastev 2007). Indeed, in Russia, sovereign democracy has become the official ideological catchword for the present Russian political system.

The meaning of sovereign democracy has been debated among intellectuals, but the general notions seem quite simple: sovereignty is a precondition for democracy, which makes sense in a state centred world. The concept is ideologically potent and originated in the aftermath of the Ukrainian 'Orange revolution' where democracy was 'defended against hostile foreign meddling' (meaning the West, Europe and the United States) (Little 2008). Sovereign democracy is not a goal but a means to realize the Russian national idea, to reach elite consensus on the basic rules of the political and economic rules to replace the struggles among different elites. It is thus instrumental, 'it has mobilization objectives' (Okara 2007; Konnander 2008: 28, 31). The notion of sovereign democracy suggests that Russia has its own way, which is different from 'real democracy': sovereign democracy 'protects the state and those in power', meaning in short that 'We'll do it our way' (Goble 2008; Amsterdam 2006).[33]

So, finally, what is the status of Russian electoral democracy? Is it fair to talk of Russia as an electoral democracy at all? The short answer is 'no', since there are simply too many deviations from any resemblances of a 'Western' model for electoral democracy. Furthermore, these deviations are of a nature that let us question the use of 'democracy' at all with respect to the Russia of today. The main reasons for the latter stem from the inconsistencies between the Russian Constitution itself and the realities of Russian politics today, it is also precisely because of this, a deviation from any European-like model for 'free and fair' elections. The formality with which the Russian political leadership today points at the existence of elections as such and at the fact that there is more than one party or one presidential candidate is somewhat beside the point. After all, there were elections also in the Soviet Union and people did vote and the extension of the ballot to include several 'not so real alternatives' to the party or president is but a technicality. The description by Lucas of the present Russian political system is to the point: 'Opposition parties are allowed to exist, albeit on the fringes of the political system. However, they cannot demonstrate easily. They have no access to the media. In a free, law-governed country, the executive power is checked and balanced from all sides: by elected representatives, by the media, by public organizations, and by the judiciary. All these – almost everything that could constrain the power of the Kremlin – are broken or co-opted. So, too, are the most fundamental political rights: free speech and free association of individuals. These are guaranteed by the Russian constitution; they flourished during the 1990s. But under Putin, they have shrivelled' (Lucas 2008: 58). In comparison to the West, says Lucas, '/f/or all the abuses and sleaze that disfigure the advanced industrialized countries of the West, honest judges and due process are the ultimate backstop, upholding the law and transcending politics. Not in Russia' (Lucas 2008: 72). Moreover, it is precisely this lack of 'checks and balances' that is the difference between real and fake democracy and real and fake electoral democracy: there should be a real alternative political force and there should be more than one centre of power. Otherwise, democracy is but a sorry excuse for authoritarianism and possibly also for totalitarianism.

Notes

1 Putin used the 'dictatorship of law' to create order out of the chaos of the Yeltsin era (Putin 8 July 2000).
2 In other CIS countries, countermeasures against colour revolutions were even more hash, especially in Uzbekistan, Belarus and Azerbaijan, but tough countermeasures were taken also in Kazakhstan and Tajikistan. In the majority of post-communist states, the democratization process lost momentum and either resulted in partially democratic systems or in authoritarian regimes. Today, there are three categories of regimes in the CIS region: democratic, semi-democratic and autocratic regimes (Ekiert *et al.* 2008: 7, 11).
3 In fact, democracy means much more than elections, political parties and individual rights, a stable electoral system with institutionalized parties and alterations in power (Rose-Ackerman 2007: 31–32). Nevertheless, even concentrating on elections only, multiple dimensions are needed to establish the quality of any elections, and there is a need to consider both 'pre-election day, election day, post-election day issues'. One obvious dimension is 'fair and technically sound' elections and that voter lists and vote counting could be trusted, another that 'electoral campaigns should not be unduly biased in favour of particular parties and candidates'. In general, however, the focus tends to be on the former (Hartlyn *et al.* 2008: 76–77).
4 The ODIHR was founded in 1992 but originally established in 1990 as the Office for Free Elections.
5 The IDEA Standards provide the general framework and basis for internationally recognized electoral standards and emanates from several declarations and conventions, including the 1948 Universal Declaration of Human Rights and the 1990 Document of the Copenhagen Meeting of the Conference on the Human Dimension of the Conference for Security and Co-operation in Europe (CSCE). Such declarations and conventions have an especially evident role once a country is a signatory to such documents.
6 The Constitution comprises only the fundamental electoral rights, such as the formal rights to vote and be elected, institutions to be elected and their terms of office (IDEA 2002: 14). Particularly important are 'the laws governing media, registration of political parties, citizenship, national registers, identity documents, campaign finance and (relevant) criminal provisions' (IDEA 2002: 16).
7 Transparency means that voter registers are public documents (IDEA 2002: 46).
8 The 'right of all individuals and groups to establish, in full freedom, their own political parties or other political organizations with legal guarantees to enable them to compete with each other on a basis of equitable treatment before the law' (IDEA 2002: 49–50).
9 The number of votes cast in the five parliamentary (State Duma) elections since the demise of the USSR is: 55 per cent voted in 1993, 64 per cent voted in 1995, 62 per cent

voted in 1999, 56 per cent voted in 2003 and 63 per cent voted in 2007 (of the total number of voters). The number of votes cast in the presidential election is: 70 per cent in 1996, first round, 69 per cent in 1996, second round, 69 per cent in 2000, 64 per cent in 2004 and 70 per cent in 2008. Generally speaking, voter turnout is higher in presidential elections than in parliamentary elections. Participation in the first (not so) democratic elections to the Supreme Soviet in 1989 was 90 per cent, and in comparison with this figure, later Duma elections draw the attention of much fewer voters.

10 The 'vote against all' category was also used in presidential elections in 1996 (both in the first round and in the second, where only two candidates are named), in 2000, in 2004, but not in 2008.

11 The Duma changed the election law in May 2005, and substituted the combined party elections and single mandate elections into a proportional party system (Federal Law 2005).

12 When Putin had expressed his support for Unity in the parliamentary elections, Unity managed to become the second-largest party. When Primakov's party lost, it was difficult for him to run credibly for president a few months later (and he withdrew).

13 This has had some comic effects. For one thing, since Putin has not been a member of a Duma party, he has had to collect signatures. At the same time, the Communist Party and the LDP have generally presented their candidates without any problem – sometimes their own party leader, sometimes someone else.

14 It is also true, of course, that Duma members have great advantages in presidential elections in comparison with non-members (Remington 1999: 170).

15 Yeltsin's decision to resign prematurely only added to this advantage. This was the predominant judgment by political commentators in Russia. See, for example, *Nezavisimaya gazeta, Segodnya* and *Izvestiya*, the early January editions, 2000. From the interview book with Putin, released on the Net only a week before the March 2000 elections, it seems that Putin himself was informed about Yeltsin's decision to resign only about a week before the resignation (Putin 2000a).

16 There are limits, however: state-owned television and radio channels cannot sell more than two hours of paid advertising per day, and state-owned newspapers cannot sell more than ten per cent of their space.

17 Articles 47, 49 and 50 of the Federal Law (1999) on Election of the President of the Russian Federation.

18 ORT is fully owned by the state, and RTR is 51 per cent state-owned (while the rest was owned by Boris Berezovsky, who was closely allied to Yeltsin in 1996 and less of an ally to Putin in 2000). NTV, the only privately owned channel was controlled by Media Most, owned by another media tycoon, Vladimir Guzinsky, in 2000. In 1996 and 2000, ORT and RTR sided with Yeltsin and Putin, respectively, and NTV supported Yeltsin in 1996 and Fatherland – All Russia (Otechestvo–Vsya Rossiya) in 2000. In 2004, Berezovsky and Guzinsky had been 'neutralized' and all three national channels supported Putin. The government and the media empires of the oligarchs also largely control the print media. Until 2000, Berezovsky controlled the company that published *Kommersant*, and Media Most oligarch Guzinsky controlled *Segodnya* until it was closed. Several of the Russian newspapers have also supported political parties, for example *Pravda* (which supported the leftist parties), and *Izvestiya* (which supported rightist parties). Newspapers had, however, largely lost their importance for public opinion in Russia already by the mid 1990s, and the larger newspapers are distributed only in the large cities (White *et al.* 1997: 250).

19 The OCSE report referred to an investigation of the time given to candidates between 12 and 26 February on Channel One: 2 hours and 38 minutes to Putin, and 22 minutes to all the others. The report also said that the fact that Putin did not participate in debates was not countered by critical mass media, so the net effect was all in Putin's favour (Moscow Times 2004a).

20 Putin's 12 February meeting with his election agents, where ORT and RTR aired Putin's 29-minute speech live (RFE/RL 17 and 23 February and 2 March 2004). (For an analysis of the speech, see McGregor 2004).

21 The Communist Party made an investigation into the airing of information/analytical TV programmes from 12 February to 10 March in all federal and leading regional networks. The findings showed that Putin was shown 1,584 times, compared with 275 times for former Union of Rightist Forces co-leader Irina Khakamada, 264 times for Duma Deputy Sergey Glazev, 242 times for Kharitonov, and 182 times for Liberal Democratic Party of Russia (LDPR) candidate Oleg Malyshkin. The OSCE evaluated the reports in terms of positive and negative coverage (see RFE/RL Russia Votes). Furthermore, only on NTV and TV-Tsentr did Putin receive any coverage that could be characterized as negative in tone (RFE/RL 12 March 2004. See also 'Top Channels', Moscow Times 2004b).

22 Even with respect to the advertising by the candidates, there were complaints. For example, Glazev accused state-controlled ORT and RTR of broadcasting only a portion of his campaign advertisements, and the CEC agreed with Glazev. The RTR explained to the CEC that they had technical difficulties (*RFE/RL Newsline* 11 March 2004). Rybkin complained that all of Russia's four leading television networks refused to air his campaign videos sent from London during free or paid airtime (RFE/RL 1 March 2004).

23 As Åslund notes in his majestic work on the development of democracy and market reform in the post-Communist space, several of today's authoritarian CIS states started off as semi-democratic states, but they are today captured or oligarchic (Åslund 2007: 214).

24 'Electoral autocracy' under Yeltsin was exchanged for 'bureaucratic authoritarianism' under Putin, and she has characterized Russia as a 'consolidated bureaucracy' (Shevtsova 2004: 8). Alexander Lukin calls the Russian political system today a 'clan democracy' (Lukin, 2004a: 11). Peter Rutland suggests that Russia, rather, is a custarchy, the backbone of which are men in 'epaulettes' from the KGB/FSB (Rutland 2004). Ekiert *et al.* (2008: 11, 13) refer to the same system as 'competetive authoritarianism, which may very well end up with authoritarianism'.

25 The tool has been a general trait in all post-Soviet states and it has been a tool that has promoted centrism (Dawisha and Deets 2006: 700,703).

26 The background to this paragraph has been derived from http://www.heritage.org/research/worldwidefreedom/bg2088.cfm

27 Nashi – the quasi nationalist youth group tied to the Putin regime and with an ideology close to that of Unified Russia, is also based on strong nationalist feelings (Hanson 2008: 77). Nashi was created in 2005 and has rather typical paramilitary elements and is based on the writings of Vladislav Surkov, Putin's own political advisor. The basic objective was to oppose the creation of groupings like those involved in colour revolutions.

28 'They can sell you a party programme, create a new party, sell you bogus opinion polls and exit polls, get them placed prominently in the press – and, most important, sell prime television time' (Council for the National Interest 2006).

29 Another analyst claims that Putin's Russia constitutes 'the gravest setback for democracy in the post-Cold War era', that the 'essence of Putinism is simply the accumulation of power by the leader and his associates' (Puddington 2008).

30 Putin's own comments about democracy sound almost like statements taken out of a textbook on democracy: 'I think a lot about how life in Russia should develop after 2008 and in the longer term and I see no instrument capable of stabilizing the country other than democracy and a multiparty system. We cannot build Russia's future by tying its many millions of citizens to just one person or group of people. We will not be able to build anything lasting unless we put in place a real and effectively functioning

multiparty system and develop a civil society that will protect society and the state from mistakes and wrong actions on the part of those in power. There is no other road we can take and there is no question of inventing some kind of homegrown local-style democracy' (Putin 14 September 2007). On the other hand, in the spring of 2008 Putin said that the West would have no influence over elections (Putin 30 January 2008).

31 In Russia, as elsewhere in the CIS, the new parliamentary system was designed to promote reform parties although it did in fact produce a legislature filled with opponents. The election rules themselves were designed to help parties of their own reformist orientation (Dawisha and Deets 2006: 698).

32 More specifically, 'the dominant strain in Russian political thought throughout history has been a conservatism that insisted on strong, centralized authority, unrestrained either by law or by parliament' (Richard Pipes 2005: 1, cited in Åslund 2007: 237).

33 Putin and Medvedev themselves rarely use 'sovereign democracy' in their speeches, but on one of the very few occasions when Putin commented on the term sovereign democracy, he actually more or less denied its usefulness: 'I think that sovereign democracy is a debatable term. It creates some kind of confusion. Sovereignty is something that refers to the quality of our relations with the outside world, while democracy refers to our inner state, the inner substance of our society'. However, he also noted that the discussion on the term was useful, although he himself did not want to take part in it (Putin 14 September 2007). Medvedev has also criticized the notion of sovereign democracy, pointing out in an interview for 'Ekspert' that sovereignty and democracy are different conceptual categories and therefore difficult to fuse: 'If you take the word 'democracy' and start attaching qualifiers to it, that would seem a little odd. It would lead one to think that we're talking about some other, non-traditional, type of democracy' (Medvedev 24 July 2006).

4 The theory and practice of reciprocity in EU–Russia relations

Tatiana Romanova

Introduction

Reciprocity has recently become the buzzword in EU–Russia relations, already full of contradictions, ambiguities and mutual suspicion. The term 'reciprocity' is frequently employed by both Russian and EU politicians and officials. Representatives on both sides accuse each other of violations of this fundamental principle. Mutual irritation is additionally nurtured by the current conceptual crisis in EU–Russia relations, which is due to the lack of vision on the part of both Russia and the EU about how to construct their future relations.

A close examination of Moscow's and Brussels's statements and their mutual critique, however, reveals that Russia and the EU understand reciprocity in different ways, and, hence, presuppose different consequences. Therefore, there is little wonder that the expectations of both sides fail to come true.

Reciprocity, as such, is one of the most difficult, ambiguous and obscure terms in international relations. In fact, it is not an objective category, nor is it as transparent as it initially seems. Already in 1986, Robert Keohane rightly argued that this term 'appears in so many different literatures. Each school of thought defines reciprocity in accordance with its own theoretical purposes, with little regard for its other definitions and little comprehension of the conceptual progress that other disciplines may have made' (Keohane 1986). The multiplicity of this literature, however, serves us to distinguish different facets of reciprocity and to identify the core of the confrontation about this concept in today's EU–Russia relations.

Current political science writing identifies five dichotomies of reciprocity (specific vs. diffuse; simultaneous vs. sequential; restrictive vs. open; negative vs. positive; and homomorphic vs. heteromorphic). This typology is instrumental in the analysis of various facets of reciprocity. It appears that in most of the cases the EU and Russia, when approaching the concept of reciprocity, converge in their understanding of reciprocity in the last three dichotomies (negative vs. positive; restrictive vs. open; homomorphic vs. heteromorphic), but drastically diverge when it comes to the first two (simultaneous vs. sequential; and specific vs. diffuse).

This lack of a shared understanding prevents fruitful discussions on the energy cooperation, the construction of a common economic space, details of a new basic

EU–Russia agreement, and a regime for foreign investments, to give just a few examples. The EU's and Russia's understanding of reciprocity are conditioned by their past and present experience, their notion of the world as well as by their self-perception and relative strengths and weaknesses. However, given the fact that reciprocity is not an objective and a cast-in-iron category, there are several windows of opportunity that can be explored to make the EU and Russia converge in their understanding of reciprocity.

Hence, the present chapter starts with an outline of the concept of reciprocity, divergent approaches to it as well as five dichotomies that describe different facets of reciprocity. Then I go through some recent examples in EU–Russia relations where both parties applied this concept, but clearly understood it in different ways. Finally, I compare the findings about the EU and Russian position to see whether something can be done to improve the understanding between the parties and to help them overcome the current dead end in their relations.

Theorizing reciprocity

The word 'reciprocity' comes from the Latin word 'reciprocus', which means alternating. A decomposition of this word into syllables helps to identify two prefixes 're' (i.e. back) and 'pro' (i.e. forward). Thus, the notion of a backward and forward movement, of mutuality and circular dependence underlies the original meaning (Carter 2002). It also follows that both sides are the cause and the result of each other. Similarly, in international relations theory (IR theory) it has been mentioned that reciprocity implies 'interdependence and the wish to maintain this interdependence' (Rhodes 1989: 279–80).

In IR theory, R. Keohane has so far provided the deepest analysis of the reciprocity phenomenon. He defined it as 'exchanges of roughly equivalent values in which the actions of each party are contingent on the prior actions of the others in such a way that good is returned for good, and bad for bad' (Keohane 1986: 8). In designing this definition, he identified two fundamental features of reciprocity for IR theory. They are contingency and equivalence among either actors or the concessions that they make to each other.

Taking up the problem of equivalence, Keohane rightly argued that by far not all international relations are reciprocal, because there is very little equality in today's world and among today's actors. The necessity of equality in reciprocity is also stressed by other IR theorists (see, for example, Wight 1977: 135; Byers n.a.) as well as by specialists in international private law (see, for example, Lunts 1984: 75). The latter state that reciprocity is a specific case of sovereign equality among the states. Equivalence can be provided either between the relevant actors or between their benefits (Keohane 1986). The equality among the relevant actors is by definition difficult in today's relations, given the wide discrepancy among the states, their economic standing, political strength and military capabilities. Therefore, today's international relations are more prone to the equality of benefits. However, a deep and all-embracing reciprocity would require equality between the actors in question.

Another fundamental feature of reciprocity identified by Keohane, is contingency, which could be narrowed down to the readiness to reciprocate (Keohane 1986). Two observations are important in this regard. First, the question is who sets the framework and conditions for reciprocating, and who decides what reciprocity is and whether the taken measures are sufficient and balance the efforts of the other side. Second, it is natural for every human being and every social entity to overestimate their own efforts and to underestimate the behaviour of others. Clearly, there are no objective criteria to evaluate it, and, hence, the process is quite biased (Keohane 1986: 10).

It then follows that equivalence and contingency in reciprocity raise multiple questions. More importantly, however, this discussion demonstrates that reciprocity is not an objective category and is open to competing interpretations.

J. Goldstein, in his recent book on the principles of IR theory (Goldstein 2008: 4–9), classified reciprocity as one of the ways of solving the 'collective goods problem', which itself occurs because each nation is sovereign. In the situation of missing central authority, J.Goldstein sees three 'principles' for agreeing on some basic rules of the game that are common for most of the social sciences. One is dominance, when a state or a group of states forces the others to contribute to the common good. The second principle is reciprocity, which presupposes 'rewarding behaviour that contributes to the group and punishing behaviour that pursues self-interest at the expense of the group' (Goldstein 2008: 4–9). Goldstein goes on to define reciprocity as a 'linchpin of cooperation' in IR. The third and final principle for solving the problem of the collective good is identity. Contrary to the dominance and reciprocity solutions, which focus on the self-interest of the players, the identity principle presupposes that participants 'care about the interests of others in the community enough to sacrifice their own interests to benefit others' (Goldstein 2008: 4–9).

Thus, J. Goldstein posits reciprocity as a solution that is more moderate and acceptable than the dominance but at the same time hinges on self-interest. It also converges with the idea of R. Keohane, who observed that reciprocity 'seems to be the most effective strategy for maintaining cooperation among egoists' (Keohane 1984: 214).

The concept of reciprocity emerged in international trade, which historically was the first form of external economic relations. Initially, reciprocity was applied to the tonnage of ships and goods that was allowed from each signatory to the other, as well as to the conditions for access to markets. However, already in the eighteenth century, reciprocity was used to provide 'equivalent treatment of foreign nationals ... Since the latter part of the nineteenth century, the meaning of reciprocity has been enlarged to include equal access to raw materials, the protection of foreign investments, aviation overflights and landing rights, treatment of tourists, and a host of financial matters involving such items as exchange controls and debt payments' (US Foreign Policy Encyclopedia 2009). Thus, the sphere of application of reciprocity has been consistently enlarged in line with the evolution of international economic relations.

Later, reciprocity entered the realities of political relations. According to the US Foreign Policy Encyclopedia, reciprocity is a fundamental characteristic of diplomatic negotiations; it is 'a process of exchange between nations, a negotiating tool whereby nations bargain with each other for equivalent treatment' (US Foreign Policy Encyclopedia 2009). The encyclopedia goes on to stress that reciprocity could be understood as a combination of diplomacy and an idealized concept; in other words, it is both an instrument for achieving a certain degree of cooperation among sovereign states and a set of beliefs that underlies this instrument.

Historically, reciprocity in IR evolved from being solely an instrument to becoming a ruling concept. Initially, it was used to denote a 'short-term variation in foreign policy behavior' of two superpowers (the Soviet Union and the United States) but was later classified as a norm, which 'establishes an underlying level of expected behavior which serves to determine the pattern of foreign policy interactions among contemporary superpowers' (see Rajmaira and Ward 1990). Interestingly, some authors argue that the more the countries interact the further they move from a short-term engagement, a reaction, to the establishment of long-term conditions for cooperation.

Along the same lines, some concepts of international private law define reciprocity as a link between two countries (in the form of some agreements), on the basis of which they treat each other's laws and judgments, as well as a principle of actions (*jus cogens*). Here, one can also clearly see an instrument of cooperation and a set of beliefs that underlie this principle. Sociologists also draw a distinction 'between (1) reciprocity as a pattern of mutually contingent exchange of gratifications, (2) the existential or folk belief in reciprocity, and (3) the generalized moral norm of reciprocity' (Gouldner 1960). Here, the difference between an act of reciprocating and reciprocity as a set of norms can also be discerned.

This leads us to the first, and, probably, the most important dichotomy in reciprocity. Keohane defined it as a specific vs. diffuse reciprocity. Specific reciprocity describes 'situations in which specified partners exchange items of equivalent value in a strictly delimited sequence. If any obligations exist, they are clearly specified in terms of rights and duties of particular actors' (Keohane 1986: 4). In contrast, diffuse reciprocity means that 'one's partners may be viewed as a group rather than as particular actors, and the sequence of events is less narrowly bounded' (Keohane 1986: 4). In this latter version, obligations and common and accepted standards of behaviour are of the highest importance.

Keohane further pinpoints that 'specific reciprocity requires bilateral balancing between particular actors; diffuse reciprocity emphasizes an overall balance within a group' (Keohane 1986: 7). Thus, it follows that in specific reciprocity actors do not perceive themselves as members of the same group. Rather, they emphasize differences and cooperate despite this divergence. In the case of diffuse reciprocity, partners accentuate their similarity and equality, and the notion of common identity is of the utmost importance for them. They cooperate because of this common identity, a shared set of rules and a sense of community. Naturally, cases of specific reciprocity are much less predictable than those of diffuse reciprocity. Specific reciprocity is also based on self-interest, whereas the diffuse one hinges on the sense of obligation to the group.

The divergence between specific and diffuse reciprocity can be rendered even more vivid by drawing on some literature from social anthropology. In 1965, anthropologist M. Sahlins conducted a study of gift exchanges and identified three types of reciprocity. According to this work, generalized reciprocity refers to the situations when one does not expect an immediate return to their gifts. However, the level of trust is so high that they know that they can rely on this person in the future. Balanced reciprocity describes the situations when one expects to get something in return immediately. A purchase in a supermarket is a good example of balanced reciprocity: it is a situation where one gets something and immediately pays for it. Finally, there are instances of negative reciprocity, which take place when one tries to acquire a valuable item for a small price. This could be done through cheating, the use of force or bargaining in unequal circumstances (Sahlins 1972: 194). Negative reciprocity cannot, however, be practised on a long-term and consistent basis within one society, because it will eventually disrupt it.

Interestingly, Sahlins concludes that generalized reciprocity mostly occurs among members of one family or close communities – it is a continuous chain of gives and takes. Balanced reciprocity characterizes relations within one tribe, whereas negative reciprocity refers to the relations between different tribes (See Figure 4.1).

There are numerous parallels between the specific reciprocity, defined by Keohane, and balanced reciprocity as described by Sahlins. Similarly, diffuse reciprocity almost resembles generalized reciprocity in Sahlins typology of the exchange of gifts. It means that diffuse reciprocity requires a strong sense of identity, and of shared community as well as certain commonality of views. It would also require a certain degree of equality between the actors in question. On the other hand, specific reciprocity can exist between different communities, and, as the notion of balanced reciprocity implies, it is characterized by equivalence of benefits in every given moment. It seems that terms of economic exchange are more applicable in this situation (Blau 1964: 93–97).

This brings us to the second dichotomy in reciprocity – the differentiation between simultaneous and sequential types. Simultaneous reciprocity presupposes that all parties involved exchange benefits at the same time, whereas sequential reciprocity allows for a time lag in this exchange of benefits among the parties. Sequential reciprocity requires more trust than simultaneous reciprocity, as the delay for one of the sides in the acquisition of benefits can be significant. Keohane rightly argued that 'sequential reciprocity promotes long-term coopera-tion much more effectively than simultaneous exchange does. Conversely, when a simultaneous exchange takes place, it often reflects a breakdown of confidence' (Keohane 1986: 22).

The difference between simultaneous and sequential reciprocity is of impor-tance for the first dichotomy in reciprocity (into diffuse and specific types).

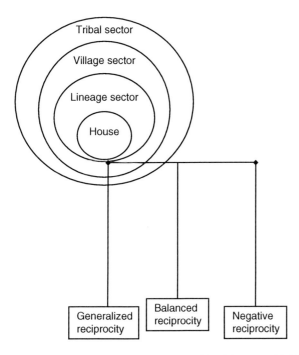

Figure 4.1 Reciprocity and Kinship residential sectors (*Sahlins 1972: 194*).

Clearly, diffuse reciprocity is characterized by more sequentiality, whereas specific (balanced) reciprocity is mostly about simultaneous exchanges of gifts/ favours/benefits/evils.

According to the *Britannica Concise Encyclopedia*, reciprocity means 'granting of mutual concessions on tariffs, quotas, or other commercial restrictions' in international trade (*Britannica Concise* 2009). The encyclopedia goes on to stress that reciprocity 'implies that these concessions are neither intended nor expected to be generalized to other countries with which the contracting parties have commercial treaties' (*Britannica Concise* 2009). Thus, the exclusivity of the club is important in this definition as well as the fact that reciprocity is not universal and relies on the reciprocal actions of the parties involved in a specific deal.

Similarly, a law dictionary defines reciprocity as 'a relationship between persons, states, or countries whereby favors or privileges granted by one are returned by the other' (*Law Dictionary* 2003). The dictionary goes on to specify that reciprocity 'does not involve a vested right that would exist without it', meaning that it creates an additional value that would not otherwise exist. Thus, the give-and-take idea is emphasized in numerous definitions.

At this point, it is essential to stress that reciprocity can characterize both the relations between two (or more) states and 'a systemic pattern of action' (Keohane 1986). In other words, it could either be limited to a number of participants or could be turned into a systemic pattern of cooperation among all the states, provided they all agree to reciprocate.

This leads us to the third dichotomy in reciprocity that is also stressed in the US Foreign Policy Encyclopedia: restrictive and open reciprocity. The former means 'agreement between two countries and can involve privileges (or different types of treatment) that are denied to other parties, or that must be specifically bargained for by third parties' (US Foreign Policy Encyclopedia 2009). The latter can be either bilateral or multilateral and presupposes that 'concessions granted to one nation are automatically extended to all others that have signed most-favored-nation agreements with the granting nation' (*Britannica Concise* 2009). Restrictive reciprocity is most frequently a feature of a protectionist regime, while open reciprocity characterizes a liberal system. Contemporary international relations evolved in such a way that reciprocity was initially practiced in its restricted form and then actors gradually moved to open reciprocity.

Furthermore, most IR (and other social sciences) specialists agree that reciprocity can be positive and negative.[1] In other words, one can do good for good or bad for bad ('An eye for an eye, a tooth for a tooth', as the Old Testament states). This thought is also implied in the quotation from Goldstein mentioned earlier. In the case of positive reciprocity, the interdependence between the actors will be positive (i.e. the general direction of global trade talks), in the case of the negative one, we will witness a negative interdependence (like an arms race, for example). Both positive and negative reciprocity promote cooperation. However, the positive one ('a carrot') seems to be more constructive and rewarding, as well as not threatening the sovereignty of the players in question, whereas the negative one ('a stick') is alienating, propelling a conflictual mode of international relations and is frequently perceived as undermining the sovereignty of states. The negative reciprocity can lead to 'a downward spiral as each side punishes what it believes to be negative acts by the other' (Goldstein 2008: 4–9).

Last but not least, reciprocity can be homomorphic or heteromorphic (Gouldner 1960: 170). In the first case, one partner answers to the concessions of the other with the very same steps. In the case of heteromorphic reciprocity, two sides exchange different benefits. The homomorphic version is frequently a characteristic of restricted reciprocity, whereas the more open reciprocity becomes more heteromorphic. Moreover, homomorphic reciprocity often characterizes cooperation between similar or identical actors, whereas the heteromorphic one is a feature of interrelations among divergent actors and is more frequent if the actors practise diffuse reciprocity.

To sum up, reciprocity is about circular movement and interdependence among the actors. Therefore, reciprocity is characterized by contingency. Equivalence between actors or their benefits is another fundamental feature of reciprocity. Given today's international relations, it is much easier to get equivalence of benefits than equivalence of actors, which significantly restrains the practice of reciprocity. Second, reciprocity is a process of exchange, which could be understood in two ways: as an exchange of benefits or evils per se and as a fundamental principle of cooperation. This leads us to the most important dichotomy in

Table 4.1 Typology of reciprocity

Specific	Diffuse
Simultaneous	Sequential
Restrictive	Open
Negative	Positive
Homomorphic	Heteromorphic

reciprocity – specific (*quid pro quo*) vs. diffuse (where the reward does not follow immediately and in a precise way). Third, four more dichotomies of reciprocity have been detected in the literature, which have shed light on other aspects of this phenomenon (simultaneous vs. sequential; restrictive vs. open; positive vs. negative; and homomorphic vs. heteromorphic). These findings are summarized in Table 4.1.

Clearly, the types, which are located in the left-hand column, characterize more distanced and even hostile interactions; they also denote states which are drastically different (with a possible exception of the pair homomorphic–heteromorphic). A closer interaction among actors who are alike would be characterized by the types of reciprocity located in the right-hand column. However, the 'left' and 'right' types are not mutually exclusive. Rather, they should be imagined as different poles of the five continuums with states' positions on reciprocity being located in different parts of these continuums (see Figure 4.2). Moreover, the positions on the continuum are not given once and for all. Rather, they should be imagined as being dynamic. The experience of cooperation can shift the exercised reciprocity to both the right-hand and left-hand column poles.[2]

Having clarified the term of reciprocity and its different facets, let us turn to some recent cases in EU–Russia relations and apply the outlined theoretical framework for reciprocity to them.

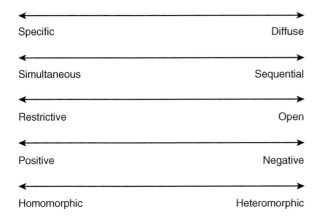

Figure 4.2 Typology of reciprocity.

EU–Russia relations: some recent discussions on reciprocity

The recent history of EU–Russia relations provides several examples of the difference in how Moscow and Brussels interpret reciprocity. I will, firstly, take up the case of energy cooperation, where the nature of cooperation has been the bone of contention since the early 1990s. I will then discuss how divergence in the EU's and Russia's interpretations of reciprocity manifested itself in their wider economic relations (at the time of negotiations on the common economic space, during the exchange of views on the new EU–Russia basic agreement, and, last but not least, in the context of a recent discussion on investments).

Since the early 1990s, the EU has been trying to ensure that its companies get access to the exploration of Russia's oil and natural gas resources. Therefore, it has asked for stable and predictable conditions in the energy sector of Russia (and in Russian economy, in general). Initially, the EU wanted Russia to ratify the Energy Charter Treaty, which contains provisions for cooperation in the field of energy, modelled in the early 1990s after the EU's *acquis communautaire* (Romanova 2007). Through reforming Russia's energy sector on the basis of the provisions of the Energy Charter Treaty, the EU tried to establish compatible market conditions in the EU and Russia throughout the energy sector. When the ratification of the Energy Charter Treaty came to a stall in the State Duma, the EU and Russia initiated an energy dialogue. It has been conducted since 2000 and devoted a significant amount of effort to the improvement of the investment climate. Notably, the European Commission insisted on the improvement of the investment climate in energy through legal stability (including the grandfather clause) and market reforms in Russia, designed after the EU's *acquis communautaire*. The Commission expected the eventual establishment of a Pan-European energy market.

Thus, the EU's strategy could be summarized as an effort to level the playing field, and to establish stable and predictable conditions for all companies operating throughout the EU and Russian energy market, for their investments in exploration, production, transportation and distribution. The EU and Russian players would then have a choice of investing in Russia and the EU. Their decision would be made on the basis of the market consideration, whereas the role of the public authorities would be limited to providing a good legislative framework.

Similarly, the EU has argued that synchronization of the EU's and Russia's electricity networks that would unleash an increased flow of Russian electricity to the EU, can only occur provided Russia has the same market opening conditions and energy saving and safety provisions. This strategy was additionally justified by the fact that, otherwise, Russian electricity would dump the one that is produced in the EU. The European Commission's Director General for Energy and Transport, F. Lamoureux, famously argued in 2003 that the problem of reciprocity had to be sorted out before electricity synchronization took place (Finasovye izvestiia 2003).

In fact, the EU has been arguing for the diffuse, sequential and, mostly, heteromorphic reciprocity. It promoted the general conditions that would allow energy companies to decide on specific deals without much public involvement and the exchange of simultaneous assets. The EU's approach has also been mostly about positive reciprocity: it is looking to reward the compliance of a partner (in this particular case, Russia) with clearly defined rules and benefits. However, Brussels stresses that this reciprocity is indeed a restrictive one: it is provided only to those who guarantee similar treatment to EU companies.

This EU view on reciprocity was made even more acute in the so-called third energy liberalization package, presented by the European Commission in September 2007 (EU Commission 2007a). In the clause, which was immediately branded 'Gazprom's clause', the European Commission argued that foreign investments into the EU's energy sector should be enabled by a special agreement between the EU and a third country in question. This agreement was meant to guarantee that the market conditions in the third country were roughly similar to those of the EU, and that the EU's companies had access to the market of that country on the basis of decent market conditions. In turn, the EU's partner should try to limit investments coming from the EU – its companies could be constrained in their activities in the EU's market.

Russia initially did its best to just honour its commitments and supply obligations and conducted the discussion on the possible reform of its energy sector. However, little progress has been achieved in this reform. At the turn of the millennium, Russia started defining its own vision of reciprocity. One of the first illustrations was provided by the negotiations between Gazprom and two German energy companies (E.ON and Wintershall) about the North European Gas Pipeline (currently Nord Stream) project and the development of the Yuzhno–Russkoye gas field. The focus of these discussions was on the swapping of assets, and Russia was ready to offer a share in Russian natural gas deposits in exchange for access to the EU's distribution networks, which Russian players believed to be the most profitable in the gas supply chain.

Yet another indication was the decision on the exploration of the Shtokman gas condensate field. Discussions around this gas deposit in the Barents Sea, with probable reserves of 3.7 trillion cubic metres, went on for more than 10 years. The latest stage of the debate started in September 2005 when a shortlist of candidates for a consortium to develop the field was announced.[3] Gazprom was in dire need for a partner because Shtokman required access to advanced technology, great financial resources and a retail guarantee. Nevertheless, the Russian gas monopoly dragged its feet on the establishment of the consortium. On 9 October 2006, Gazprom announced that it did not need any partner to develop the Shtokman field, because none of the candidates offered a good stake in exchange for the share in Shtokman. The agreement about the consortium was eventually found in 2007 when French Total and Norwegian Statoil and Norsk Hydro became Gazprom's partners offering Gazprom access to their assets.

It is noteworthy that then Russian President V. Putin underlined in 2007 that the EU is the party that has a problem with reciprocity because Russian investments in the EU are ten times smaller than the EU's investments in Russia (Putin 26 October 2007). Drawing on this, V. Khristenko, who until very recently was the top Russian official responsible for the EU–Russia energy dialogue, stressed on

17 October 2007 that Russia is against the 'simplistic' efforts to reach reciprocity, which is about Russia's ratification of the Energy Charter Treaty. He claimed, instead, that the volumes of Russian investments in the EU and the EU's capital flows to Russia are drastically different (55 bn. Euros of the EU's investments in Russia compared with 8 bn. Euros invested by Russian companies in the EU) (Lobjakas 2007). Hence, reciprocity will take place when the levels of investments are more or less the same.

Thus, attempting to define its version of reciprocity Russia has been increasingly talking about assets swapping and specific deals. The underlying idea, as it looks, is that of fair profit distribution between the EU and Russian market players (because the perception is that most of the profit is gained in distribution, while the largest share of investments is made in exploration and production of oil and gas). However, doing so, Russia, in fact, insists on specific, simultaneous and mostly heteromorphic reciprocity. Like the EU's version of reciprocity, the Russian one is mostly positive: it is ready to reward the concessions of the partner. Lastly, Russian reciprocity is also restrictive, i.e. it is only provided to the partner, who agrees to very specific deals, in particular, to the swapping of assets.

A recent interim report on the energy dialogue gives a good illustration of an effort to integrate the EU's and Russia's versions on reciprocity. It reads that 'mutual participation in the assets of energy companies plays an important role in ensuring secure energy supplies. Fair and reciprocal access to markets will contribute to such a development' (EU Commission 2006a:2). It demonstrates both Russia's preference for the swapping of assets and clearly defined benefits for Russian and EU companies (i.e. specific and simultaneous reciprocity), and EU's interest in the mutual opening of the markets and in creating stable, transparent and equal market conditions (i.e. diffuse and sequential reciprocity). It is also an excellent illustration of the efforts on both sides to polish a mutually acceptable formula instead of converging on the essence of reciprocity.

K. Barysch put it even more crudely, saying that 'In fact, the EU and Russia mean different things when they talk about reciprocity. For Europeans, reciprocity means a mutually agreed legal framework that facilitates two-way energy investment. For Russia, reciprocity means top-level talks to identify assets of similar market value, and then swap these assets. Since the EU is not making headway on the Energy Charter Treaty, while Gazprom has been acquiring more and more downstream assets in Europe, it looks like the Russian idea of reciprocity is prevailing at present' (Barysch 2007). Again, in terms of the present theoretical discussion, the difference is between specific and simultaneous reciprocity, on the part of Russia, and a diffuse and sequential one, on the part of the EU.

These divergent interpretations of reciprocity, which emerged in the field of energy cooperation, eventually spilled over to the wider EU–Russia economic cooperation. In 2003–5, Russia and the EU debated the possibility to create a common economic space. EU's representatives argued for regulatory approximation between Russia and the European Union (i.e. for Russia bringing its legislation closer to the EU's along the lines of article 55 of the Partnership and Cooperation Agreement), and for the gradual establishment of the four freedoms.

Olli Rehn, current Commissioner for Economic and Monetary Affairs, summed up this position by saying that the,

> ... ultimate goal of the Common Economic Space is to create an open and integrated market between the EU and Russia, to promote trade, investment and the competitiveness of our economic operators. However, in order to promote economic integration, it is not enough to liberalize trade. The essential efforts must be geared towards the promotion of compatible regulatory frameworks and the proper enforcement of rules.
>
> (Rehn 2004)

Thus, in the EU's view, the Common Economic Space was about the gradual liberalization of trade and about the regulatory convergence in *all* sectors of the economy. In other words, we witness here another illustration of the diffuse and sequential reciprocity: the essence is to remove the distortion in legislation and to allow the market players to reap the benefits.

Russia's chief negotiator, V. Khristenko, on the other hand, stressed that the work should be based on two pillars, which are soft legal harmonization and deep economic cooperation in *some* clearly defined spheres. According to his visions, full harmonization of Russian and EU economic and legal systems was the issue in the long-term perspective. Therefore, in the short-term, cooperation should take place in separate sectors, prepared for intensive cooperation and integration. He did not exclude harmonization of the most fundamental economic norms (property rights, contract law, competition, non-discrimination, stability of tax law, transparency, etc.); however, in all other relations, regulatory convergence would take place gradually under the influence of the practical needs in integration processes (Khristenko 2004). Khristenko also specifically insisted on the fact that energy had to be treated separately from other spheres of cooperation.

Hence, Russia again underlined its preference for specific and simultaneous reciprocity. Not being competitive with the EU in most of the sectors, the Russian side preferred to insulate most of its economy, so as to deprive the EU's companies from taking advantage and driving Russian competitors out of the market.

The multiplicity of sectoral dialogues launched in recent years proves that the parties are striving to find a balance between a general approach and specific asset swaps in different sectors (i.e. between diffuse and specific reciprocity). Currently, the parties are, with varying success, holding dialogues on investments, intellectual property rights, transports, state procurement, environmental protection, competition, macroeconomics, financial policy, telecommunications and information society, industrial policy, entrepreneurial activities, agriculture and regional cooperation.

The discussion on the renewal of the legal basis for EU–Russia relations provides us with another example of the divergence in their interpretation of reciprocity. In 2005, the EU and Russia started their preliminary discussions in view of the fact that the initial period of validity of the 1994 Partnership and Cooperation Agreement was coming to the end. In this exchange of views, the

EU has been arguing for a long and detailed agreement, which would include all possible spheres of economy as well as political and security cooperation. This vision is also outlined in its mandate for the negotiations with Russia (EU Commission 2006b). Russian representatives of the Ministry of Foreign Affairs,[4] on the other hand, insist that the agreement should be short and concise; that it should concentrate on the basic principles of cooperation and its institutional framework with sectoral cooperation being left to separate agreements (or protocols).

Again, we are witnessing a wish on the part of the European Union to decide on the level-playing field in economic cooperation as a whole (with particular attention to energy as a prime area of cooperation between the EU and Russia), while Russia insists on the specific treatment of each sector that would let it guarantee equality (in terms of benefits). Hence, Russia and the EU again agree that reciprocity should be restrictive and positive, but differ on whether it should be specific and simultaneous (Moscow's vision) or diffuse and sequential (the EU's preferences).

Finally, the most recent EU–Russia argument, where reciprocity is involved, touches upon investments. Russia has been trying to foreclose some sectors of its economy (or at least some companies) from foreign investments (including those from the EU). The law on strategic sectors, which eventually entered into force in 2008 (Federal Law 2008), defines 42 spheres as strategic (including nuclear energy, aviation, exploration of natural resources as well as some mass media). According to this new legal act, all foreign investments, which lead to the acquisition of 50 per cent stakes in an enterprise from a strategic sphere, should be approved by the state. The threshold for approval is decreased to 25 per cent in case the shares are purchased by a state or an international organisation, and to 5 per cent in the oil and gas sector. This law means that the state can examine all the deals involving strategic assets and propose the best solutions for them. Thus, a more legalized form for Russian-style reciprocity (specific and sequential) has been developed.

The EU, on the other hand, gave a more official form to its principle of reciprocity. In 2007, the European Commission introduced a communication on globalization, saying that the EU's partners, who want their enterprises to have access to the EU's markets, have to establish rules similar to those that are adopted within the EU (EU Commission 2007b). Hence, the essence of the EU's reciprocity roughly boils down to legal approximation. Moreover, this principle foresees sanctions in the form of restrictions in their access to the EU's market in case third countries refuse to guarantee a certain level of legal approximation. The European Council, which took place in December 2007, confirmed this idea by adopting the Declaration on Globalisation (European Council 2007). In particular, it says that the 'European Union will press for increasingly open markets which should lead to reciprocal benefits' (EU Declaration 2007).

Furthermore, the EU initiated the discussion about the activities of the so-called sovereign funds in the EU. Both the European Commission and the European Council stressed the necessity to guarantee transparency of ownership structure and interests of these funds as well as independence of their management. In February 2008, the European Commission came up with the proposal of a common approach in this sphere (EU Commission 2008a). These discussions can potentially create a leverage to limit Russian investments in the EU if the EU's views on reciprocity are not taken into account.

Thus, the EU perseveres in its reading of reciprocity, which should be diffuse and sequential. This latest example (internal discussions on the limits to be placed on foreign investments) deserves a closer look. Both the EU and Russia clearly shift here to a more negative version of reciprocity compared with past experience. Both Moscow and Brussels stress their ability to punish the other side by preventing relevant investments rather than underlining that 'good' interpretation of reciprocity will be rewarded.

The EU and Russia decided on 16 October 2007 to establish a specific expert group to deal with the problem of reciprocity and its divergent interpretations. The group was established within the framework of the EU–Russia investment dialogue, and consists of politicians, governmental officials, business representatives and academia. The results of the deliberations of this group are unknown at present.

The four outlined examples allow me to conclude that both Moscow and Brussels view reciprocity as a way to solve the problem of the collective good and to improve their relations.

However, it is clear from the discussions about energy cooperation, common economic space, new treaty and investments, outlined earlier, that the disagreement originates from the profound difference in how the EU and Russia interpret two dimensions of reciprocity, the dichotomies of specific-diffuse and of simultaneous-sequential. To visualize the difference, the EU and Russian visions have been marked on the five reciprocity continuums (see Figure 4.3).

The recent discussions in Russia and the EU demonstrate that both Moscow and Brussels converge in their reading of at least three facets of reciprocity. Moscow and Brussels look for a rather heteromorphic reciprocity. More importantly, both are ready to reward the behaviour of the partner that matches their vision of what the relations should be and punish the one that goes against their vision (i.e. they have a roughly similar vision of the positive–negative dichotomy of reciprocity). In addition to this, none of the parties are prepared to sacrifice their interests and to grant concessions unilaterally, which means that they share the vision on the dichotomy restrictive–open reciprocity as well. While this is a positive achievement in conceptual terms, in practice, it means that the parties can enter a downward spiral unless they find a compromise about the specific–diffuse and simultaneous–sequential reciprocity.

The next part explores the reasons for the EU's and Russia's visions and looks at whether their difference can be bridged in the short-term and long-term perspectives, thus avoiding the downward spiral. In doing this, I will mainly look at the two dichotomies of reciprocity where the positions of Moscow and Brussels are manifestly different (i.e. specific vs. diffuse; and simultaneous vs. sequential).

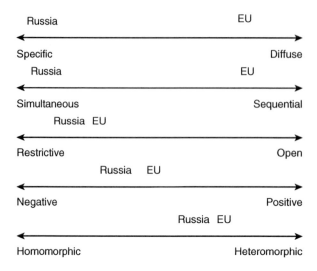

Figure 4.3 Different types of reciprocity in EU–Russian relations.[1]

[1] The location of each political entity represents my own assessment. Its main purpose is to visualize the difference between the EU and Russia, and their position in respect to the two poles of each dichotomy and to each other.

Explaining the difference, looking for solutions

Why do the parties diverge so widely on two dichotomies of reciprocity (specific–diffuse, and simultaneous–sequential)?

First, there is a fundamental difference in how the EU and Russia understand the world and their position in it. It has become conventional to argue that the EU lives in a postmodern world and views the situation in the world arena through Kantian lenses of eventual perpetual peace (Kagan 2003). It is natural for the EU, therefore, to support the widest possible cooperation on the basis of shared values. The idea of an eventual *community* of states that share the same values underlies both the process of European integration and the EU's policy towards many third countries. It is also oriented towards constructing transnational relations, which make participating states increasingly integrated. The EU's preference for diffuse and sequential types of reciprocity is, therefore, a natural result of its own specificity, its past experience and ideology.

It should be stressed, however, that the EU's approach to reciprocity cannot be classified as 100 per cent diffuse. It is definitely closer to the diffuse pole but difficult negotiations on institutional changes, on financial perspectives and some other issues of the EU's internal development supply multiple examples of interest-based interaction. These relations are driven by a specific reciprocity rather than a diffuse one. This tendency is even more pronounced in the EU's relations with third countries. However, the point that is of primary importance for us is that today's EU is more predisposed to the diffuse and sequential reciprocity than Russia.

Russia, on the other hand, views the international arena in a very modern way. Its favourite concept is that of the balance of power, which is reminiscent of the international relations in the nineteenth–early twentieth century. Russia views itself as a pole in the multi-polar world, characterized by interest-based interaction and the necessity to balance the other poles to ensure that none of them becomes too strong. Shared values and beliefs are of secondary importance. This concept is further reinforced by Russia's deep-entrenched respect for sovereignty and its suspicion of transnational interaction, which leads to the socialization and eventual construction of the feeling of a shared community. This set of beliefs naturally predisposes Russia to specific and simultaneous reciprocity, to short-term actions and gains rather than long-term engagement.

If one here remembers the meaning which social anthropology attaches to different types of reciprocity, an interesting conclusion emerges: the EU treats Russia as if it were a part of the same tribe (community), while Russia explicitly sets the boundary, insisting that partners come from different tribes and this should be reflected in their interaction.

Second, the approach of each of the partners is predetermined by their strengths and self-perception: '*market-based and rules-based in the EU; state-controlled in Russia*' (Barysch 2007). P. Sutela rightly argues that the way the EU's and Russia's approach to reciprocity reflects the difference between 'liberal and authoritarian capitalisms' (Sutela 2007). 'For the EU, the more market-oriented partner, it is primarily a matter of commonly agreed access to markets and investments. After rules have been established, let the best competitor win. For Russia, the more state-oriented partner, it is primarily a matter of asset swaps assumedly of similar value' (Sutela 2007). However, it deserves some further precision.

The logic of the European integration is that of gradual legal approximation. Initially, the EU applied it only to its members. More recently, the approach was broadened to neighbouring countries and, to a much less extent, to the developing states. Hence, it is not surprising that legal approximation (or levelling the playing field) is the basis for its approach to reciprocity. Furthermore, inherent in the process of European integration is the elimination of any discrimination of economic players coming from different member states. Thus, the EU tries to guarantee them equality of opportunities, limiting the interference of public authorities to this function, and letting private actors take a lead.

However, there is no guarantee that the legislation that has been designed in the EU and for its member states is by definition good for all partners of the European Union. When drafting the legislation, the EU institutions and its member states took into account their specificity and needs but not the (potential) concerns of third countries. It is, therefore, at least partly hypocritical to argue that this legislation is by definition good for any third party and to insist on its adoption as a condition for deeper relations. An effort to impose this legislation on other countries, in fact, can be viewed as neo-imperialism[5] and legislative expansionism.

As a result, the EU's approach to reciprocity can also be criticized from the neo-Marxist positions as an effort to impose its legislation and hence deprive the periphery of any independence (in terms of the legislative development).

Russia, on the other hand, has always been characterized by the tradition of state intervention in all spheres of the economy. Today we witness a certain deprivatization of some sectors (not only that of oil and natural gas but also of the automobile industry, banking, etc.), and, at the same time, a growing assertiveness of the public authorities. The point of departure for most of these activities is that the state is the key-driver, the subject of modernization (see Pain and Volkogonova 2008), as opposed to civil society. The weakness of the latter[6] gives the state a *carte blanche* in pursuing this attitude. It is further strengthened by today's Russian lack of belief in self-organization and in the ability of the private business to respect the interests of society if left uncontrolled. This attitude contrasts vividly with the EU's deeply entrenched belief in the self-organization of the society and in the good behaviour of the business world, provided the essential legislation is in place.

The Russian position can be criticized by drawing on neo-liberal thinking. Moreover, the danger of the Russian attitude is over-reliance on bureaucracy, which is a very fertile ground for corruption and overall lack of effectiveness.

Third, diffuse reciprocity in EU–Russia economic relations is hampered by the inherent inequality of the two sides. Keohane stressed that one fundamental of reciprocity is the equivalence in either benefits or actors. Equivalence in benefits is much easier but it provides grounds for specific and simultaneous reciprocity rather than a diffuse and sequential one. The latter requires equivalence in actors. However, this is precisely what EU–Russia relations currently lack.

The EU argues for the level playing field. However, Brussels suggests that its own legal norms and rules are taken as the basis for this process. This notion was present already in the 1994 Partnership and Cooperation Agreement between the EU and Russia. Its article 55 stated that 'the Parties recognize that an important condition for strengthening the economic links between Russia and the Community is the approximation of legislation. Russia shall endeavour to ensure that its legislation will be gradually made compatible with that of the Community' (EU/RF 1994). This idea was later reaffirmed in other documents.

In this situation, Russia becomes a dependent partner. The very fact that Russia has to take on board this legislation, which was designed without its participation, is alien to the great power ideas of Russia and to its search for equality. Moreover, unlike the acceding countries, it will never take part in the further development of the EU's legislation, nor will it be able to reap all the benefits of the EU's membership. Thus, legal approximation on the EU's terms makes diffuse reciprocity highly improbable in EU–Russia relations, because it makes the situation of inequality cast in iron. Russia's insistence on the specific (equality in volumes of investments, sector-by-sector approach in common economic space and new agreements), and simultaneous reciprocity become the only remedy in these circumstances.

In this light, Russia's insistence on equality in designing new legislation for EU–Russia energy and other economic relations, on mutual legal convergence (as opposed to just copying the EU's norms in case of legal approximation) gets a completely different meaning. If developed, it can be an avenue to a more diffuse and sequential reciprocity.

To sum up, the difference in how the EU and Russia see themselves and the world around them, the specificity of the actors, and insufficient equality and equivalence leads to the situation when Brussels and Moscow understand reciprocity differently while insisting that they speak about the same thing.

Is there any remedy to the present conceptual divergence? The answer is at least threefold. In the first place, the two sides should take all possible measures to prevent the situation when mutual disappointment produces a spiral of negative reciprocity. On several occasions, the EU, misunderstanding Russian stances, threatened to adopt a similar (or even stricter) behaviour. In February 2004, Commission spokesperson for external affairs Diego de Ojeda argued that the EU should learn from Russia how to link different issues and achieve them 'on a basis of reciprocity' (Lobjakas 2004). In January 2007, A. Merkel stressed that the EU can limit Russian investments should Russia arbitrarily treat the EU's capital flows (Russian Taxpayer's Union 2007). Recent debates about investments in strategic sectors and the possibility for the EU to limit activities of foreign enterprises on its territory give yet another example of the potential growth of negative reciprocity.

However, such an approach would only further complicate EU–Russia relations. Rather, reaction, which is based on positive reciprocity, would be much more constructive in this situation. It is worth remembering at this instance that economists have traditionally argued that throwing stones in your own harbour when your neighbours do the same, would not lead to the flourishing of your economy.

Furthermore, EU–Russia relations are characterized by the overall lack of trust, which is driven by some 'in Russia who think that the European Union and the West tried to exploit the transition period of relative weakness when the Soviet Union was being unwound and the modern Russia rebuilt. Also, at the other end of the telescope there are those in Western Europe who feel Russia was heading in a broadly European direction, but now has chosen a different path, and, furthermore, believes it can live without integration in the global economy' (Mandelson 2008). The resulting politicization and securitization of the relations could already be discerned on both sides. The spiral of negative reciprocity would only complicate the current relations.

Second, it is worth remembering that specific and diffuse reciprocities do not exclude each other. Rather, the relations between the two types are dialectical. This is mainly due to the fact that experience of specific reciprocity creates a culture of cooperation and trust. 'In the long run, reciprocity based on self-interest can generate trust based on mutual experience as a result of the "recurrent and gradually expanding character" of process of social exchange. That is, by engaging successfully in specific reciprocity over a period of time, governments may create suitable conditions for the operation of diffuse reciprocity' (Keohane 1986: 21).

The most constructive solution for EU–Russia relations, when Moscow supports specific reciprocity and the EU is looking for the diffuse type, seems to start with the specific version and then gradually moves in the direction of diffuse reciprocity. The habit of cooperation and growing trust should also help Russia shift from the simultaneous type of reciprocity to the sequential one. This approach should be viewed as trust creation and not as a victory of the Russian vision.

In this sense, the idea of the former chancellor of Germany, Gerhard Schroder, who argued for opening the European market to Russian investments in the same fashion as Russia was doing the same with its energy reserves to foreign investors, is worth noting. He equated the Russian version of reciprocity to the foreign-policy doctrine of 'convergence through interlocking', which he championed while being in office. Conceptually, it presupposes a bottom-up (inductive) approach, going from small cases of successful cooperation to a bigger concept (cf. Schroder 2007).

Making a reference to social anthropology again, it takes some time to become one tribe; it could be constructed only through day-to-day interaction that beefs up confidence and trust. However, once the perception of commonality and shared destiny is here, the transfer to diffuse (or generalized, to use Sahlin's terminology) reciprocity becomes unavoidable.

Third, and finally, movement in the direction of diffuse and sequential reciprocity in EU–Russia relations requires a certain institutional adaptation. There is a need to bypass the notion of legal approximation as a spreading of the EU's *acquis* on Russia. Russia would like to have a key say in the development of its own regulative norms and legislation, and legitimately so. On the other hand, a certain legal convergence between Russia and the EU is needed to enable greater economic cooperation and to implement in practice the diffuse reciprocity. However, it is difficult to imagine the EU letting Russia interfere in its decision-making process or changing the results of its deliberations, because they are not satisfactory for Russia.

This contradiction could be solved by extensive consultations within different international fora where both Russia and the EU are equal partners. The discussions that took place, for example, in the framework of the International Maritime Organization on the single-hull tankers are a good example of this strategy. From this point of view, the earliest possible accession of Russia to the WTO (World Trade Organization) is essential.

Furthermore, the EU and Russia need to develop regular consultations between the middle-ranking and lower-ranking officials and regulators about new legislative drafts and the modification of existing norms. It is a trans-governmental type of cooperation, which is essential for developing a habit of working together[7] and for the overall trust-building between Brussels and Moscow. Moreover, this cooperation can alleviate some problems of legal approximation/convergence at an early stage before they produce any negative results or become politicized problems of EU–Russia relations.

Finally, the EU and Russia need judicial authorities which would examine conflicts and disputes in their relations and would provide judgments on the basis of legal argumentation. This would prevent politicization of some problems and would also create a certain level of confidence and trust, and an eventual spirit of commonality that is a necessity for diffuse and sequential reciprocity.

The EU–Russia negotiations on the change of their legal basis, which were launched in June 2008 (EU Commission 2008c), provide an excellent opportunity for creating institutional prerequisites for diffuse and sequential reciprocity. Some institutions need only a slight correction. Others (such as judicial authorities and trans-governmental dialogues) are to be created from scratch. Trans-governmental cooperation would also require more delegation to lower levels in Russia.

In the short-term, these institutions would ease the practice of specific and simultaneous reciprocity. In the long-term, they would lay the ground for the practice of more diffuse and sequential reciprocity.

This threefold scheme (an emphasis on positive reciprocity; gradual movement from small instances of cooperation to a bigger project, from specific reciprocity to the diffuse one, and from simultaneous to the sequential one; and institution building for diffuse reciprocity) will help to alleviate the problem of a divergent interpretation of reciprocity, which haunts EU–Russia relations today.

Conclusion

The seemingly easy and transparent notion of reciprocity turns out, at a closer look, to be quite a complex one. Multiple definitions of this phenomenon help to discern its basic features like the notion of secular movement and interdependence, equivalence and contingency. Furthermore, the evolution of the concept in trade and international relations as well as its analysis in international law and social anthropology reveal that there are numerous facets of reciprocity. This chapter has limited itself to five dichotomies (specific–diffuse; simultaneous–sequential; restrictive–open; negative–positive; and homomorphic–heteromorphic). The analysis also reveals that actors can understand reciprocity differently, based on their experience, world perception and present strengths. Therefore, they locate themselves in different parts of the five reciprocity continuums.

When applied to the recent cases of EU–Russia cooperation (discussion on the modalities of cooperation in the field of energy, concepts of common economic space, drafts of a new EU–Russia agreement, and the exchange of views about how to regulate investments), the typology becomes an interesting tool for analysis. The EU and Russia converge in the way they treat each other when it comes to three dichotomies (negative–positive; restrictive–open; homomorphic–heteromorphic). However, they fundamentally differ when it comes to the last two dichotomies (simultaneous–sequential; specific –diffuse).

The discrepancy in the EU's and Russia's vision of reciprocity comes from their divergent past experience, their different concepts of contemporary international relations and their role in the world as well as from their inequality, which at present rule out the diffuse reciprocity.

However, the EU and Russia view reciprocity as the only means to solve the problem of the collective good and accuse each other of not applying it instead of paying attention to the fact that they interpret it differently. The danger of this situation – in the current circumstances of drafting a new agreement – is enormous. The parties can eventually end up with preparing a nicely worded future agreement, which would not be implementable due to the lack of shared understanding of the basic (yet not objective) categories. Furthermore, their misunderstanding can produce a dangerous spiral of negative reciprocity.

However, the situation is not hopeless. Certain steps can be taken to enable EU–Russia converge in how they view reciprocity. One is to stick to positive reciprocity. Another is to realize the interdependence between specific and diffuse reciprocity and to enable a gradual movement from bottom-up, from small cases of cooperation to a general principle; from a specific reciprocity to the diffuse one, and from a simultaneous to a sequential one. The third step is to modify institutions of cooperation and to use the current negotiations on the new EU–Russia agreement as the arena for this change. This modification should create the basis for an enhanced practice of reciprocity.

The overall aim of the present EU–Russia relations should be to agree on the basic concepts that underlie their cooperation (and not solely reciprocity). Only such a consensus will, in the future, prevent conceptual deadlocks and will enable greater progress in EU–Russia relations.

Notes

1 See, for example, Goldstein. In sociology, this typology was very well formulated by Gouldner (1960: 172).
2 Note, for example, that Keohane once wrote that 'diffuse reciprocity, in the absence of strong norms of obligation, exposes its practitioners to the threat of exploitation. In the absence of strong norms of obligation, specific reciprocity may provide an antidote to the abuse of diffuse reciprocity' (Keohane 1986: 24).
3 It included Norwegian Statoil and Norsk Hydro, ChevronTexaco and ConocoPhillips of the United States, and Total of France.
4 Discussions at conferences held in St. Petersburg, Moscow, Ljubljana, and Madrid in 2006–8.
5 To be fair, the EU is not the only example of this kind of attitude. In the twentieth century, the United States applied a similar strategy towards Japan.
6 See Bova's contribution to this volume.
7 On the implementation of this idea in the EU–Russia energy cooperation, see Romanova (2009).

5 Beyond the paradigm of integration in EU–Russia relations

Sovereignty and the politics of resentment

Sergei Prozorov

Introduction

For the most part of the post-communist period, EU–Russia relations have developed in a singularly paradoxical manner. On the one hand, the idea of Russia's progressive 'integration into Europe' has remained virtually unchallengeable in both Russia and the EU, serving as the background assumption of any assessment of EU–Russia relations. On the other hand, there is an abundance of empirical evidence of the increasingly conflictual character of EU–Russia relations from 1995 onwards. We need only recall the sharp divergences between Russia and the EU over the two Chechen wars, the Kosovo crisis, the authoritarian regime consolidation in Russia, the Yukos case, the 'colour revolutions' in post-Soviet states, etc. There appears a wide gap between the unchallengeable goal of integration on the level of policy rhetoric and the manifestly problematic state of EU–Russia relations that may well be considered not the exception but the rule throughout the post-communist period. For example, in 2004, the year when EU–Russia relations sharply deteriorated over the issues of Yukos, Chechnya, Kaliningrad and concerns over Russian democracy, President Putin stated in his address to the Russian Federal Assembly that European integration was not only a matter of economic policy but also a 'spiritual question' for Russia (Putin 2004). It therefore appears that the rhetoric of integration has acquired a life of its own, entirely detached from the actual state of Russia–Europe relations.

According to Sergei Sokolov (2007), the 20th EU–Russia Summit in Portugal in October 2007 demonstrated an extreme degree of alienation between the two parties, which precludes the formation of a meaningful agenda of cooperation and downgrades EU–Russia interaction to the recycling of vacuous declarations, insofar as neither of the parties wishes to abandon the rhetoric of partnership and recognize the impasse in EU–Russia relations. In Sokolov's argument, the key conclusion drawn from this summit by both the practitioners and the analysts of EU–Russia relations is the need to decrease the frequency of such summits in the future so as to avoid a disgraceful biannual demonstration of the ineffectiveness of 'strategic partnership'. In October 2008, this suggestion was echoed in the EU's review of its relations with Russia before the Nice Summit (Barber 2008). In a more radical reassessment of EU–Russia relations, Boris Mezhuev (2007) has argued that we may presently observe the collapse of all 'pan-European' institutions, in which Russia previously participated (OSCE, the Council of Europe, the CFE Treaty, etc.), while Russia's relations with the EU are marked by the failure of the two parties to agree on almost anything whatsoever. While such pessimistic diagnoses of EU–Russia relations have recently become ever more pronounced, the vacuity of the partnership and the prevalence of conflict issues in EU–Russia relations is not at all a new phenomenon. As Sokolov notes, 'for the last four years at least the Russian Federation and the EU have been unable and even unwilling to formulate the long-term goal(s) of their interaction' (Sokolov 2007).

From this perspective, the currently perceived decline of EU–Russia relations, whose intensification is often dated to the Litvinenko assassination in November 2006 and which reached its climax with the Russia–Georgia war of August 2008, should not be viewed as an aberration in the otherwise cooperative pattern of interaction, but a somewhat belated *adaptation* of foreign policy rhetoric to the actual situation in EU–Russia relations. The contemporary crisis must therefore be understood against the background of the patterns of EU–Russia relations that prevailed throughout the postcommunist period. In this chapter, we shall pursue this line of reasoning by analysing the current state of EU–Russia relations in terms of the dynamic of the interface of the logics of sovereignty and integration in the policies of both parties. In the following section, we shall present the key principles of the conflict theory developed in our *Understanding Conflict between Russia and the EU* (Prozorov 2006). We shall then analyse a number of statements on EU–Russia relations in the contemporary Russian political discourse in order to demonstrate the abandonment of the integrationist paradigm in Russia's orientation towards Europe, conditioned by the more fundamental process of the destitution of the very figure of Europe in the symbolic order of contemporary Russian politics. Finally, we shall address the implications of this destitution for the EU–Russia interaction and discuss the potential of remobilizing the 'idea of Europe' in the wider domain of social practices in contemporary Russia.

Exclusion and self-exclusion in the EU–Russia conflict discourse

In *Understanding Conflict*, we have presented an interpretive framework for the study of EU–Russia conflicts, developed on the basis of the empirical analysis of the narrative structure of the conflict discourse of the two parties.[1] This framework is grounded in the distinction between structural and interactional determinants of the outcomes of the EU–Russia interface.

Structural determinants are policy logics, opted for by the conflict parties in their relation to each other in a concrete encounter. These logics may be *sovereign*, i.e. establishing a link between territory and identity and excluding the other from

Table 5.1 Interpretive framework of conflict analysis

Deployed logic	Response by other party	
	Equivalence	*Dissent*
Sovereign Logic	Mutual Delimitation	Exclusion
Integrationist Logic	Transnational Integration	Self-Exclusion

the domestic space, or *integrationist*, i.e. advocating the transformation of the pluralistic structure of international relations into integrated 'common spaces', whose ultimate ideal outcome would be the unification of the world at large. Interactional determinants are the results of the encounter of the two parties' logics in each concrete case. One party's move may either be matched by the other's deployment of the same logic (resulting in a situation of *equivalence*) or challenged by the deployment of the opposed logic (resulting in a state of *dissent*). Thus, we end up with four possible outcomes of the EU–Russia interface, presented in Table 5.1: sovereign or integrationist equivalences (non-conflictual outcomes) and two dissensual outcomes, whereby the logics of the two parties clash. We may observe either the exclusion of one party by the other from its sovereign domain or the self-exclusion of this party from the integrationist designs proposed by the other. It is easy to see that these dissensual patterns are mirror images of each other, exclusion of the EU by Russia being equivalent to its self-exclusion from the process of European integration and the self-exclusion of the EU from Russia's integrationist proposals being equivalent to its exclusion of Russia from its own sovereign space (see Prozorov 2006, Chapter 4).

Against the facile argument about the EU as the champion of the 'integrationist' logic and Russia as the stubborn defender of state sovereignty, this approach demonstrates that *both the principles of sovereignty and international integration are at work in the policies of both Russia and the EU* without coinciding in particular cases (see Prozorov 2007 for a detailed analysis). For instance, in the case of Russia's recurrent proposals for a visa-free regime between Russia and the EU, it is evidently the latter that is deploying the conventional instruments of sovereignty in its insistence on the uniform and stringent visa regime for Russian visitors to Europe. The same logic could be observed in the tendency to view socioeconomic developments in Russia through a 'security lens', which resulted in the inflated images of 'new security threats' allegedly emanating from Russia, from infectious diseases to organized crime. In such cases, we may speak of the pattern of the EU's *exclusion* of Russia from the European space, which manifestly contradicts the EU's own integrationist ambitions (see Browning 2003). Conversely, in the cases of the EU's admittedly modest attempts at influencing the sociopolitical situation in Russia through, e.g., technical assistance and policy advice programmes, support to non-governmental organizations or oppositional public figures, it was Russia that has regularly invoked the claims of sovereign equality and non-interference against the expansion of the EU's 'normative power' into the Russian political space (see Bordachev 2003). In these cases, which have become increasingly accentuated during President Putin's second term, we may speak of Russia's *self-exclusion* from the space of European politics.

These two conflict patterns have arguably dominated the development of EU–Russia relations since the mid-1990s, the main tendency being the gradual abandonment by Russia of the position of the complainant over unwarranted exclusion by the EU in favour of a more assertive 'self-exclusive' orientation that devalues concrete moves towards greater integration between Russia and the EU without entirely dispensing with the ideal of integration as such. The key motif in Russia's policy towards the EU in the second term of the Putin presidency has been the demand for symmetric, non-hierarchical interaction, particularly in the sphere of norms and values, in which the EU has been held to exert hegemonic influence on Russia. Russia's unwillingness to maintain the 'subject-object' pattern of relations with the EU was particularly evident in its refusal to participate in the European Neighbourhood Policy (ENP) that succeeded the TACIS (Technical Assistance to the Commonwealth of Independent States) programme, which was the key EU instrument of managing the post-Soviet transformation (see Prozorov 2004), opting instead for a bilateral framework of the four Common Spaces, whose ineffectiveness is now evident to most observers. In this manner, Russia has upgraded its symbolic status in relations with the EU even though the political and economic benefits of such a decision remain dubious (cf. Browning and Joenniemi 2008).

The contemporary crisis in EU–Russia relations may be understood as the radicalization of the logic of self-exclusion on the part of Russia. This radicalization is most evident in the political sphere, in which the increasingly active European criticism of the authoritarian tendencies of the Putin presidency is either ignored by the Russian side or reciprocated by arrogant reprisals and crude 'look at yourself!' arguments. The latter form the substance of the proverbial doctrine of *sovereign democracy*, originally developed in the amateurish theorizing of the Deputy Chair of the Presidential Administration Vladislav Surkov and elaborated in the writings of numerous apologists (see Chadaev 2006; Yuriev 2007). As numerous critics have noted (see e.g. Magun 2006; Morozov 2008), the notion of 'sovereign democracy' ultimately comes down to the first term devouring the semantic content of the second, so that 'democracy' begins to denote whatever the sovereign wants it to. Moreover, taking into consideration the conventional definition of democracy as 'popular sovereignty', the term 'sovereign democracy' either becomes a classic case of a pleonasm or implies the expropriation of the sovereignty of the people by another sovereign figure. Whatever its conceptual deficiencies, the discourse of sovereign democracy resonates perfectly with the self-exclusive orientation of Russia with respect to Europe, insofar as it allows to dismiss all European criticism of the anti-democratic tendencies of the present regime while retaining 'democracy' as a mode of the regime's self-identification. In this manner, Russia reserves for itself the sovereign right to define both the *content* of the concept of democracy and its own *correspondence* to this concept.

As Russia no longer participates in the EU's technical assistance programmes and does not depend on external financial aid, the EU is deprived of the instruments of conditionality in dealing with Russia and its capacity to influence the course of political developments is strongly undermined. Faced with the self-exclusive orientation on the part of Russia, the EU can either continue, rather half-heartedly, with its project of the expansion of European norms and values to Russia or *reciprocate* Russia's self-exclusion with its *own* self-exclusive project. The first option logically reproduces a conflictual pattern of the EU–Russia interaction (inclusion vs. self-exclusion) of the kind that prevailed at least since 2004. In contrast, the second scenario would entail the recognition by the EU of the existence of clear limits to further integration with Russia and thus of Russia's *legitimate difference* from Europe in sociopolitical terms. This pattern of development that we have termed '*mutual delimitation*' is also conditioned by the abandonment of the ideal of Russia's 'integration into Europe', whose unchallenged status was mentioned earlier. The equivalence of sovereign logics is arguably the *least* conflictual mode of EU–Russia relations, whose actualization is hindered by the predominance of the telos of integration in the policies of both parties. The exit from the dissensual mode of interface into a situation of mutual delimitation requires a *voluntary* self-exclusion of both parties from each other's sovereign spaces, a mutual 'getting over' each other. The phrase 'interaction without integration', often used to characterize current EU–Russia relations, may then be reformulated as a positive programme for the non-conflictual development of these relations in the future. Opting for this pattern requires a frank admission that in the last decade Russia moved *further away from (rather than closer to) Europe* and its political system is ever more divergent from the normative standards operative in the today's EU. At the same time, as Russia must renounce its perennial ambition to 'enter Europe', the EU must logically renounce its ambitions to govern the post-communist transformation in Russia, which is arguably no longer driven by the European ideal and should not be expected to be modelled on the European experience. In this manner, the two parties would bring their grand policy rhetoric in line with the actual state of their relations, thereby bridging the gap that was easily noticeable since the early 1990s.

Why is this pattern of interaction least conflictual? As we have argued with respect both to its conceptual principle and empirical actualization, the logic of integration is inherently unstable insofar as it subsumes the pluralism of the international terrain under an identity which can never attain the universality it attests to and thus remains particularistic (Prozorov 2006: Chapter 7). What is at stake in the integrationist ideal is the ambition to resolve international conflicts through the *domestication of the international,* which would destroy the existing inter-subjective pluralism of the international society and the establishment of a hierarchical order along the lines of the 'domestic' statist model. It is evident that such a project will encounter resistance from the subjects, whose inclusion in this hierarchical order places them in the subordinate position of the 'integrated' as opposed to the 'integrators'. In contrast, mutual delimitation is constituted by the recognition of legitimate difference and the symmetric structure of sovereign equality. Yet, despite its *conflict-mitigating* potential, mutual delimitation in itself does not do anything to *foster cooperation*. Cooperative interaction requires not only that the inter-subjective pluralism of the international society be maintained but also that the exclusive link between territory and identity, central to the logic of sovereignty, be *loosened*, making possible the formation of extraterritorial spaces of pluralistic interaction. In contrast to the domestication of the international in integrationist designs, in this case, we might rather speak of the *internationalization of the domestic,* the opening of domestic political orders to the pluralism of the international domain. Thus, mutual delimitation offers minimal conditions of possibility of a more ambitious project of cooperation that we have termed 'common European pluralism' (see Prozorov 2006: Chapter 7).

While this scenario might appear disappointingly unambitious, at least in comparisons to the grand designs for Russia's progressive integration into Europe as a newborn liberal democracy, its actual implementation in EU–Russia relations is nonetheless remarkably complicated due to the persistence in the Russian discourse of the integrationist paradigm in the negated form. As we shall demonstrate in the following section, Russia's avowed self-exclusion from Europe does not lead to a simple removal of the idea of Europe from the symbolic order of Russia's self-constitution as a subject but its paradoxical persistence therein in a destitute mode. Thus, rather than lead to mutual delimitation, the denigration of the ideal of integration has resulted in what we shall call a politics of resentment, in which the EU–Russia conflict persists despite the reciprocal abandonment of integrationist designs and what is communicated in the conflict discourse is nothing but the mutual resentment between the two parties. Let us now analyse this paradoxical politics in the context of the most recent intensification of the EU–Russia conflict in the aftermath of the Russia–Georgia war of August 2008.

On speaking of Europe with a clear conscience (and its impossibility)

At first glance, the state of EU–Russia relations in the aftermath of the Russia–Georgia war of August 2008 appears to conform to the structural pattern of mutual delimitation, which brings in the puzzle of why we nonetheless observe the intensification of the existing conflictual dispositions rather than their mitigation. The Russian invasion of Georgia, provoked by its military offensive in South Ossetia (allegedly itself provoked by the Russian military build-up in the region) and the subsequent recognition by Russia of South Ossetia and Abkhazia not only marks the lowest point in the history of EU–Russia relations but allegedly also destabilizes the overall post-Cold War world order, in whose context EU–Russia relations have unfolded. According to Alexander Stubb, Finnish foreign minister and, as chair of the OSCE, an active participant in the resolution of this crisis, we may even speak of a 'post-080808 world' (Stubb 2008). Even if this statement might appear to be an exaggeration, given the relatively low intensity of this war, compared even with the two Chechen wars, and the ultimately inconsequential character of Russia's unilateral recognition of the Georgian provinces, it is undeniable that the Georgian crisis has serious systemic implications at least for Russia's relations with Europe. Indeed, for the first time since 1991, the rhetoric of integration seems to be abandoned by both parties, particularly Russia, whose leaders' statements during and after the conflict did not exclude suspending relations with EU and NATO and proclaimed, in the widely reported phrase of President Medvedev, that Russia 'did not fear a new Cold War'.

Before addressing the implications of this war for EU–Russia relations, we must first emphasise that in a strict sense, the war itself was not a conflict *between* Russia and the EU but a conflict over a third party (Georgia), whose own sovereignty was at stake in it. While within the pattern of mutual delimitation a relation between two parties A and B is likely be both stable and non-conflictual, this argument does not cover the relations between A and B with regard to party C, with which at least one of the parties does not have a relation of mutual delimitation. The classical example of such a conflict setting from the European states' system is the existence of the so-called 'amity lines', which delimited the space of bracketed and limited conflict between European states from the open space of unrestrained pursuit of, e.g., colonial possessions (Schmitt 2003). As every regulated system, the structure of mutual delimitation must either be global and thus closed (which precludes the existence of the open space of unlimited conflict) or spatially circumscribed and thus surrounded by the Outside, where its principles do not apply. In the case of EU–Russia relations, such an exterior is arguably exemplified by the post-Soviet space of the CIS, which has been increasingly torn between contending hegemonic projects since the early post-Cold War period, particularly starting from Putin's second term. Thus, conflicts between Russia and the EU over the influence on these states, such as the conflict over the Orange Revolution in the Ukraine in 2004 (see Trenin 2005) and the subsequent Russia–Ukraine 'gas wars' may well persist even if the relations between Russia and the EU are characterised by mutual delimitation, just as long as at least one of the parties does not enjoy a similar delimitative arrangement with the 'third party' in question, as is certainly the case with Russia's relation to the post-Soviet states, the legitimacy of whose sovereignty and territorial integrity has been eroding under Putin's reign. The logical conclusion is that it is only the *universalization* of mutual delimitation in interstate relations, whose contours are at least hinted at by the postwar development of the United Nations that precludes the resumption of conflictual dispositions. As long as the sovereignty of post-Soviet States remains contested by Russia, which engages in rather heavy-handed attempts to prevent the integration of these states within European and transatlantic structures, the region will remain prone to conflictual dispositions that will at the very least *involve* both Russia and the EU.

However, the significance of the Georgian conflict arguably exceeds that of a circumscribed dispute about the status of post-Soviet secessionist territories and has direct implications for EU–Russia relations. By practicing military intervention in the absence of the UN sanction and unilaterally recognizing breakaway regions, Russia has laid a claim to regional hegemony, which logically entails abandoning the entire EU-oriented integrationist agenda. As the Russian conservative philosopher, Mikhail Remizov, whose work on Europe we shall address next, has succinctly put it: the hegemon has nowhere *to be integrated* into, it is itself that which *integrates* (Remizov 2003). As soon as a claim to regional hegemony is advanced in practice, any further invocation of the rhetoric of EU–Russia integration cannot but appear hollow, since there no longer exists a space, either conceptual nor geographic, in which such a project of integration is conceivable.

Of course, as numerous apologists of the Russian government never fail to remind us, such an intervention without UN sanction and the unilateral recognition of the independence of a breakaway province have already been practiced in the post-Cold War period, most notably by the United States and most of the European countries in the case of Kosovo. While there are certainly plausible analogies to be drawn between Russia's intervention on behalf on Abkhazia and South Ossetia and the West's intervention in Yugoslavia on behalf of Kosovo, it is not entirely clear what these analogies are supposed to prove. It is impossible to deny that the Kosovo intervention violated the principles of the UN Charter and exemplified the exercise of international hegemony by the leading Western powers. What the comparisons of this case with the Georgian crisis establish is that Russia's actions in summer 2008 were *also* a violation of international law (thus delegitimizing Russia's own invocations of international law as quintessentially hypocritical) and an exercise or at least a claim to hegemony in the post-Soviet space. The only difference is that by committing this violation and attempting this hegemony, Russia *excludes* itself from the Western hegemonic order, which it previously sought to be integrated in.[2] It is in the framework of this *concrete order*, irreducible to the formal principles of international law, that the Kosovo operation *is* legitimate (because it fortifies the hegemony in question) and the Russian actions are *not* (because they obviously challenge it). Thus, it is doubly absurd to suggest, as many Russian defenders of the intervention do, that the Western intervention in Kosovo somehow gives Russia the 'right' to do the same in South Ossetia and Abkhazia. Firstly, it would be quite preposterous to claim that a crime committed by one party somehow legitimized the repetition of this crime by the other. In terms of international law, which is admittedly ever less relevant in the context of the contemporary global 'state of exception', both the Kosovo intervention and the invasion of Georgia are *equally* illegal, if this is any consolation to anyone. Secondly, once we move away from the abstract maxims of international law towards the consideration of the concrete hegemonic order, the absurdity of the logic of justifying the invasion with the example of Kosovo is even more evident, as the defence of and the challenge to the current global hegemony are obviously not equivalent events and would never be given an identical normative assessment. This is less a matter of the proverbial 'double standards' that Russia has routinely invoked in its criticism of Western hegemony than the problem of the application of a single normative standard, immanent to the current hegemonic order, to two distinct events that, irrespectively of their formal similarity, serve completely different functions in this order. What the reference to Kosovo in the Russian discourse attempts to legitimize its intervention points to is then nothing more than a self-exclusion of Russia from the Western hegemonic order and the consequent abandonment of the integrationist paradigm. In this sense, the Georgian war is not merely about Georgia but rather disrupts the entire post-Cold War course of relations between Russia and the West, including Europe.

Let us now elaborate this abandonment of the integrationist logic by analysing a number of samples of this logic from the contemporary Russian discourse. As we have discussed at length elsewhere (Prozorov 2006: Chapter 2), the radical dissociation from the process of European integration has been advocated in the 'left-conservative' discourse since the late 1990s. For Remizov (2001, 2002a), the century-old discourse on Russia and Europe has achieved little more than condemning Russia to the endless enactment of the struggle for recognition in the manner of the bondsman in the Hegelian dialectic. Ironically, the entire Russian political spectrum in the post-communist period has been tied to the figure of Europe, be it the early liberal–democratic optimism of the 'Common European Home', the desire of liberal–conservatives to 'abduct Europe' by disassociating it from the United States and thus claiming the mythical role of Zeus for Russia or the perception of Russia as the 'true' Europe as opposed to the degenerate Europe of today, characteristic of the nationalist rhetoric, etc. (see Prozorov 2008, 2009 for a detailed discussion). This wild oscillation of positions that nonetheless all refer to Europe as a Big Other, whose recognition is necessary for the validation of one's subjectivity, is for Remizov a symptom of hysteria that must be ceased by a simple dissociation of Russia from Europe as such:

> Up to this moment European politics has been an existential zone for us, an area of fateful deeds, in which we fought not so much for our interests, but for the formation of our identity. […] The 'abduction of Europe' resembles an erotic game with a succession of sadistic and masochistic phases. First we impose ourselves on it in order to define ourselves through its frightened glance and then reject our selves to be defined by it through a condescending glance. [Thus,] the very abduction of Europe is twisted inside out and is presented as a return to it.
>
> (Remizov 2001)

Remizov's strategy is distinct from the more familiar geopolitical constructions of, e.g., Alexander Dugin (2000) or Gennady Zyuganov (2004), prevalent in the Russian 'national–patriotic' discourse of the 1990s, which sought to conjure a figure of a 'true' (usually Christian, conservative and anti-American) Europe, with which Russia could positively identify and cooperate. This strategy of distinguishing between 'true' and 'false' Europe has been central to the century-old Russian discourse on Europe, which has ensured that irrespectively of its numerous conflicting definitions, Europe remains the constitutive Other, whose recognition is to be sought for Russian identity to be validated (Neumann 1996). In this historical discourse, the question of being inside or outside of Europe (defining the positions of respectively 'Westernisers' and 'Slavophiles') is complicated by the fragmentation of the figure of Europe itself into a 'true' Europe (variably conceived as conservative, liberal or socialist), with which various political forces identify positively, and a 'false' Europe, the object of negative identification (see Morozov 2003; Prozorov 2008). Against this logic of the fragmentation of Europe, Remizov is considerably more attuned to the realities of contemporary European thought and practice with the consequence of abandoning all attempts at finding there the figure of Europe that would accord with the self-understanding of Russian conservatives: 'The internal bifurcation of Europe, its self-alienation, the abstraction of universal substance from the singularity of historical existence is what makes possible the phenomenon of non-European ideologies of Europeanism'. Thus, if 'Europe' became a disease for Russia, isn't this because it has already become a disease for itself?' (Remizov 2002a: 47) For Remizov, there is little sense in Russia's attempt to secure its identity by articulating it with the contemporary Europe, which for him is precisely the epitome of non-identity, abstract universalism and limitless dissemination: 'To be rootless, one does not have to be European' (Remizov 2002a: 47). Thus, neither affirming Russia's Europeanness nor denying it, Remizov attempts to remove the question of Europe from the Russian political discourse as such, opting for a purely *autopoietic* constitution of Russia's identity in terms of its difference from its exterior (Luhmann 1995: 34–37). Russia must neither join nor confront Europe; instead, it must simply '*get over*' it (Remizov 2002b).

Originally belonging to the fringes of the political spectrum, this strategy of dissociation has largely lost its ideological specificity and gradually begun to dominate the mainstream of the Russian discourse on Europe. In a reassessment of Russia–Europe relations on the eve of the widely criticized parliamentary elections of 2007, Boris Mezhuev (2007) argues for the need for Russia to 'divorce Europe': 'Russia is not merely a complex and problematic part of Europe; Russia is already quite simply not European, as year by year it outgrows the framework of any pan-European institution'. In Mezhuev's argument, any attempt by Russia to assert its Europeanness is doomed to fail, insofar as it is the EU that sets the rules of the game and the criteria for recognition in the field of European politics. The only effect of Russia's trying to 'force' its way into Europe would be the intensification of EU–Russia conflicts that have no other substance than the Hegelian struggle for recognition that is purely symbolic yet no less lethal for this reason:

> In any domestic-political configuration, Russia will increasingly stick out from every pan-European construction, losing its positions in the united Europe, while desperately trying to split the Union in order to deal with individual European states. Instead, it would face systemic retaliation from the European bureaucracy and pro-integration forces. This will create a perfect background for the strengthening of the 'new', Central Europe, with all its anti-Russian complexes, within the EU. We will enter an unnecessary conflict, just like two irritated spouses, who could have improved their relations simply by having a civilized divorce years ago.
>
> (Mezhuev 2007)

Mezhuev is clearly aware that after the 2004 enlargement EU–Russia relations can and will only get worse, as the EU is less and less conceivable as a privileged club of Great Powers, with which Russia could envision a nineteenth-century style partnership. While the dwindling geopolitical discourse constructed a mirage of 'continental' Europe that was cast as a 'natural' partner of Russia, contemporary conservatism is increasingly bereft of such idle fantasies and renounces the very logic of 'true' and 'false' Europe as resigning Russia to a perpetually frustrated search for its own traces in the Other. Instead, the 'question of Europe' is simply removed from the Russian political agenda in the strictly sovereignty-based vision of foreign policy. This apparently negative gesture is nonetheless of profound significance for the future development of EU–Russia relations, since it targets not merely the practical implementation of the policy of 'strategic partnership', whose problematic status is self-evident, but also the overall telos of 'integration into Europe', which, as a 'spiritual question', has been virtually uncontestable for most of the post-communist period. Such authors as Remizov and Mezhuev confront precisely this 'spiritual' or existential dimension, considering it a symptom of political immaturity that leads to a discursive self-entrapment in the infinite struggle for recognition of Russia's 'Europeanness'.

The aftermath of the 2008 war on Georgia has witnessed a veritable explosion of similar narratives of 'divorce' or 'dissociation' in the mainstream political discourse. A highly representative and influential sample of such a narrative is an article by Alexei Chesnakov, a political philosopher and a former top-level official of the presidential administration, tellingly entitled 'On Europe, Without Pangs of Conscience' (Chesnakov 2008). Chesnakov begins with recycling the key arguments, made during the previous decade by such conservative politicians and commentators as Natalia Narochnitskaya, Dmitry Rogozin and Mikhail Remizov: the ambiguity or non-existence of a 'European identity', the loss by Europe of its Christian cultural heritage, the undemocratic character of the 'European bureaucracy', etc. Insofar as European identity evades all determination, the very question of 'whether to include oneself in Europe or not' should not even be raised. This does not mean that Russia should isolate itself from Europe – on the contrary, Chesnakov views cooperation with the 'political and bureaucratic' Europe as a possible course of action among others. What is effaced in this discourse is not the contingent possibility of cooperation but the identitarian background, formerly dominated by the integrationist telos:

> We should keep our hands untied for any decision. If tomorrow we need to leave some European structures in order to save the lives of our citizens or even to make their existence more comfortable, we must have this possibility. If, conversely, we need to enter European structures for these very egoistic national purposes, we must use this opportunity. […] Both the government and the people of Russia must hold their own interests as a priority, not the interests of Sarkozy or Kaczynski. We must think and decide for ourselves. We have neither a historical nor a moral right to humiliate ourselves by fulfilling

new conditions or succumb to the nagging of Euro-bureaucrats. After all, every nation must have a national dignity.

(Chesnakov 2008)

The logic of reasoning in this fragment is evidently suspect. How can one be so sure that one can enter and exit 'European structures' at will, especially if one is explicitly doing it for 'egoistic purposes'? Yet, what is important about this argument, as well as its precursors discussed in the foregoing section, is the modification it introduces to the status of Europe in the Russian political discourse. Irrespective of whether on the policy level one advocates a 'civilized divorce' or a ceaseless movement 'in and out' of European structures, the symbolic significance of Europe for Russia's identity declines drastically in comparison with the 1990s, in which Europe was the primary reference point of Russia's self-constitution as a polity. None of three authors discussed in the foregoing section are at all interested in delimiting true Europe from a false one or finding somewhere in the contemporary Europe a potential ally of Russia. In this manner, Europe is deprived of its hegemonic status in Russia's self-understanding, which has been ensured by the ceaseless operation of the true–false distinction. While the content of these binary poles was historically variable, the figure of Europe itself invariably maintained its symbolic authority in the manner of the master-signifier that quilts the floating signifiers of the political field into a hegemonic articulation (Laclau 2005).

It is arguably this discursive structure that is undergoing a fundamental shift in the contemporary politics of EU–Russia relations. In fact, none of these statements exhibit any interest in what Europe is as such, simply taking its existence as 'non-identical', 'ambiguous' or 'rootless' for granted in the very act of dissociation from it. While previously the question of what Europe was and where Russia stood in relation to it gave rise to grand ideological battles, it now appears to be removed from the agenda. While Europe traditionally played a *constitutive* role in defining Russia's identity, its symbolic status in the contemporary discourse is increasingly *destitute*. This destitution entails that the questions of Europe's identity and its recognition of Russia become irrelevant, since Europe's status in the symbolic order of Russia's identity is no longer constitutive. This demotion in status is especially evident in Chesnakov's repeated pejorative references to Euro-bureaucrats and his somewhat infantile enthusiasm over the widely reported telling off of the British foreign minister by his Russian counterpart in September 2008 in the context of the management of the Georgian crisis ('shutting up Millibands'). Similarly, Mezhuev's metaphor of divorce clearly indicates a change in the status of Europe in the symbolic order of Russian politics: if the divorce is initiated by us, we can hardly be expected to seek recognition from an eventual ex-spouse. While conflicts over recognition usually form part of the marital relationship, the decision on divorce entails that recognition is no longer at stake. Indeed, the pragmatic intention of Mezhuev's discourse is precisely to *avoid* a conflict (struggle for recognition) that has become 'unnecessary', since its sole substance consists in 'mutual irritation'. It is precisely the reduction of the problem of mutual *recognition*, that has animated conflict narratives of exclusion and self-exclusion, to the mundane fact of mutual *irritation* that marks the destitution of the figure of Europe in the Russian symbolic order. The politics of Russia's struggle for recognition by Europe, which may be traced back to the early modern period, appears to give way to what we may term a politics of *resentment* in the Nietzschean sense, a defensive reaction that conceals one's inferiority before the other by its self-righteous and hyperbolic denigration (Nietzsche 1998).

This attitude of resentment is much more pronounced in the Russian discourse, since Russia has never played a significant constitutive role in the European symbolic order (cf. Neumann 1998, Chapters 3, 6). As we have discussed elsewhere (Prozorov 2006, Chapter 1; 2008), the politics of EU–Russia relations in the post-communist period has been manifestly asymmetric not merely because of Russia's economic weakness and political instability during much of that period, but also primarily because in its relations with Europe Russia has been engaging in a project of existential self-questioning and self-formation, while Europe was largely engaged in a more outward-oriented project of the promotion of liberalization and democratization in Russia that did not affect its own self-formation, other than by strengthening its own reputation as a champion of liberalism and democracy. Thus, the current impasse in EU–Russia relations does not entail a similar destitution of Russia in the European discourse, since Russia never occupied the locus of the constitutive other in the first place. Nonetheless, Russia's claim to 'alternative hegemony' and its irritated dismissal of the symbolic role of Europe in turn renders problematic any integrationist orientation on the part of the EU, making the EU's sovereign self-exclusion from the Russian political space the only meaningful option. The substance of EU–Russia relations is rapidly being deprived of all political content and reduced to pragmatic aspects of sector-specific cooperation, while the overall rhetoric of the EU policy on Russia is marked by ever-more pronounced resentment about the resurgence of authoritarianism, the politicization of economic praxis, the policy of destabilizing the post-Soviet space, etc. Having failed in its project of liberalization and democratization, the EU recoils from any further integration, limiting its cooperation agenda to what serves its own 'egoistic national purposes'.

Notwithstanding the appearance of a reciprocal self-exclusion of both Russia and the EU from the integrationist project, it would be incorrect to interpret this politics of resentment along the lines of the pattern of mutual delimitation, let alone the more ambitious design of common European pluralism. Rather than recognize each other's *legitimate difference*, both parties treat the rejection of each other's integrationist moves as an illegitimate refusal of due recognition and, rather than 'get over' each other, persist in a state of *mutual resentment*. Thus, both Russia and the EU abandon integrationist designs as a matter of necessity, not policy reorientation, hence the continuation of the conflict discourse. Yet, what is communicated in this discourse is no longer any concrete grievance or the Hegelian struggle for 'pure prestige' but simply the two sides' resentment towards each other akin to the mutual 'irritation' of spouses in Mezhuev's metaphor. On the part of Russia, this discourse communicates an exhausted and vacuous struggle for the recognition, whose very desirability is denied in it, since the status of Europe in the Russian symbolic order is utterly destitute. On the part of the EU, which never sought Russia's recognition in the first place, this discourse communicates the disappointment of earlier integrationist illusions and the pragmatic adaptation of policy goals to the symbolic demotion of Europe in the Russian discourse.

It is precisely this communication of resentment that distinguishes the current politics of EU–Russia relations from mutual delimitation. If the two sides really got over each other, then what is the need for this ceaseless discourse on 'divorce', 'not succumbing to the nagging of Euro-bureaucrats', etc? What is the purpose of speaking 'on Europe without pangs of conscience', if its symbolic status for Russia's identity is so destitute? After all, Russia enjoys cooperative relations with, e.g., China and India without any need to validate its 'belonging' to either of the two civilizations, the absence of anything like a 'common identity' never hindering the articulation of common interests. The very persistence of the Russian discourse on Europe in the mode of communicating resentment and advocating dissociation demonstrates that Russia has *not* really got over Europe but remains stuck in the conflictual setting devoid of any substance. As long as dissociation from Europe continues to be communicated in discourse, it logically remains inaccessible to the speaking subject. In its sheer repetitiveness, Russia's discourse of resentment rather functions in an auto-suggestive manner that unsuccessfully seeks to gain the speaker's adherence to the content of his own enunciation. The three statements that we have analysed in the foregoing section are all marked by the non-coincidence of the *subject of enunciation* and the *subject of the enunciated content*.

This distinction, developed in the context of Lacanian psychoanalysis (Lacan 2001: 94–96, 182–84, 330–32) as part of the proverbial 'decentring of the subject', emphasizes the heterogeneity between the speaking subject who utters the enunciation and the subject, articulated or implied in the content of the enunciated statement itself. In psychoanalysis, the attention to this distinction assists in the interpretation of the analysand's discourse by tracing the inconsistencies between the two aspects of enunciation. For instance, the discourse of the obsessional neurotic is characterized by the dissimulation of his desire on the level of enunciation that is concealed by determinately clinging to the truth on the level of the enunciated content. Conversely, the hysteric reveals the truth of his desire in the guise of lying or concealing it in the positive content of his enunciations. Does not the latter pattern succinctly sum up the Russian discourse of resentment, which indeed negates its very subject of the enunciated (the destitution of Europe in the Russian symbolic order) on the level of the act of enunciation itself (the endless production of discourse about this destitution)? Moreover, to *whom* is this discourse addressed, if Europe is no longer part of the 'Big Other', whose recognition this discourse seeks? We must therefore wholeheartedly agree with Remizov's afore-discussed characterization of the Russian historical discourse on Europe as hysterical, with a proviso that his *own* statements, as well as the statements of other practitioners of the politics of resentment, do not in any way break with this symptom, but rather exemplify it most starkly. While the content of the discourse of resentment seems to demonstrate an unequivocal abandonment of the integrationist paradigm, its enunciative modality rather points to a more complex and contradictory manoeuvring that hysterically affirms what it denies in the very act of denying it. It seems that it is impossible to speak of Europe with a clear conscience after all, nor is it possible to simply stop speaking of (or to) it as such.

Singularity and the 'Wider Europe': beyond the conflictual impasse

The maintenance of the conflict discourse in the mode of the politics of resentment that we have analysed in this chapter evidently marks an impasse in EU–Russia relations. While the integrationist paradigm has been abandoned by both sides, this abandonment has not led to the two sides' reciprocal self-exclusion from each other's domains and the recognition of legitimate difference that would permit cooperation in the common pluralistic space. Rather than simply be abandoned in favour of a pluralistic 'interaction without integration', the integrationist ideal is at least in the Russian discourse subjected to an active *negation* that entails the forced destitution of the figure of Europe in the symbolic order of Russia's identity, which does not bring the conflict discourse to a halt but rather uses to communicate nothing but resentment.

Is there any way out of this impasse? It is evidently conditioned by our commitment to the binary opposition between inclusion and exclusion, whereby the recognition of Russia's Europeanness is only accessible through its integration into the EU institutions and the absence of such integration presumably invalidates Russia's European identity as such. Thus, Russia is from the outset engaged in the struggle for institutional recognition that leaves it with only three possible scenarios: its *inclusion* into Europe by means of accession to the EU (the granting of recognition), its *exclusion* by the EU (the withholding of recognition) and its *self-exclusion* from Europe (the denial of the very need for recognition). During the post-communist period, the relations of Russia and the EU moved from the brief enthusiasm over the first scenario to the protracted oscillation between the second and the third scenarios to the current domination of the third. The space for the current politics of resentment is thus conceptually preconstituted by the understanding of European identity in the manner that renders indistinct the difference between the existential properties of the subject and its participation in a certain institutional order.

Perhaps, then, the best way out of this impasse is the development of a conceptual apparatus that would be better fitted to address the distinct constellations of Europe–Russia relations than the binary logic that equates identity with inclusion and help us elucidate a possibility of participation in European politics for a non-member of the EU that does not come down to the struggle for recognition through membership. We have attempted to develop such a conceptual apparatus on the basis of Alain Badiou's set-theoretical ontology (Prozorov 2008), which introduces a distinction between *belonging* (presenting one's status as an *element* of a certain situation) and *inclusion* (representing one's status as a *subset* of this situation). This distinction fractures the unitary notion of identity into two dimensions of belonging and inclusion and draws our attention to the non-coincidence between the two aspects. Badiou distinguishes between normal sets (that both belong to a situation and are included in it), excrescent sets (that are included in the situation despite not belonging to it) and singular sets (that belong to the situation without being included in it). Thus, Russia's standing in relation to the contemporary Europe is finally freed from all ambiguity: Russia belongs to Europe without being included in it, is its element without being its subset, is presented in Europe without being represented in it. One of the consequences of deploying this approach is the understanding of Russia as a potential 'evental site' of Europe (Badiou 2005: 173–90), an *absolutely singular* element, present in Europe only as undecomposable, 'all of a piece'. Belonging to Europe is only axiomatic for the Russian state apparatus and leading corporations rather than all of the Russian society, which remains present in Europe only 'through its representatives'.

While this position of absolute singularity has evident drawbacks that have motivated the surge of the struggle for recognition in the form of inclusion, Badiou's idea of the evental site brings forth a more positive and constructive possibility of participation in European politics that has barely been addressed by the existing readings of EU–Russia relations. What our theory in *Understanding Conflict* was unable to grasp due to its grounding in the actual policy discourses of Russia and the EU, which remain structured according to the binary logic of inclusion and exclusion, is a possibility of an unrepresented singular element becoming the source of radical novelty that dislocates the very order, from which it is excluded. In this manner, the issue of Russia's presence in Europe without representation, which is at first glance highly particularistic and of no consequence to the overall European political process, may be reconceptualised as having universal implications for European politics at large. An eventual site is the locus of a possible eruption of an event that disrupts the entire situation, precisely because it introduces into it the elements that were formerly absent from it.

This is not to say that the hypothetical event would introduce something specifically Russian into the European situation – in contrast, Badiou's theory is extremely hostile to any form of particularism. Any political event worthy of the name is characterized by a manifest universality that subtracts itself from any particularistic predicates of identity and addresses itself to anyone and no one in particular. Indeed, for all their obvious differences, the October Revolution of 1917 and the process of Perestroika in the Soviet Union both exemplify events that have radically restructured the overall European situation without being expressions of Russian particularism and, furthermore, renouncing particularism as such in their explicitly universalist political orientations. While it is certainly difficult to observe any signs of fidelity to these events in the contemporary Russia, whose political mainstream is marked by the denigration of all forms of universalism in favour of a crude valorization of sovereign particularism, it would be premature to exclude from the outset the possibility of the 'resurrection' of these events in new forms of political praxis (Badiou 2006: 75).

Even if we bracket off the discussion of the nature of this hypothetical event or the resurrection of past events, the understanding of Russia as a singular element of Europe and its potential evental site has serious consequences for rethinking EU–Russia relations since it permits us to move beyond the reiteration of the integrationist paradigm in the absence of any substance to it and the hysterical discourse that resentfully communicates to Europe its destitution *qua* Big Other. What this approach offers us is the possibility of a re-engagement with the idea of Europe *as such* from the perspective of singularity, which is the site of the production of any possible novelty.

It is evident that the ceaseless oscillation between the affirmation of the integrationist ideal and its hysterical denial is utterly incapable of any discursive innovation, insofar as the repertoire of possible positions is established from the outset. The limited political imagination, guiding the discourse on Europe–Russia relations, envisions as the most radical, if not outright unthinkable, scenario of their development the eventual accession of Russia to the EU, which is presented as an event of an almost eschatological significance. In fact, there would be nothing evental whatsoever in such a development, which would rather exemplify a radical depoliticization and normalization of Europe–Russia relations, which would henceforth be reduced to negotiations and bargaining over the adoption of *acquis communautaire*. On the other hand, it is certainly not a sign of a vibrant intellectual culture that the Russian political discourse has in recent years been reduced to drooling over (or an equally pointless bashing of) the slogan of sovereign democracy that is at best (when not dismissed as an embarrassing pleonasm) approachable as a hysterical communication to Europe (or the West more generally) of Russia's own dissociation from it, which, by retaining the very concept of democracy, betrays its continuing dependence on the 'Big Other' it disavows. Focusing on Russia's relation to Europe taken as an unproblematic given, neither of the two positions is capable of questioning what contemporary Europe actually is, let alone reconstruct its meaning. Rather than move from the lack of institutionalization of Russia's belonging to Europe to the hysterical reduction of the sheer fact of this belonging to a destitute symbolic status, we may proceed from the affirmation of the singular status of Russia in Europe to the pursuit of possibilities for reconstituting our idea of Europe as such.

In this manner, the Russian discourse may participate in the wider European discussion of the politics of European identity and the reactivation of the progressive value of the 'idea of Europe'. The identification of European identity with the institutional fact of the EU membership has been strongly criticized in the European politico-philosophical discourse by such authors as Jacques Derrida (1992) and Etienne Balibar (2003), whose works alert us to the dangers of an essentialist understanding of Europeanness. In his remarkable Nietzschean reading of the concept of Europe, Stephan Elbe (2003) has similarly suggested that the contemporary version of Nietzsche's 'good European' disposition must be characterized by the suspension of all attempts to strictly delimit and ground

'European identity' that only serve to replicate the exclusionary mode of identity constitution, which the 'European project' historically sought to transcend, on a higher level of the integrated unity of the EU. In Elbe's criticism, such designs are characterized by what Nietzsche called 'incomplete' or 'imperfect' *nihilism* – the attempt to deal with the presently problematic status of formerly secure foundations (e.g. nationhood) by inventing new ones (e.g. 'European integration'). For those whom Nietzsche has called 'good Europeans', Europe must always be a 'Wider Europe', i.e. a Europe that is always wider than any of its positive determinations, a Europe, whose elements are far greater than its parts and which presents far more than is currently represented in it. In this 'wider Europe', the unrepresented singularities (e.g. non-member-states, sub- or transnational regions, international social movements, immigrant workers and refugees) are no longer confined within a debilitating choice between normalizing their singularity in a frequently hopeless quest for institutional inclusion and the valorization of one's particularity in a politics of resentful entrenchment. Instead, they may directly participate in European politics as the forces of novelty that by definition ksubtracts itself from all institutional determination yet conditions any meaningful transformation on the level of institutions (cf. Prozorov 2008: 200–202).

Thus, it is precisely the *absence* of institutional inclusion that authorizes an innovative discourse on the future of Europe that does not seek to cleanse one's conscience by downgrading its symbolic status but rather affirms as the substance of the 'idea of Europe' the very *impossibility of a clear conscience*, which is the condition for the permanent renewal of political praxis, which otherwise degenerates into nihilistic administration or, worse, administered annihilation. Indeed, to attempt to speak of Europe 'without pangs of conscience' is say nothing at all about Europe, irrespectively of the volume of the produced discourse, since in the absence of this dimension Europe, as a political concept, is perfectly vacuous. This is clearly not to say that contemporary European politics is an adequate embodiment of the ethico-political imperative that Europe stands or stood for – on the contrary, it frequently testifies to its obscurity and perversion, often illustriously exemplified by some of Russia's European 'allies' or 'advocates'. Yet, this very obscurity does not authorize a forced dissociation from Europe but rather invites new efforts at rearticulating what the contemporary content of the European ethico-political imperative may be.

Given the contemporary condition of authoritarian depoliticization in Russia, it would be highly unlikely if such a project of rearticulation were undertaken in the official and semi-official discourses. This is not only because the Russian state has no use for re-engaging with the idea of Europe as long as it maintains its anti-hegemonic orientation or pursues its August 2008 claim to a regional counter-hegemony. More importantly, any re-engagement with the idea of Europe from the perspective of singularity, i.e. belonging to Europe without inclusion in it, has serious implications not only internationally but also domestically, serving as Pandora's box for the contemporary political order. If belonging to Europe is posited axiomatically, we are not merely enabled to participate in the pan-European discourse on the meaning of this belonging but also empowered to question or contest the domestic political order as deficient from the perspective of what we assert as the European ethico-political imperative. Overcoming the politics of resentment and re-engaging with the idea of Europe from the standpoint of singularity will thus ultimately also reactivate the domestic political discourse within Russia, whose degradation both coincided with and is best illustrated by the advent of 'sovereign democracy' on the planes of both theory and praxis. The revival of Europe as no longer a destitute 'big Other' but rather a permanently open ethico-political imperative may ultimately inspire the formation in Russia of a public sphere that is genuinely democratic and *for this reason* truly sovereign.

Notes

1 This framework has been developed on the basis of the empirical analysis of Russian and European foreign policy discourses during 2000–2004, including statements of political parties and parliamentary factions, reporting and commentary in print and electronic media and the 'second-order' discourses of academic and expert communities. A similar analysis was undertaken on the macro-regional level of Russia's Northwestern Federal District, which has been the site of the most intensive EU–Russia cooperation since the early 1990s, particularly in the framework of the EU's Northern Dimension.
2 See Morozov (2008) for a detailed discussion of the Russian criticism of the Western hegemonic order in the discourse of 'sovereign democracy'. While Europe is evidently part of this hegemonic order, it has largely been spared the negative connotations of 'hegemony' and 'unipolarity' that the West (and, more specifically, the United States) are endowed with in the Russian discourse, which made it possible to be simultaneously 'anti-Western' and 'pro-European'. The symbolic destitution of Europe in the contemporary Russian discourse, analysed elsewhere, may be understood in terms of a belated identification of Europe with the West, with all the negative consequences that ensue from this move.

Part 2

Moscow, Brussels and the big three

6　The return of history

Hard security issues in the Russia–Europe relationship

Yuri E. Fedorov

Introduction

Russia's intrusion into Georgia in August 2008 sparked a new round of debate about Russia and its relations with Europe. Many believe that Moscow's militant words are matched by no less militant deeds, that aggressiveness and neoimperial ambitions result from systemic features of the Russian society gravely poisoned by morbid mental syndromes, and thus that a new period of political-military confrontation between Russia and Europe is in sight. If so, hard security issues will become central elements in the Russia–Europe relationship.

Others still argue that Russia is not hopeless, not yet anyway; that the war in the Caucasus was an unpleasant but essentially local incident caused by Georgia's recklessness; that the strategic implications of the Russian intervention should not be exaggerated; and that any prediction of a new Cold War is a product of archaic thinking. This philosophy proceeds from the view that the West has neither the muscle nor the willpower to influence Kremlin's policy, and thus that Europe and the United States should resign and allow Russia's dominance in the post-Soviet space. However, it is doubtful whether the 'policy of appeasement' of Russia will actually strengthen European security and mitigate the hard security problems in Russia–Europe relations.

In order to assess such notions we will look into the principal hard security issues that have emerged in the last few years: the suspension of the Treaty on Conventional Armed Forces in Europe (the CFE Treaty) and the debates about the conventional balance in Europe; the military-political crisis that may result from Russian 'countermeasures' to the American ballistic missile defence in Europe; and obviously also the roots of the war in the Caucasus and its consequences.

Russia's CFE Treaty suspension

On 12 December 2007 Russia 'suspended' its participation in the CFE Treaty until NATO member states ratified the modified 'adapted CFE Treaty' and accepted conditions 'necessary for restoring the viability of the CFE Treaty' (Putin 14 July 2007). This was a sign of warning that Russia was returning to the practices of the Cold War and also a gross violation of the Treaty which lacks a suspension clause. The CFE Treaty concluded in 1990 'had been a landmark achievement that had appeared to signal the end of East-West military confrontation' (Rumer 2007: 9). Moscow justified its 'suspending' the Treaty by 'exceptional circumstances that affect the security of the Russian Federation'.[1] This poses two questions. Are Moscow's arguments profound? Has the strategic landscape in Europe changed so drastically as to enforce Russia's withdrawal from the CFE Treaty?

The CFE Treaty: main points and adaptation

The CFE Treaty divided Europe into four geographical zones, in each of which equal limits were established for the treaty limited equipment (TLE) that belonged to the states, which, at the moment of signing, were members of NATO and the Warsaw Pact.[2] During the 1990s, the total number of weapons on the continent was reduced by more than half. Regular exchanges of detailed information on armed forces and on-site inspections made it impossible to prepare for major surprise offensives unnoticed. This made the military situation predictable and thus much more stable than before.

Yet, an essential modification of the CFE Treaty became necessary as the demise of the Warsaw Pact made the very principle of equivalency between two groups of states obsolete. An Agreement on adaptation of the CFE Treaty was signed in Istanbul in 1999. Instead of zonal limits, it established national and territorial ceilings for each state.[3] National ceilings limit the TLE that belongs to a country, while the territorial ceiling limits the total numbers of land force TLEs stationed on this country's territory.[4] Thus land force armaments held by foreign troops in this state are limited by the difference between its national and territorial ceilings. This was the principal distinction as opposed to the CFE Treaty, which merely demanded that the total numbers of armaments owned by a group of states in a particular zone should not exceed a certain ceiling.

For the NATO member states bordering on Russia the national and territorial ceilings coincide.[5] Hence, the balance of conventional armaments between Russia and NATO will be retained, which is certainly in the interest of Russia. Strangely enough, by 'suspending' participation in the CFE Treaty the Kremlin destroys the only legally binding instrument limiting NATO troops close to Russia's borders.

Ratification of the adapted CFE Treaty: who is guilty of delay?

Moscow has accused NATO member states of making ratification of the adapted Treaty conditional upon Russia's complying with the commitments to withdraw its troops from Georgia and Moldova which it agreed to in Istanbul in 1999. Since 2004, Moscow has named its own commitments 'unsubstantiated conditions'.

In 2000–2004, Russia reduced troops and armaments in the Transdniestrian region in Moldova and retrieved some ammunition stocks from there. Yet, since 2004, Moscow has ceased withdrawing the remaining troops and ammunition from Transdniester. As for Georgia, right up to 2006 Russia refused to evacuate its military bases from Georgian soil. Yet, to everyone's surprise, in the beginning of 2006, Moscow changed its stance, and in March 2006, a Russia–Georgia agreement on dismantling the two Russian bases in Batumi and Akhalkalaki by 2008 was reached. The withdrawal of Russian troops was completed ahead of schedule and since the end of 2007, the only remaining issue was the Russian military base in Gudauta, Abkhazia. Georgia wanted the Russian forces to withdraw, while Moscow insisted that the base was used by the Russian 'peacekeepers' in Abkhazia.

The ratification of the adapted CFE Treaty was thus blocked by Russia's refusal to withdraw its remaining forces from Moldova and settle the issue of the Gudauta base with Georgia. Moscow insisted that the Istanbul obligations were of a political character and did not carry legal force; that the Russian obligations relating to the CFE Treaty had been fulfilled; and that Russia's commitment to withdraw troops from Moldova do not include any rigid timetable.

These arguments are beyond criticism. The Istanbul Summit Declaration signed by Russia established the exact deadline of withdrawal of the Russian troops from Moldova (OSCE 1999: 50).[6] Besides, the CFE Treaty and its adapted variant, both ratified by Russia, stipulated that foreign troops could only be present in the territory of a state party to the Treaty on condition of explicit consent of the latter.[7] This means that Russia in a legally binding way has agreed that it could station troops in the territory of other CFE Treaty states only after a clearly stated agreement of the latter. Moldova and Georgia definitely disagreed on the presence of Russian troops. Furthermore, the political character of an obligation does not exempt the state that has recognized it from the need to fulfil it.

Moscow also justifies retaining its force in Transdniestria by the need to protect the storing of Russian munitions and by the obstacles to withdraw the munitions posed by the Tiraspol regime. This suggests that the solution to a problem of strategic importance to Russia would depend on the position of a small and nasty separatist clique. This hardly suits Russia's great power ambitions.

Group of state-parties to the CFE Treaty and members of military alliance

The Kremlin has announced that Bulgaria, Hungary, Poland, Romania, Slovakia and the Czech Republic failed 'to make the necessary changes in the composition of group of states party to the Treaty on the accession of these countries to NATO'. This led to the 'exceeding of the TLE limits by parties to the CFE Treaty that belong to NATO'. In this light, Russia demanded a reduction of the TLE of NATO countries 'in order to compensate for the widening of the NATO alliance'. This demand is based on equating a 'group of states parties to the CFE Treaty' with a military alliance. However, membership in an alliance was not qualified by the Treaty as a necessary condition of membership in a 'group of states parties to the Treaty'. The preamble to the Treaty says 'that they (i.e. the state parties to the Treaty – Yu.F.) have the right to be or not to be a party to treaties of alliance'. Finally, if the Russian interpretation of the relationship between a 'group of states' and a military alliance is correct, then Russia, Belarus, Ukraine, Moldova, Kazakhstan and the three South Caucasian states, which now form one of two groups of state parties to the CFE Treaty should be regarded as members of a military alliance, which they are not.

Russia–NATO military balance in Europe

The principal question, however, is not whether NATO expansion should automatically result in changes in the groups of states or not, but whether the military balance in Europe is so threatening to Russia that it needs to withdraw from the CFE Treaty.

After two rounds of expansion, NATO member states, taken together, have an advantage over Russia in conventional armaments. However, due to massive military build-down, the actual amounts of TLE of the NATO member states are visibly smaller than NATO's quotas on the TLE established in 1990. Besides, a comparison of armed forces of all NATO member states and Russia would only make sense if all troops of all NATO member states in Europe were deployed on Russia's borders when a conflict arises. Yet, one should not imagine that all NATO forces would be transported to a region of a hypothetical conflict such as the Caucasus or the South Baltic region. Therefore, it would only make some sense to compare the actual armed forces of Russia and those of the NATO member states located in the relative proximity to Russian territory, adding the US troops in Germany and Turkey.

In regions geographically close to Russia, NATO member states hold 10–20 per cent more of heavy ground-force armaments than Russia; the numbers of attack helicopters is approximately equal while Russia has a definite advantage in combat

Table 6.1 The TLE for Russia and NATO countries (as of January 1, 2005). (Treaty on Conventional Armed Forces in Europe, Article IV, para 1.)*

	RUSSIA		NATO		
	1992 Ceilings	2005 Holdings	1990 Ceilings	2005 Total holdings	2005 Holdings near Russia**
Battle tanks	6,350	5,088	20,000	15,313	6,622
ACVs	11,280	9,671	30,000	27,433	10,055
Artillery	6,315	6,061	20,000	16,296	7,074
Attack helicopters	855	484	2,000	1,361	530
Combat aircraft	3,416	2,152	6,810	4,322	1,292

* 'Treaty on Conventional Armed Forces in Europe, Article IV, para 1; http://first.sipri.org/dan/cfe_country_list.php?year=2005.
** The Czech Republic, Poland, Hungary, Germany, Turkey and the US troops in Europe. The accession of Baltic States to NATO has not significantly changed the balance of forces between Russia and NATO.

aircraft. In such conditions, hypothetically offensive operations by NATO against Russia are meaningless.

The Baltic issue

Justifying the 'suspension' of the CFE Treaty, Moscow insists that the accession of Latvia, Lithuania and Estonia to NATO radically changed the military balance in the Baltic region, which 'has adverse effects on Russia's ability to implement its political commitments to military containment in the northwestern part of the Russian Federation'. The Kremlin demands that the three newly independent Baltic States 'return to the negotiating table' and join the CFE Treaty with a view to eliminate the zone in 'which there are no restrictions on the deployment of conventional forces, including other countries' forces' (Information on the Decree 2007).

The three Baltic States never joined the CFE Treaty. Therefore, there are no legal restrictions on the deployment of foreign troops in their territories. In practice, however, the accession of the Baltic States to NATO did not change the balance of forces in the Baltic region whatsoever. Only four combat aircrafts from NATO countries are stationed there on a permanent basis. The size of the Estonian, Latvian and Lithuanian armed forces taken together is less than 24,000; the three countries have about 250 armoured combat vehicles and 550 artillery pieces; no one possesses combat aircraft or attack helicopters; the three Latvian tanks, obsolete T-55s, are only good for training purposes (IISS 2008: 117–18, 135–37). This minimal military potential cannot have 'adverse effect on the Russian ability' to implement military containment near the Baltic region.

By demanding that Latvia, Lithuania and Estonia join the CFE Treaty, Moscow overlooks the fact that this Treaty does not envisage the expansion of membership. Only the states that signed the Treaty in 1990 or their assignees may be parties to it. Latvia, Lithuania and Estonia do not have this opportunity, as not one of them is an assignee to the former USSR. They cannot 'return to the negotiating table', simply because they never were at such a table. In turn, the adapted CFE Treaty has a clause on accepting new members. Thus, if it enters into force the three newly independent Baltic States will accede to it as they have officially declared. Therefore, if Russia wanted to limit the deployment of foreign troops in the Baltic States, it should accelerate the ratification of the adapted CFE Treaty instead of destroying it.

American bases in Bulgaria and Romania

Moscow's list of 'exceptional circumstances', meaning extraordinary threats to its military security, includes the deployment of American forces in Bulgaria and Romania. Actually, due to the large-scale reorganization of US forces abroad most of the seventy thousand American military personnel stationed in Europe will be moved to the United States, while about 5,000 American armed forces personnel would be stationed in Bulgaria and Romania.[8] In this light, the Russian 'concerns'

about American troops in Bulgaria and Romania were manifestations either of
paranoid mentality or, more probably, cynical capitalizing on insufficient factual
knowledge among the general public. 5,000 or even 6,000 American soldiers
stationed more than two thousand kilometres away from Russian borders and
separated by the vast territory of Ukraine cannot present any threat to Russia's
security.

The flank zone issue

The adapted CFE Treaty retains sublimits on the TLE in the flank zone for
Russia and Ukraine.[9] The 'flank zone' limitations exasperated Russian military
and they demanded the abolishment of restrictions, something that was hardly
possible. The establishment of a flank zone was mainly the result of Turkey's and
Norway's attempt to limit Soviet, then Russian, capacity to concentrate troops
close to their borders. However, in May 1996 the NATO member states agreed
to alter the geographical demarcation of the flank zone in Russia. As a result, the
military capabilities of Russia, including its capabilities in the South, have grown
significantly.

As a rule, Russia has explained its aversion to flank restrictions by the need
to accumulate large forces in the North Caucasus due to the threat of massive
extremist activities there. The North Caucasus is indeed unstable, yet regular land-
force units with a large number of tanks, artillery and other heavy armaments are
useless in combating urban guerrillas and are of little use in fighting small and
mobile partisan groups in mountain terrain. However, the military build-up in the
North Caucasus that became possible after the 'suspension' of the CFE Treaty
was part of the preparations for the attack on Georgia.

'Suspension' as an element of escalation strategy

The Russian arguments supporting its 'suspension' of the CFE Treaty were either
futile (like the claim that the three Baltic States are to return to the negotiating
table), or based on arbitrary interpretation of some clauses of the Treaty, or have
little in common with the actual strategic situation in Europe. No exceptional
strategic circumstances justifying the Russian 'suspension' of the CFE Treaty
have emerged in the few years since Russia ratified the adapted CFE Treaty in
2004; and practically all Russian concerns may be removed by implementing

Table 6.2 The TLE ceilings for Russia's active units in the flank zone*

	Battle tanks	Armoured combat vehicles	Artillery
The CFE Treaty	700	580	1,280
The adapted CFE Treaty	1,300	2,140	1,680

* According to note (1) to the Protocol on territorial ceilings for conventional armaments and equipment
limited by the Treaty on Conventional Armed Forces in Europe.

the adapted Treaty. This suggests a question: what were the reasons for Russia's withdrawal from the CFE Treaty? In fact, Moscow was most probably not unduly worried about the balance of land-force armaments near Russia's borders. To assemble a massive grouping of NATO ground forces with a few thousand tanks near Russia's borders, say in the Baltic region, is a highly unlikely scenario. Today, the key component of the military balance is the ability to deploy long-range precise delivery platforms (land-based and sea-based cruise missiles, strike aviation, etc.) able to deliver conventional weapons to accurately chosen targets. In this light, the 'suspension' was rather a signal that Russia was losing patience and implied an element of the Russian strategy of escalation challenging Europe and the United States with a dilemma: either to spur Moscow's ambitions or to face the risk of an escalating confrontation with Russia.

The war on Georgia

The aggression against Georgia was the first (yet, quite probably not the last) time of Russian use of military force against a neighbouring state. It revealed the nature and methods of Russia's international policy and created a number of thorny security problems in Russia–Europe relations. A new Russian attack on Georgia is likely, as is a growing probability of armed opposition between Russia and some other post-Soviet state, first of all Ukraine. Should such conflicts take place, Europe may get involved and with unclear consequences. In addition, the Russian recognition of the so-called independence of South Ossetia and Abkhazia and deployment of the Russian bases there make null and void a restoration of the CFE regime, because Russian troops are stationed at the breakaway territories which legally are parts of Georgia.

Many analyses of the war on Georgia begin by observing that in the evening of 7 August 2008, Georgian troops launched a massive attack on Tskhinvali. This almost automatically makes Tbilisi guilty of initiating the war. It is true that Georgia started the operation against South Ossetian militants that particular night. However, a look at the history of the conflict yields a better understanding of the origins of the war. Lilia Shevtsova from the Carnegie Endowment for International Peace notes that '/t/he war had its roots in a whole number of issues: continuing fallout from the Soviet collapse; Russia's failure as a peacekeeper – the result of a deliberate Kremlin policy to maintain the Caucasus as a black hole providing Russia with a pretext for intervention; the emergence of corrupt separatist regimes feeding parasitically on loyalty to Moscow; personal animosity between Moscow and Tbilisi; the fight to control energy transit routes; and Tbilisi inability to give the Abkhazians and South Ossetians broad autonomy within Georgia' (Shevtsova 2008: 11–12).

The background of the war

In the early 1990s, South Ossetia and Abkhazia declared their independence. This was not recognized by any member of the international community and the two self-proclaimed republics went through bloody wars with Georgian government forces. Moscow did not question Georgia's territorial integrity and joined the sanctions against Abkhazia that were introduced by the CIS on 19 January 1996. Due to the economic crisis of the 1990s and the first Chechen war, Moscow had no resources to conduct active interventionist policies in the South Caucasus. However, Russia supported the Abkhazian and South Ossetian ruling cliques and used them to pressure Tbilisi. The main Russian policy instruments in Abkhazia and South Ossetia were the Russian troops stationed there as peacekeeping forces.[10]

Soon after Vladimir Putin had become president, almost all residents of Abkhazia and South Ossetia who were not ethnic Georgians were offered Russian passports. This was then used as an excuse to invade Georgia, justified by the need to protect the life of Russian passport-holders. As the Kadyrov faction was achieving relative stability in Chechnya, Moscow intensified its military support of Abkhazia and South Ossetia. In particular, in March 2005, Sultan Sosnaliev, the then 'de-facto defence minister' of Abkhazia, admitted that Abkhazian officers were regularly trained in the Russian 'Vystrel' military training centre.[11] Russia began to supply the breakaway territories with armaments and military advisers. By 2006–7, the Abkhazian and South Ossetian armed forces had between 140 and 190 tanks, between 170 and 190 armoured fighting vehicles and between 100 and 300 artillery pieces. The Abkhazian Air force had between five to eight combat aircrafts (Elenski 2006; Tziganok 2007). The only possible source of those weapons is the Russian army. At about the same time, Russian military and security officers were appointed to leading positions in the armed forces and security forces of Abkhazia, and South Ossetia in particular.[12] This is an evident violation of the sanctions established by the CIS and Russia's obligation to remain neutral due to its official position as peacekeeper.

Commenting on the dismantling of the two Russian bases in Georgia, the Russian military expert Pavel Felgengauer wrote: 'As military facilities those bases are useless. Most weapons have been evacuated, most personnel are not fully reliable as they are recruited from local dwellers. A modest number of servicemen who arrived in Georgia from Russia are not a force but hostages. That is why an accelerated withdrawal is not a gesture of reconciliation but much more likely a precondition for starting a military solution of the Georgian problem' (Felgengauer 2006). The 'suspension' of the CFE Treaty allowed Russia to build up troops and armaments close to the Georgian border. In 2006 and 2007, the Russian armed forces held the military exercises 'The Caucasian Frontier' in the course of which strong groupings of storm-troops were being formed near the northern terminal of the Roki tunnel with a view to prepare for an armed intrusion into Georgia (News.Rin.ru 2008).

In the autumn of 2006, four Russian intelligence officers were caught red-handed by the Georgian police. At a meeting of the Russian Security Council, Putin commented: 'Our military servicemen in Georgia have been seized and thrown in prison. ... They (i.e. the Georgian government – Yu.F.) are clearly trying to pinch Russia where it hurts most, to provoke us. ... Those people think that they can feel at ease, safe and secure under the protection of their foreign sponsors, but is this really so? I would like to hear the views of the representatives of the civil ministries and the military specialists' (Putin 1 October 2006). For a start the 'representatives of the civil ministries and military specialists' suggested an economic blockade of Georgia and ethnic cleansing of Georgians living in Russia. The blockade failed.

Most probably, the final decision on the military operation was made in the early spring of 2008. In March 2008, Russia unilaterally lifted economic and military sanctions against Abkhazia, which removed the legal restrictions on the transfer of armaments, military equipment and Russian military personnel to the separatist region. At the same time, the Russian General Vasily Lunev, a graduate from the most prestigious Russian General Staff's Military Academy, was appointed Defence Minister of South Ossetia. In August 2008, Lunev played a key role in the military operations against Georgia.

In April 2008, regular reports of a Russian military build-up in Abkhazia started coming in from the Georgian government and independent sources. Russian fighters took to shooting down Georgian spy drones gathering information about Russian troops in the Abkhazian border regions. Media reported that Russia was preparing a military invasion of the Georgian-controlled Upper Kodori gorge, scheduled for 8 or 9 May.[13]

In June 2008, Russian railway troops moved into Abkhazia to restore the railway from Sukhumi to the Georgian border. During the war, Russian troops and heavy armaments were transported along this railway. Finally, in July 2008 Russian troops conducted massive exercises in areas near the Georgian border. After the exercise, the commanders of the North Caucasus military district noted the 'high quality of cooperation between the participants of the exercise, and the ability of units and formations to rapidly deploy large groups of forces far from their permanent bases' (Russian Defence Ministry 2008a).

In addition, in the summer of 2008, Russia hampered American and European shuttle diplomacy aimed at a resolution of the frozen conflicts in Georgia. In particular, Moscow led the way into a dead end discussion of the peace plan advanced by the German Foreign minister Frank-Walter Steinmeier aimed at resolving the Abkhazia problem. Moscow supported demands advanced by its client regime in Sukhumi that Georgia should withdraw its forces from the Upper Kodori gorge and assume legally binding obligation of the non-use of force against the breakaway territories as a prior condition of any talks on resolving the problem. Those demands were apparently unacceptable for Georgia.

From 3 August, a few thousand 'volunteers' were mobilized in the Russian North Caucasus and with the help of local military commissariats moved to South Ossetia (Illarionov 2008a). This was an act of aggression as the UN definition of aggression stipulates 'the sending by or on behalf of a State armed bands, groups, irregulars or mercenaries, which carry out acts of armed force against another State' is aggression (UN General Assembly 1974). The rise of violence in South Ossetia was orchestrated. On 4 August 2008, the Russian General Marat Kulakhmetov, commander of the peacekeeping forces in South Ossetia, confirmed in writing that the South Ossetians used 120-mm shells forbidden by the international agreements on South Ossetia in attacks on Georgian villages. A few days later Kulakhmetov recognized that the peacekeepers were not able to stop the South Ossetians (Civil Georgia 2008). Thus, by the beginning of August 2008 Russia had completed all preparations for the war.

The night of 7–8 August 2008

Russian officials claim that late at night of 7 August 2008, Georgian troops entered South Ossetia at around 11.30 pm and began a massive shelling of Tskhinvali, which 'erazed the city to the ground', and civilian casualties reached 2,000 on the very first night of the war. Georgia offers a completely different version: late on 7 August, President Saakashvili received intelligence that Russian troops had entered South Ossetia. He then ordered the suppression of South Ossetian militants and to stop the Russian troops marching towards Tskhinvali. Georgia has provided a radio intercept in which the South Ossetian military are heard talking about Russian troop movements through the Roki tunnel. Tbilisi says that Russian figures of civilian casualties in Tskhinvali are wildly exaggerated and that many of those killed were actually killed by Russian artillery fire and air strikes after Georgian troops had taken control of most of the city.

The time of the first Russian troop movement across the Georgian border and the number of casualties on the first night of the war are crucially important. If the Russian version is true, Russia's action can be justified as defence of its citizens, not fully permissible by international law, yet politically and morally justified. Nevertheless, if we accept the Georgian version, Russia committed an act of armed aggression against Georgia, which merely took to a fully legitimate self-defence.

The actual truth about the events of the night of 7–8 August 2008 can be established by independent international investigation only; but there is no hope that Russia and its clients in Abkhazia and South Ossetia would agree to such an investigation. At present, however, the Georgian version appears much more plausible. Tbilisi has provided a detailed chronology of events and offered verifiable evidence, while Russian authorities withhold key information. They are neither saying exactly when President Medvedev made the decision to march into Georgia, nor precisely when the Russian troops crossed into Georgia and entered South Ossetia, nor when the Russian troops joined battle in the Tskhinvali area. This suggests that Russian authorities are hiding something, which is understandable, since any discrepancy between the official and real timing would belie Russia's claims. It is impossible to conceal large troop movements – there are too many witnesses, military records that have been kept, etc.

There is evidence that either the Russian troops were already in South Ossetia on 7 August, or that they had started crossing into South Ossetia much earlier than official Russian reports suggest. In particular, in the late evening of 8 August 2008, the Information and Public Relations Service of the Russian Ground Forces informs us 'The units of the 58th Army stationed at the Tskhinvali outskirts the day before (on 7 August 2008) have by tank and artillery fire neutralized Georgian arms emplacements from which Georgian troops fired upon the city of Tskhinvali and positions of the peacekeeping force' (Russian Defence Ministry 2008b). On 3 September 2008, Krasnaya Zvezda published an interview with a Russian officer suggesting that his unit was moved to Tskhinvali on 7 August and joined battle with Georgian troops in the morning of 8 August.

The Russian version of events is doubtful due to the greatly exaggerated figures on civilian casualties emanating from the Russian government. On 9 August 2008, the Russian Foreign minister Sergey Lavrov announced that 'about fifteen hundred civilians have been killed by some count which is being verified now' (Lavrov 2008a). It seems that the Russian Foreign Ministry verified that figure quite quickly, as on 10 August 2008, the Deputy Foreign Minister Grigori Karasin was absolutely definite in saying that more than 2,000 people, most of whom were Ossetians, had been killed (Russian Foreign Ministry 2008a).

These claims contradict independent evidence. According to Human Rights Watch data gathered by interviewing doctors of the Tskhinvali city hospital (which is where the bodies of Ossetians killed during the war were brought), as of 13 August, the morgue had received 44 bodies. Some 273 people were given medical assistance (Parfitt 2008). In mid-September, Ayvar Bestayev, a surgeon at the Tskhinvali central hospital, told the Ossetian radio that 70 dead and 190 wounded were brought to the hospital on the first night of the war – that is, on the very night when about 2,000 people were killed according to Russian politicians and diplomats. A total of 270 people underwent surgery over the three days presumably since the beginning of the war (South Ossetian TV and Radio 2008). On 4 September 2008, the head of the Russian group of investigators in South Ossetia said that the group had found the bodies of 134 people in Tskhinvali. On the same day, the South Ossetian prosecutor's office said that 276 had been killed during the war (Rambler.ru 2008). It did not specify how many of the dead and wounded were civilians and how many were South Ossetian fighters. Neither did it say where exactly the bodies were found, and crucially, whether they had died on the night of 7–8 August or later during the fighting between Georgian forces, South Ossetian militants and Russian servicemen. The facts also disprove Russian claims that Tskhinvali was 'wiped off the face of earth'. In fact, satellite photos made on 19 September 2008 proved that the percentage of destroyed and severely affected buildings in the city of Tskhinvali was 5.5 per cent. This was serious damage, but it cannot be qualified as 'wiping the city of the face of the earth' (UNOSAT 2008).

Thus, Russian claims that the intervention in Georgia was a humanitarian mission, or a response to a Georgian aggression does not correspond to facts. Quite likely, the first echelon of Russian troops entered South Ossetia before the Georgian attack on Tskhinvali. Yet, even if it was not so, the whole set of events preceding the war suggests that Moscow was preparing a large-scale 'military option' against Georgia. This brings us to the question as to what the goals of the intrusion into Georgia were?

The goals of the intrusion

In essence, the war on Georgia was Moscow's reaction to the 'colour revolutions' in the post-Soviet countries. Increasing public discontent combined with a split in the ruling elite brought to power a new generation of leaders who pursue independent policies and strive for Western integration. This resulted in a decline of Russia's influence in the areas Moscow saw as its sphere of interest. Fighting the 'orange threat' in order to restore Moscow's influence became one of the key elements of the Kremlin policy in the former Soviet republics. A regime change in Georgia through political intrigues, economic pressure and finally, if such levers were inefficient, by military action was seen in Moscow as a demonstration of Russia's ability to establish its influence in the former Soviet Union. Besides, Mikhail Saakashvili, a dynamic, flamboyant and controversial pro-Western leader in Georgia, was perceived by the Kremlin as an outstanding symbol of troublesome changes in the post-Soviet countries and became a personal *bête noire* for a few figures at the very top in Moscow.

In addition, through the war on Georgia Moscow attempted to avert NATO's further eastward expansion. Most likely, the NATO Bucharest summit in April 2008 was the last straw. The heads of the NATO member states promised to sooner or later admit Ukraine and Georgia (North Atlantic Council 2008).[14] Yet, no concrete date of admittance was offered and no MAPs (Membership Action Plan) were presented. This strengthened Russia's resolve to use force to 'teach Georgia and Ukraine a lesson' before NATO decided to give them membership action plans.

The announced intention to accept Ukraine and Georgia into NATO plunged Moscow, as one could imagine, into a situation of both irritation and panic. For the past several years Russia's top circles have had the aspiration, in fact the illusion that the Russian position would be taken into account when any important issues of world politics were to be decided. In reality, however, it turned out that the one issue that Moscow sees as vital was decided upon contrary to Russia's repeated demands. Hence, the prospect of a large-scale foreign policy defeat emerged. Those in Russia who still see NATO as a material incarnation of 'the world evil' imagined enemy tank armadas deploying along the Russian borders. Some, especially those in the governmental agencies involved in foreign policymaking, were in a panic fearing responsibility for a looming strategic defeat. Others were using these events in the struggle for power, influence and budget funding.

Irritation and panic led to an irrational and aggressive reaction. Just after NATO's Bucharest Summit Sergey Lavrov, the Russian Foreign Minister, sullenly announced that 'we will do our best not to let Ukraine and Georgia become NATO members' (Lavrov 2008b). A few days later, General Yury Baluevsky, the then Chief of the General Staff, reported that in case Ukraine and Georgia joined NATO 'Russia will swiftly take actions aimed at guaranteeing its interests along its state borders. These are going to be not only military measures, but also measures of a different character' (Solovev 2008). The intriguing 'non-military measures' promised by General Baluevsky turned in fact out to be the recognition of Abkhazian and South Ossetian independence.

The results of the war

The war on Georgia resulted in Russia's military victory and political failure. The Georgian army was defeated and Russian military bases appeared in South Ossetia and Abkhazia. This improved Russia's ability to threaten Georgia with a new offensive against its main economic and political centres, including Tbilisi. Yet, Moscow did not reach its strategic goals. No regime change took place in Georgia; even if Saakashvili would be forced to retire, the next Georgian leader will not be a pro-Russian figure as there are no pro-Russian politicians in Georgia left today and no one will appear in the years ahead. Furthermore, the prospect of Georgian and Ukrainian membership in NATO was not blocked. The Ukrainian elites realize that a threat of Russian political and military intervention in the Crimea is quite possible; they think about NATO membership in a much more practical way than before; and they do not agree to prolong Russian military presence in the Crimea after 2017. In addition, no substantial differences between the United States and a few new NATO members and the 'Old Europe' were triggered.

At the same time, Russia's international standing was seriously damaged. Russia fell into isolation with respect to recognition of the two Georgian breakaway republics. Former Soviet republics are now concerned with the Russian use of military force and cannot but think about strengthening their own security by developing strategic relations with the West. Instead of discussing whether Russia is a problem or an opportunity, the international community is debating how dangerous Russia is. At the same time, Moscow's attack on Georgia was not rebuked in a proper way, which may even stimulate similar Russia's behaviour towards some other neighbouring state.

The conflict over Crimea

Since the Russian aggression against Georgia, the prospect of new armed conflicts in the post-Soviet space have became more visible. Leon Aron, the director of Russian Studies at the American Enterprise Institute, has said that '/u/ltimately, this short war (i.e. on Georgia – Yu.F.) is likely to be remembered as the beginning of a decisive shift in Russia's national priorities. The most compelling of these new priorities today seems to be recovery of the assets lost in the Soviet Union's collapse in 1991, which Vladimir Putin has called the "greatest geopolitical catastrophe of the 20th century"' (Aron 2008). Among the assets that Russia would like to get, the naval base of its Black Sea fleet in Sevastopol is of special importance.

The Russia–Ukraine agreement signed on 28 May 1997 gave Russia the right to keep its warships in Sevastopol for a period of 20 years. This agreement will automatically be extended for another five years unless one of the parties not later than a year before their term expires notifies the other party of the termination of the agreement. In 2007, Ukraine warned that the agreement would not be extended and suggested starting a discussion on the schedule for the withdrawal of the Russian fleet from Sevastopol. In the summer of 2008, the Ukrainian leadership made it clear that it will expect the Black Sea fleet to leave its base in Sevastopol in 2017 and once again invited Moscow to start discussing the issue. The Russian Foreign Ministry once again, with an arrogance that has become its trademark, rejected that proposal, stating that the issue of the duration of the fleet's presence was too premature to discuss. At the same time, it was said that Russia was interested in extending its Black Sea Fleet's presence in Sevastopol beyond 2017.

The statements by Russian Navy commanders were incredible. Instead of speeding up the construction of a new naval base in Novorossiysk they said that the fleet (or at least some part of it) would be based in the Syrian port of Tartus. The latter is completely unsuitable for a permanent naval base, since it has virtually no infrastructure, and any vessels stationed there rather than at a Black Sea port would be much more vulnerable in the event of a conflict with the United States or NATO. They also promised to bring the number of ships in the fleet back to about a hundred – an increase of about 60 per cent. This can only be done by relocating some ships from other fleets – an idea even more questionable than moving the base to Tartus. In fact, these statements were just a reflection of how dazed and confused the Russian military commanders actually were. In addition, since the summer of 2008 the Navy command has intensified lobbying in favour of building a strong aircraft carrier strike force consisting of 5–6 carrier battle groups despite the evident fact that the only shipyard able to build aircraft-carriers is situated in the Ukraine.

Most probably, such statements by Russian admirals and Foreign Ministry officials signal that they do not consider a withdrawal of the Black Sea fleet from Sevastopol as an option. Probably, they flatter themselves with the hope that in several years' time Ukraine will undergo political changes and will agree to extend the Russian naval presence in the Crimea indefinitely. Hopes like these are typical wishful thinking: common sense demands that Russia commences negotiations on the withdrawal of its fleet as soon as possible and instead starts building a new base, since this is a very expensive, laborious and lengthy process. If this is not done, then the fleet will be relocated to poorly prepared bases instead. The later Russia's construction of its future main naval base near Novorossiysk begins, the more likely that the only thing built in time will be the harbour.

This looks quite probable. In 2003, former president Putin signed a decree setting up an alternative naval base for the Black Sea Fleet in Novorossiysk. Yet, a targeted federal programme that allocated 12.3 billion roubles for the five following years for the construction of the new base started in 2007. In the summer of 2007, admiral Vladimir Masorin, then Russian Navy Commander, announced that by 2012 the construction of breakwaters and piers would be completed (Moscow News 2007). This means that the Black Sea fleet will for a long time lose its combat readiness, since the latter depends critically on a huge set of coastal logistics including airfields, command posts, communications stations, warehouses, barracks, accommodations for officers, hydrographical infrastructure, and many other things. Meanwhile, instead of speeding up construction of the new naval base in Novorossiysk, the Kremlin plans to build a fleet of aircraft carriers, which is an extremely expensive adventure. Nevertheless, it confirms that Moscow has already decided that it will not evacuate its fleet from Sevastopol.

In this light, by unleashing the intrusion into Georgia, Moscow would like to demonstrate to Kyiv that Russia has enough resources and political will to force Ukraine to accept Russia's plans regarding the Sevastopol naval base. To force Ukraine to prolong Russian naval presence in Sevastopol beyond 2017 or annex Sevastopol Russia plans to stir up discontent and disturbances in the Crimea with a view to provoke harsh measures of the Ukrainian government against pro-Russian groups and thus to get a pretext for a military intervention. Thus, prospects of a conflict between Russia and Ukraine are emerging.

This will force Europe (and the United States) to make a really difficult decision: either to oppose Russia effectively or 'swallow' its behaviour. A tough reaction will trigger an acute crisis in Europe's relations with Russia, which the European nations would prefer to avoid. Yet, European passivity would be a signal that Russia's violent behaviour, although unwelcome, does not result in essential negative consequences for Moscow. This, in turn, will stimulate similar or even more defiant Russian actions towards other post-Soviet countries and even some European states.

Prospects for a new missile crisis in Europe

Another trouble-spot may emerge in Central Europe. In the summer of 2008, the United States signed the agreements with the Czech Republic and Poland on the deployment of the American ABM components (dubbed 'the third site') in these countries. In response, Russian officials once again announced that 'countermeasures' would be implemented. Commenting the signing of the United States–Czech Republic agreement on the US missile defence, the Russian Foreign Ministry declared that 'on the pretext of an imaginary Iranian missile threat' the US administration stubbornly pursues a policy fraught with threats to Russia's strategic forces. The Ministry stated that if this agreement acquires a legally binding character Russia 'will have to respond using not diplomatic but military-technological methods. There is no doubt that the setting of elements of the US strategic arsenal in place close to the Russian territory could be used to weaken the potential of our deterrent. It is understandable that the Russian side in such circumstances will take adequate measures to compensate for the emerging potential of threats to its national security' (Russian Foreign Ministry 2008b).

Russian policymakers, military experts and political analysts have suggested that the most probable 'countermeasures' would be the stationing of theatre strike aviation and new highly accurate Iskander-M missiles, including its cruise variant, in the Kaliningrad region and, perhaps, in Belarus too if only President Lukashenka agrees to turn his country into a launching pad for Russian missiles. In his address to the Russian parliament on 5 November 2008, Medvedev announced that, with a view to effectively oppose American ABM (Anti-Ballistic Missile) elements in Europe he decided not to decommission the Strategic Rocket Force division; to deploy Iskander-M missiles in the Kaliningrad region, and to suppress new American ABM facilities by electronic countermeasures (Medvedev 5 November 2008). According to Russian mass media, the battle range of a new Iskander-M ballistic missile can be more than 500 kilometres while the cruise version of the Iskander missile is about 2000 kilometres (Voenno-promishlenny kurier 2007). Hence, not only deployment but also production of such missiles is forbidden by the INF (Intermediate range Nuclear Forces) Treaty. The most probable response to Russia's 'countermeasures' would be the reinforcement of American forces in Europe, including intermediate range missiles. Therefore, if Russia deploys new nuclear missiles near its western borders, it may trigger a new crisis in Europe similar to the missile crisis of the 1970–80s.

Moscow's allegations that the US ABM components in Europe will challenge Russia's strategic forces fail to explain in any intelligible way why this is so. For a start, Russian officials claim that there is no Iranian nuclear threat to Europe. Such optimistic statements contradict credible academic assessments and well-known facts. For instance, David Albright, one of best experts on proliferation matters, concludes that 'Iran would need approximately 6–12 months to produce enough highly enriched uranium for its first nuclear weapon' (Albright 2007). Iran has developed and recently flight-tested the 1,300-km-range single-stage liquid-fuelled ballistic missile Shahab-3, capable of reaching Israel. Iran is also implementing an aggressive missile development effort which includes developing longer range ballistic missiles, cruise missiles and a space launch capability (NTI 2008).

Besides, many Russian experts and political figures state that the American ABM facilities in Europe will not be capable of defending Central Europe against Iranian nuclear missiles. Therefore, they infer that the radar installations in the Czech Republic and the ten interceptors in Poland are designed to damage Russian strategic forces. This is an apparently defective logic: if a missile defence system is useless against primitive Iranian warheads, then it is definitely incapable of destroying Russian strategic missiles which are several generations ahead of Iranian missiles.

In addition, a professional analysis shows that the ABM facilities to be installed in Poland and the Czech Republic are simply incapable of striking intercontinental ballistic missiles (ICBMs) launched from Russia. To do that, the ABM components, above all, the ballistic missile interceptors need truly extraordinary performance characteristics, which are unachievable in the foreseeable future, beginning with their speed. In fact, the launching of an ABM can only take place after the detection of the launching of a Russian ICBM, and only if the potential margin of error has been eliminated. In addition, it is essential to determine, even in the most general term, the flight trajectory of the missile. Only then can the decision to launch an interceptor missile be made. All these steps require time even if it is just a matter of a few minutes. These short minutes may make the

ICBMs simply unattainable. Finally, it is unclear how, in fact, ten interceptors can pose a threat to Russia's strategic forces that possess several hundred land-based ICBMs, most launch sites of which are located more than two thousand kilometres from Poland.

A group of Russian and American missile experts with worldwide reputation, including General Vladimir Dvorkin, the former head of the Russian military research institute specialized in missile issues, have concluded that '/e/ven if the United States expands the system, say, by increasing the number of interceptors, it would not be able to neutralize the retaliatory capability of the Russian missile force. ... The location of the radar in the Czech Republic would not allow it to see missiles launched from any of the Russian test sites used for launches of sea-based or land-based ballistic missiles. The curvature of the Earth completely prevents this. Thus the radar cannot be used to gather intelligence on Russian missiles. ... Overall, the European system in the configuration that is proposed by the United States today cannot present a significant direct threat to the Russian strategic force' (Dvorkin *et al.* 2007). Therefore, Russia's violent objections to the ABMs in Europe are not valid.

In September 2009 the Obama administration backed down on its plans to station American ABMs in Central Europe, which for good reasons was seen as a sort of American security guarantee for Poland and the Czech Republic. This decision, as such questionable, was even more confusing since Russia has not responded to it by making adequate political concessions. Albeit the immediate prospect of a new missile crisis in Europe was removed, this American move convinced Russia that tough pressure and threats of a Cold war type are the best levers in relations with the West.

Since a deployment of the third site of US missile defences in Central Europe does not threaten Russia's security then the question is: what are really the reasons for Moscow's opposition to build this site? On the one hand, its position may be explained by the striving for consolidation of Russia's position as the power with a deciding vote in security related issues in Europe. In addition, if Russia is able to foil these plans, it will demonstrate that the United States is not a reliable ally for Europeans, which of course will diminish Europe's capacity to resist Russian political pressure and blackmail. On the other hand, the Russian approach to the US ABM facilities in Central Europe may be a result of the efforts to find grounds for escalating tensions and confrontation with the West.

The roots of Russia's aggressiveness

Moscow's assertions that the current exacerbation of tensions between Russia and Europe (as well as the United States) result from the West's disdain of Russia's strategic interests and from the provocations of clients of the West, like Saakashvili and Yushchenko, are futile. There are no 'exceptional circumstances' that justify Russia's suspension of the CFE Treaty. The Russian aggression against Georgia was not a humanitarian mission or a response to Tbilisi's provocation encouraged by Washington. A future deployment of US ABMs in Europe cannot threaten Russia's strategic missiles. So, then, what are the roots of Russia's aggressiveness?

Aggressiveness as neurotic behaviour

Some suggest that Russia's escalating bellicosity results from a mixture of transient factors, like personal attitudes of a few people at the very top suffering from wounded *amour propre*, a post-imperial syndrome, a parochial mentality and inadequate knowledge of the rest of the world. The Russian Professor Dmitry Furman describes the behaviour of the Russian ruling elites as a neurotic syndrome: 'This is a neurotic hysteria. You (i.e. the West – Yu.F.) have recognized Kosovo; we shall recognize South Ossetia and Abkhazia. You do not like us, and do not recognize us as equals; we do not need your love at all, we spit upon you; you will creep begging for gas from us. The Georgians do not like us and want to attend NATO, so they should and will be punished' (Furman 2008).

The notion of Russia's policy as neurotic behaviour is in some respects correct. The current Russian strategic thinking is a mixture of megalomania and paranoia. Russian elites are dreaming about a Russia that 'has risen from its knees'; they believe that the *période de revers*, a time of retreat, chaos and decline so typical of Yeltsin's era is now over; that Russia has acquired new muscle and wealth. In addition, the Russian elites have concluded that the West is now weak because of Iraq, Afghanistan, Iran, the new American president and the differences between the United States and Europe. Thus, they think the international situation is favourable for Russia, which is why Russia can and should reformulate its relationship with the West. Russia should be a forceful and in many cases decisive voice on international issues, particularly in areas close to Russia.

At the same time, Russian elites suffer from a paranoid syndrome. They are firmly convinced that the West, especially the United States, is afraid of a new powerful Russia and is doing what it can to resist Russia's rebirth. President Medvedev has said that '/t/oday Russia competes increasingly confidently in the economic, political and military spheres. Moreover, we must frankly acknowledge that many are not pleased with this development. Perhaps some forces in the world would like us to remain weak, and to see our country develop according to laws dictated from the outside' (Medvedev 30 September 2008). It seems that the Russian elites sometimes believe their own lies. For instance, Putin claimed that 'someone in the United States (i.e. most probably meaning the Bush administration – Yu.F.) created this conflict on purpose to worsen the situation and create an advantage in the competitive struggle for one of the presidential candidates in the United States' (Putin 29 August 2008).

Within this conceptual context, the 'suspension' of the CFE Treaty, the war on Georgia and the threats against Poland and the Czech Republic with 'counter-measures' are undertaken precisely as demonstrations of Russia's resoluteness and Western vulnerability. In a wider context, a mechanism that leads Russia's foreign policy down a dangerous blind alley has been formed. The gross exaggeration of Russia's potency leads to the advancement of admittedly unattainable goals.

The inevitable failures, of course, are explained not as results of the Kremlin's own errors, but as the result of the hostile intrigues of the West. This distorts perceptions of international realities even further and aggravates Russian suspicion towards the outside world. The inability to attain proclaimed strategic goals is perceived as a threat that has to be counteracted, if necessary by military means. This should be a source of concern, because the policy of the second world nuclear power should not be determined by inadequate perception of the global strategic realities.

The vested interests

However, Russia's foreign policy is driven not only by a damaged national mentality but also by essential interests of influential groups in Russia's top circles. The American scholar of Czech origin, Jiri Valenta, in his brilliant analysis of the Soviet decision to intervene in Czechoslovakia in 1968 wrote that '/t/he general argument of the bureaucratic–politics paradigm can be summarized as follows: Soviet foreign policy actions, like those of other states, do not result from a single actor (the government) rationally maximizing national security or any other value. Instead, these actions result from a process of political interaction ("pulling and hauling") among several actors' (Valenta 1991: 4).

This paradigm can describe the shaping of Russia's foreign policy today. There are a substantial number of pressure groups, clans and cliques within the top echelons of the governmental bureaucracy, armed forces, security and intelligence services merged with business groups. They compete for control over alluring segments of the national economy, flows of financial assets and influence on government decision-making, including foreign policy. As a result, the zigzags of Russia's policy reflect, directly and indirectly, the balance of influences between domestic actors and their coalitions that exist at each instant of time.

Since the demise of the USSR, the Russian elites have been divided into two 'super-groups' with different interests with respect to the principal national strategic goals on the international arena. The first group consists of business, political and bureaucratic clans that are deeply involved in economic relations with the outside world, primarily associated with export oriented and raw materials branches of the economy. They would like to reach a new Yalta-type agreement with the West in order to secure preferred positions for Russian business, above all the energy supplying companies, in Europe and in the former Soviet republics. To achieve this they seek Russian dominance in the post-Soviet space and strong political influence in areas of Europe close to the former Soviet Union. They would also like the West to permit authoritarianism in Russia. Yet, this part of the Russian political class is not interested in a new Cold War, which would result in a principal redistribution of the national wealth in favour of the military and defence industry at the expense of the export branches and in intrusive governmental control over the economy. Some experts believe that president Medvedev belonged to this part of the Russian elites at the very beginning of his tenure.

However, a number of influential interest groups and lobbies in Russia, generally associated with the security sector, the so called 'siloviks', are looking for a new confrontation with the West in order to justify an increase in defence expenditure and a return to a mobilized economy. In part, this is a result of the traditional Soviet paranoid mentality, in part a product of vested interests. The former high-ranking officer of the Soviet military intelligence, Vitaly Shlykov, noted that 't/he raw material complex has created a parallel economy where there is no place for the VPK (military industrial complex). The expectations of the Russian leadership that the impoverished defence industry would be able not only to survive by exporting arms but also to finance the rearmament of the Russian Army with weapons of a new generation are truly odd (Shlykov 2005). Therefore, despite the recent feeling that the country has been restored to its former *grandeur*, the Russian generals, the masters of the military–industrial complex and the heads of security services must have understood that Russia's conventional forces are far behind NATO troops in Europe, as well as American and Japanese forces in the Far East, and that Russia has not been able to take advantage of the 'revolution in military affairs'.

They are worried about the progressive degradation of Russia's armed forces, military science and industry, and the declining ability to develop and introduce new high technologies that are crucial for the fighting efficiency of the armed forces. In addition, despite the rocketing defence budget, which in nominal terms has grown by seven or eight times since Putin entered the Kremlin, the Russian military are not able to purchase modern armaments in more or less significant volumes and to intensify essential vitally important military research and development. These groups are concerned with Russia's transformation into a 'petro-state' suffering from defects typical of such states, including lack of motivation for technological modernization.

In this context, a cohort of the diehard generals, the masters of the military–industrial complex, the heads of security services together with governmental officials, politicians and academics closely associated with these circles are seeking to rechannel export revenues into the defence sector and into the restoration of the privileged status they enjoyed in the former Soviet political system. With this in view, they attempt to provoke Western behaviour that may then be construed as a violation of Russia's legitimate interests and a military threat to the Russian state – all in order to convince the Russian society that a new militarization of the country is the only way to ensure its survival.

The growing influence of the 'siloviks'

The war on Georgia provided conclusive evidence of the growing influence of the security sector on Russia's policy. Andrey Illarionov, the former economic adviser to the Russian president, concluded that the war 'was a spectacular provocation that had been long prepared and successfully executed by the Russian "siloviks" that almost duplicates the "incursion of Basayev into Dagestan" and the beginning of the second Chechnya War in 1999' (Illarionov 2008b).

During the war, Moscow step by step destroyed prospects for a political resolution of the conflict and escalated enmity and tension in its relations with the West. James Sherr, the well-known British analyst, wrote that 'Russia was given the perfect opportunity to use military force on a limited and responsible scale. ... It did not. Had it done so – had it ejected Georgian forces from South Ossetia and stopped at the border – it would have won a convincing military, political and psychological victory. The West would have felt humiliated, it would have been apprehensive about Moscow's long-term game plan, but it would have been in no position to object or make demands. Russia would have secured the moral high ground, and the west would have to live with it' (Sherr 2008: 8). However, instead of stopping at the border with Georgia proper after defeating the Georgian troops in South Ossetia, the Russian Army invaded the so called 'internal areas' of Georgia and the Upper Kodori gorge in Abkhazia that was controlled by Georgian forces.

The same behavioural pattern was repeated after the signing of the Medvedev–Sarkozy plan on 12 August 2008. This plan opened up good diplomatic and political prospects for Russia. Had Moscow implemented the plan, Russia would have emerged as a responsible and reasonable player on the international arena. It would have helped to minimize the damage done to Moscow's relations with the West and to improve its international moral standing. At the same time, Moscow could demonstrate that it had the military capacity and political will to be a dominant force in the post-Soviet space and that it could defend its interests by all available means, including force.

Yet Moscow did not use that opportunity. In defiance of the Medvedev–Sarkozy arrangement, Russian troops were not withdrawn 'to the line where they were stationed prior to the start of hostilities', i.e. to the territory of the Russian Federation, and two 'security zones' deep in Georgia proper have been arbitrarily established, including the outpost close to the port of Poti, which allowed for the control of cargo transfers to and from the main Georgian sea terminal. The Russian military wasted no time in correcting the President. The Defence Ministry said that it was just a pullback of forces, not a complete withdrawal that Russia had agreed to. This meant that a large part of the Russian troops that entered South Ossetia and Abkhazia still remain there. This caused a huge diplomatic scandal; the West took Russia's actions as a slap in the face and started discussing sanctions, some of which could have been quite painful.

On 26 August 2008, Russia recognized the independence of Abkhazia and South Ossetia, then signed agreements on friendship, cooperation and mutual assistance with them, and announced its intentions to build military bases there. In violation of the agreement on troop withdrawal to the pre-conflict positions, Russian troops remain in South Ossetia and Abkhazia, where new military bases will be built. Moreover, following Russia's recognition of the independence of the two breakaway Georgian territories, Article 6 of the Medvedev–Sarkozy peace plan to launch an international discussion on the stability and security of the region becomes pointless. The prospect of returning Russia's relations with the West to the *status quo ante*, which could have defused the tension between Russia and the West, was erased.

Finally, just after the second Medvedev–Sarkozy arrangement was reached at the beginning of September 2008, Moscow announced the joint Russia–Venezuela naval exercises and sent two strategic bombers to Venezuela with the only possible aim to demonstrate its readiness for a military confrontation with the United States. At the same time, Russia's defence budget for the coming year was enlarged by almost 30 per cent in comparison to 2008. The scenario of the 'Stability-2008' military exercise held in September 2008, the largest since the demise of the Soviet Union, proves that the Russian armed forces are preparing for a military conflict with Western armies with the use of nuclear weapons.

Medvedev's doctrine

Another evidence of the growing role of the generals in shaping Russia's strategy is found in Medvedev's presentation of something looking like a strategic doctrine. Just after the recognition of Abkhazian and South Ossetian independence he was quite open and brave enough to state that '/w/e are not afraid of anything, including the prospects of a new Cold War. ... In such a situation everything depends on the stance of our partners in the international community and our partners in the West. If they want to maintain good relations with Russia, they will understand the reason for our decision, and the situation will remain calm. If they choose a confrontational scenario, well, we have lived in different conditions, and we can do so again' (Medvedev 26 August 2008). To put it differently, Medvedev actually said that if the West wishes to avoid a new Cold War, it should agree with Russia's annexation of the two Georgian territories (which was exactly what he meant by 'understanding' Russia's reasons) and put on a show of bravado having said that Russia was not afraid of a new confrontation with the West. The aforesaid statement was not a casual one: the views on Russia's foreign policy and security advanced by Medvedev since the war on Georgia, 'the Medvedev doctrine', may, in a certain sense, be summarized in the following way:

*   After the war on Georgia, the world has changed and Russia needs to revise its approaches to global and regional security.
*   There are regions in which Russia has 'privileged interests'; countries to which Russia shares 'special historical relations', and with which Russia is 'bound together as friends and good neighbours', and with which Russia will 'build friendly ties' (Medvedev 31 August 2008) (The war on Georgia revealed the methods by which Moscow plans to 'build friendly ties' with countries with which Russia is 'bound together as friends and good neighbours'. Also, 'special historical relations', 'good neighbours', 'friendly ties' are used in the Russian political jargon with reference not only to the former Soviet republics but also to the former Soviet bloc members.)
*   A strategic confrontation with the West is in the coming. Medvedev has stated that 'the determination with which Russia was forced to stand up for ordinary people, for those who had Russian passports, carrying out its obligations under international mandate, was never going to satisfy the large number of

forces that believe that only they have the power to influence the climate on our planet, that only they are capable of taking meaningful action. I would go even further: they will not forgive us for this' (Medvedev 30 September 2008).

- There are powerful foreign actors that are encroaching upon Russia's rich raw material recourses. 'Russia', the president said, 'can either be big and strong, or it will cease to exist. This morsel will prove too tasty to resist, our lands and our natural resources will attract too many envious glances, all our capacities will be sought after. The world has not become any easier, but another force has emerged that is capable of maintaining order in the world. And perhaps that is the principal lesson of the Caucasian War' (Medvedev 30 September 2008).
- In this context, it is of vital importance for Russia to modernize, 'completely remodel' its armed forces, generally improve its combat preparation and training. For this, Russia needs to create new organisational structures, develop new types of weapons and put them on combat duty and develop social support systems for personnel quicker than it has done in the past. In particular, Russia has to strengthen its nuclear deterrence forces, establish a unified system of airspace defence and increase its status as a great maritime power.
- The construction of new aircraft carriers is 'obviously the most important area for the future development of the Navy' and there is an 'absolute need to restore the navy's aircraft carriers' (Medvedev 11 October 2008). This ignores the apparent fact that there are no shipyards able to build such warships, and that the construction of shipyards and building aircraft carriers is an extremely expensive task; and that the principal mission of aircraft carriers is power projection into distant areas of the World Oceans which in no case can be of immediate importance to Russia.
- The modernization of the armed forces including building new nuclear submarines will be implemented regardless of any economic crisis (Medvedev 25 September 2008).

None of these ideas are new, all of them were developed by the diehard military, political and academic circles long before the war on Georgia. Yet, now they have been systematized and presented by the president. Medvedev has thus turned into a mouthpiece of the security sector. Quite likely, he seeks support from the military command with a view to turn the officer corps of the armed forces into the core of his own power base. His remark that Russia 'cannot rest on what has been established previously, with all due respect to the traditions and forms of organisation that existed before' (Medvedev 25 September 2008) was a plain hint that during the previous reign the Russian Army was lacking proper care despite the rocketing rise of export revenues. If a political alliance between Medvedev and his generals is established, it will help the president to improve his position vis-à-vis his prime minister; but at the same time, it will turn him into a puppet in the hand of the generals.

The future trajectories of Russia–Europe strategic relationship

As a consequence of Russia's war on Georgia, Europe has arrived at a watershed in its relations with Russia. Future trajectories of the Russia–Europe strategic relationship depend on several factors, above all on the balance of influence of different pressure groups in and around the Kremlin; Russia's economic situation; and Western responses to Russia's militant actions. Today, we can neither exclude an attempt to make a strategic deal with Russia nor a new period of political–military opposition.

Enduring status quo

The important question is how sustainable the current state of the Russia–Europe relationship is, or, to put it differently, whether an enduring status quo is plausible. For many in Europe, above all for those who are comfortable with events and do not go beyond rhetorical annoyance with Russian aggressiveness, a continuation of the situation after the EU monitors have been deployed in Georgia is acceptable: Russian troops are withdrawn from Georgia proper and Georgia's security is strengthened by the presence of the EU monitoring team; Russia, hopefully, refrains from a new aggression and escalation of hostilities; recognition of Abkhazian and South Ossetian independence does not threaten European strategic or economic interests. In this context, NATO may not hasten admittance of Georgia and Ukraine but will obviously hardly refrain from further eastward expansion in principle.

Yet, for Moscow, a continuation of the current situation means that it does not reach its basic goals, including a regime change in Georgia, a guaranteed refusal to invite former Soviet republics to the North Atlantic alliance, a continuation of Russian naval presence in Sevastopol beyond 2017, etc. In addition, the 'siloviks' do not find a convincing pretext to accelerate the militarization of the country all the way to an actual transition to a mobilized economy. They may push Moscow to escalate further belligerence, especially because its intrusion into Georgia was not reprimanded by effective sanctions. Thus, the status quo between Russia and the West is essentially unsustainable and it may evolve into a new round of confrontation fairly easily. It may be triggered by Russian moves like the annexation of the Crimea, or at least of Sevastopol; the implementation of 'countermeasures' against Poland and the Czech Republic with a view to force them to refrain from deployment of the ABM components, or to transform Central Europe into a *de facto* 'neutral belt' between Russia and other parts of Europe; the cutting off of NATO's transit to Afghanistan via Russian territory. Sooner or later this behaviour will force the West to respond with harsh measures.

A deal with Russia

Europeans may yield to the temptation of striking a deal with the Kremlin. Most European states would like to avoid a crisis in their relations with Russia, as it will add more difficult problems to the already long list of strategic challenges to the community of democratic nations. A deal might include a continuation of NATO's non-military transit to Afghanistan, political guarantees of Ukraine's territorial integrity and of the non-use of 'energy weapons'. In return, Russia may demand from the West the dismissal of Saakashvili, freeze Georgia's and Ukraine's NATO membership, refuse the hypothetical deployment of Western (above all American) troops in Georgia; a ban on arms supplies to Georgia and a refusal to support Ukraine in terminating Russia's naval presence in the Crimea.

If such a deal is concluded, Russia will achieve some of its basic goals – a prevention of further NATO expansion eastward and retaining its naval base in Sevastopol beyond 2017. This will be a strategic defeat for the West not only because the Kremlin capitalizes upon its armed aggression against Georgia, which is morally depraved, but also because Moscow will definitely perceive the consent of the West in a deal as proof of its military weakness and lack of political will to oppose Russia's further expansion. Most probably, such Western weakness will encourage the Russian top echelons to undertake new hostile and provocative steps, which will sooner or later trigger Western responses.

Thus, making deals with Russia will not prevent a new Cold War just as the arrangements in Yalta and Potsdam had not prevented the first Cold War, and the Munich agreement of 1938 had not prevented World War II. Stephen Sestanovich, Ambassador-at-large for the former Soviet Union from 1997 to 2001, wrote: 'Let's make a deal approach to diplomacy has a tempting simplicity to it. ... Diplomats are widely thought to be negotiating such deals all the time, but it is in fact very rare that any large problem is solved because representatives of two great powers trade completely unrelated assets. The "grand bargains" favoured by amateur diplomats are almost never consummated' (Sestanovich 2008).

Western sanctions

The other trajectory starts with severe punitive sanctions on Russia by the West. Sanctions may include a denial of bank credits, a reduction of economic relations, to diminish energy import from and the export of food and medicines for the Russian populace and of luxury goods for the masters of the country; a refusal of political dialogue and semi-isolation of the Kremlin, including its expulsion from the G8 and some other international bodies, etc. At the moment, this trajectory looks quite unlikely. The financial crisis, the wars in Iraq and Afghanistan, the Iranian nuclear problem and similar issues divert the attention of European leaders from Russia. Yet, punishment of Russia may be inevitable – or quite likely – if Moscow undertakes new aggressive actions, for instance, if it annexes the Crimean peninsula. Imposing severe sanctions upon Russia will most probably result in a new Cold War type of confrontation between Russia and the West. Russian 'siloviks' will interpret sanctions as a proof of their theory of the West's unshakable hostility to Russia to justify a principal and severe rise of its defence budget, then a transition to a mobilized economy, and unleash a full-scale military–political confrontation with Europe and the United States.

A 'new Cold War'

In starting a new Cold War, Russia will deploy nuclear weapons and their delivery vehicles in the Kaliningrad region and perhaps also in Belarus, withdraw from the INF Treaty, undertake subversive actions against Ukraine and Georgia and against oil pipelines running through Georgia, cut off NATO's transit to Afghanistan, and hamper Western efforts to stop the Iranian nuclear programme. The escalating military–political confrontation with the West will result in a new arms race. If so, there will be a decline in the Russian standard of living, a rapid growth of popular dissatisfaction, and essential changes in the economic and political systems due to the transition to a mobilized economy, which, in turn, will be strongly resisted by those groups in the elite and society that are flourishing in the export-oriented sectors of the economy. In the aggregate, a deep political crisis will follow in Russia, which, in turn, may evolve either into a democratic 'colour' revolution, or into the establishment of a fascist military dictatorship, or into disintegration of the country. But before such a crisis results in a democratic revolution, if it ever takes place, Russia may cause a few dangerous conflicts in the strategic space 'between Russia and Europe', above all in the Black Sea region, and provoke a military crisis in relations with the West if Russia implements 'countermeasures' to the construction of the third site of the American BMD in Poland and the Czech Republic. In addition, a disintegration of the second world nuclear power and the largest supplier of energy to Europe will in itself be a serious challenge to Europe.

Russia's return to normality

We can not totally exclude that Russia will return to normality, meaning that Russian troops are withdrawn from Abkhazia and South Ossetia, that Moscow *de facto or de jure* retracts its recognition of these territories, that talks about the status of Abkhazia and South Ossetia and/or security and stability arrangements in those territories start. This would be the most welcomed option for Europe, and for Russia itself. Yet, this trajectory is quite unlikely at the moment, as it would mean that Moscow recognizes a failure of its current strategy and minimizes the political influence of the 'siloviks'. This may take place only if Russia is confronted by an economic catastrophe and vitally needs large-scale economic assistance from the West.

Conclusions

Despite the drastic fall of the Russian industrial production and stock market, a swift depletion of foreign currency reserves, and the drop in oil prices, many Europeans are still mesmerized by Russia's oil and gas bonanza and its nuclear weapons, the pandemonium of luxury cars in Moscow, and the impertinent lifestyle of the *beau-monde*. Europeans are frightened, while many Russians are delighted by a vision, really a *chimera* of a potent empire arising once again in Eurasia.

In truth, Russia is by no means an 'energy superpower', it is a petro-state unable to modernize. Lilia Shevtsova concludes that '/c/urrent trends give no ground for optimism. The ruling class and Russian society itself are drifting downstream with no thought of where the drift will take them. … Numerous weaknesses are disguised by demonstrations of strength (which are often imitations of strength or strength that soon turns into a weakness), make it difficult to generate the energy needed to change things' (Shevtsova 2007: 317–18) Putting it differently, something is rotting in a state camouflaged by the splendid showcases of Moscow's downtown. Russia's aggression against Georgia was a demonstration of strength that is turning into weakness.

It was also a signal that Russia's foreign policy is increasingly loaded with traditional Soviet motivations: morbid suspiciousness, an imperial syndrome; attempts at playing the United States off against Europe, a desire to preserve Central and Eastern Europe as zones of probable expansion, and such issues. Russian international behaviour is for the most part decided by the 'siloviks' faction which wittingly provokes Russia's defiant and aggressive international behaviour with a view to restoring a mobilized economy and its own priv-ileged status in a political system typical of Soviet days. In this way, hard security problems emerge as central long-term elements in the Russia–Europe relationship.

Faced with the Russian challenge which evolves into a threat, Europe should refute its abortive policy of 'engaging with Russia' and at the same time avoid a new Cold War. For this, Europe needs to reassess its relationship with Russia and find the right balance between cooperation, containment and deterrence. Europe and the West as a whole should under no circumstances give up the deployment of the US ABM elements in Poland and the Czech Republic. Further, Georgia's security should be assured either by affiliation with NATO, or by other means. Otherwise, the West will lose face while the militant factions in Moscow will be encouraged to further escalating demands and exerting pressure on Europe and the newly independent states. At the same time, Europe should use all sorts of arms control and economic levers in order to minimize an escalation of military opposition in Europe. However, if the second Cold War arrives, the West should win it.

Notes

1 In this paper, Russian allegations and demands are set forth as they were formulated by the official Information on the Decree 'On Suspending the Russian Federation's Participation in the Treaty on Conventional Armed Forces in Europe and Related International Agreements'.
2 The CFE Treaty's area of application includes the entire territory of the European continent 'from the Atlantic to the Ural Mountains', as well as the vast majority of Turkey's territory, with the exception of a small region on the border with Syria. The area of application was divided into 'Central Europe', 'expanded Central Europe', the 'flank zone' and the 'rear zone'. Ceilings for combat aviation were established for the entire area of application only. The ability to relocate air forces over large distances

in short periods rendered zonal restrictions for air forces meaningless. In each of the groups of state parties, quotas for the TLE were distributed in such a way that the total number of armaments in each zone would be less than the ceilings established for this zone.

3 For the United States and Canada, national limits were only established to limit armaments of the two countries in the Treaty's area of application.

4 Air forces are not limited by territorial ceilings. Military aircraft and attack helicopters may travel throughout the entire area of application of the Treaty.

5 Except Norway, where seven foreign ACVs (armoured combat vehicles) and 66 pieces of artillery can be stationed.

6 The Declaration says that 'We welcome the commitment by the Russian Federation to complete withdrawal of the Russian forces from the territory of Moldova by the end of 2002. We also welcome the willingness of the Republic of Moldova and of the OSCE to facilitate this process, within their respective abilities, by the agreed deadline' (OSCE 1999: 50).

7 Para 5, Article IV of the CFE Treaty states that 'no State Party stations conventional armed forces on the territory of another State Party without the agreement of that State Party'. Article II of the Agreement of Adaptation stipulates 'Conventional armaments and equipment of a State Party in the categories limited by the Treaty shall only be present on the territory of another State Party in conformity with international law, the explicit consent of the host State Party, or a relevant resolution of the United Nations Security Council. Explicit consent must be provided in advance, and must continue to be in effect'. See Agreement on Adaptation of the Treaty on Conventional Armed Forces in Europe (OSCE 1999: 131).

8 About 2,500 American servicemen are to be deployed in Bulgaria and about 2,300 in Romania. The troops are deployed on a rotational principle (see Crawley 2005).

9 There are sublimits for tanks, armoured combat vehicles and artillery in the Russian Leningrad and North Caucasus military districts with the exclusion of some areas in both of them. In Ukraine, there are sublimits for land TLEs in the Odessa oblast.

10 Officially, the Russian military contingent was stationed in Abkhazia as the CIS Collective peacekeeping force under the mandate of the CIS Council of the heads of states. In 2003, the CIS Council of the heads of states decided that those forces were to stay in Abkhazia 'until one of the parties to the conflict demands ending of the peace-keeping mission'. The Russian troops in South Ossetia was part of trilateral Collective peacekeeping force consisting of the three battalions – Russian, Georgian and Ossetian – established in accordance with the Russia–Georgia accord of 1992, also known as the Dagomys agreement.

11 'Vystrel' is the training centre for foreign military cadres, mainly from developing countries, established in the former Soviet Union (Civil Georgia 2005).

12 Since 2004 the Russian Lieutenant-General Anatoly Zaitsev, former deputy commander of one of the Russian military districts in Siberia was the first deputy of the Minister for Defence and since 2005 the head of the General Staff of Abkhazia. Russian officers and individuals occupying commanding positions in South Ossetia include: Yuri Morozov, Prime Minister since 2005, occupied various positions in Bashkiria; Mikhail Minzayev, Minister for Internal Affairs since 2005, a colonel in the Russian police; Anatoly Barankevich, Secretary of the National Security Council of South Ossetia since 2006, a colonel in the Russian armed forces; Boris Attoyev, Chairman of the Committee for State Security since 2006, previously held various positions in the Soviet KGB and in the central apparatus of the FSB in Moscow; Vasily Lunev, Minister for Defence of South Ossetia since March 2008, a general in the Russian Army; Vladimir Kotoyev, head of the Government Protection Service since 2007, a colonel in the Russian Army; Oleg Chebotariev, head of the State Border Protection Services since 2005, a colonel in the Russian FSB. See: APSNYPRESS 2005: 4–5.

13 Yuliya Latynina, a well-informed source on Caucasian affairs, reported that '/o/n May 8–9, Russian paratroopers were supposed to enter Upper Kodori. ... That was cancelled after hectic shuttle diplomacy ... On May 21, we had the election in Tbilisi. Had a war broken out, the mess would have been such that anything could happen, including such an impossible thing as a victory of the Georgian opposition' (Latynina 2008).

14 The declaration adopted at the Bucharest summit states that 'NATO welcomes Ukraine's and Georgia's Euro-Atlantic aspirations for membership in NATO. We agreed today that these countries will become members of NATO. ... Today we make clear that we support these countries' applications for MAP. Therefore, we will now begin a period of intensive engagement with both at a high political level to address the questions still outstanding pertaining to their MAP applications' (Bucharest Summit Declaration).

7 Russia–EU relations

The economic dimension

Boris Frumkin

Russia–EU economic relations during Putin's second presidency (2004–7) and the first year of Medvedev's presidential term (2008) remained the main driving force of the bilateral partnership and developed reasonably steadily. Political coolness emerging from time to time on the level of the EU or some member states (the United Kingdom, Poland and the Baltics) had in fact only a formal impact on the trade and economic ties with Russia (for instance, hampering the renewal of its legal–institutional framework due to the suspension of negotiations on new basic partnership agreement). Their influence on the dynamics and structure of the economic cooperation was negligible. On the contrary, there was reticence and disagreement on important economic problems (at the sectoral level, in large projects and at big company levels) that can reflect negatively on the political interrelations. Meanwhile, striving for a diversification of economic relations in key spheres from both sides (EU attempts at replacing the supply of oil and gas from Russia with other sources and a connected Russian quest for an alternative to the European energy markets) have practically weakened bilateral political ties.

Presently, these processes do not have a strictly negative and irreversible character. Moreover, the global financial–economic crisis has for the time being confirmed the balance of interests predominantly over the balance of values in Russia–EU interrelations. '... We should not try to hide certain differences which exist between us such as Georgia and human rights', underlined the Commissioner for external relations and European neighbourhood policy B. Ferrero-Waldner, 'but we should equally not let these prevent us from hard-headed engagement on matters of mutual interest' (Ferrero-Waldner 2009).

Nevertheless, the crisis revealed uncertainty of the future of Russia–EU 'positive economic interdependence', its conditionality upon prospects of crisis and especially post-crisis development of the Russian economy and its potential participation in the recovery of an aggregate economy of the EU. For Russia, this is the first crisis as a national market economy and an organic part of the world global economy, while the most important items are its internal systemic problems, only being revealed and aggravated by external difficulties.

Russian economy in 2004–8: a boom without development

This period of Russian economic history may be divided into two different subperiods. The first one (from 2004 to the middle of 2008) was characterized as being stable and more rapid than economic growth in the EU and created the impression of an economic miracle among the politicians and a large part of the population. The second one (from the second half of 2008, without any clear prospects yet) was featured by being more impetuous and harder than the EU's expansion of the financial–economic crisis and the shock of sobering up the political leadership and society. This harsh macroeconomic jump broke the previous development trend, though the main socioeconomic indicators grew in 2008 (see Table 7.1).

As is clear from the table, despite the negative change of the second half of 2008, Russia ended that year in a rather good macroeconomic situation. It remained the world's tenth largest economy with approximately $1.7 tn. GDP (almost 1.3 times more than the 12 new EU states together), $384 bn. in international reserve assets, with a budget surplus and balance of payments. In January 2009, it still had about $203 bn. in two sovereign wealth funds: Reserve ($125 bn.) to support budgetary expenditures in case of oil prices falling and National Welfare (about $78 bn.) to assist pensions funding and infrastructure development.

During these five years, Russia enjoyed economic growth averaging 7 per cent (higher than in the 12 new EU member states, except the Baltics before the 'hard landing' of their economy at the end of 2007). Over these years, the fixed capital investment growth had averaged more than 14 per cent and real disposable money incomes – about 10 per cent. The share of loss-making organizations in the economy has fallen by 1.6 times up to a 23.4 per cent., profitability of assets, on the other hand, has grown by almost 1.8 times up to 10.5 per cent in 2007. The scale of poverty and unemployment was declining steadily in 2004–7: the share of citizens with incomes below subsistence minimum decreased 1.5 times up to 13.4 per cent of the population, the unemployment rate – 1.4 times up to 5.6 per cent of the economically active population. Average life expectancy in 2008 reached almost 68 years – an increase of three years.

The middle classes continued to expand: the share of citizens with per capita money incomes equal or above the average level increased by almost 1.2 times up to 46 per cent of the population, including the growth of share for those with incomes twice and more above the average level by 1.5 times up to 10 per cent. The number of 'well-off' Russians (earning disposing 1.3 to 13 million rubles in cash or in bank deposits) has increased by 2.5 times up to 1.2 million (0.8 per cent of the population) with total disposable assets of 4.2 tn. rubles (about $169 bn.). The number of Russian billionaires reached 110 persons. The share of private property in enterprises and organizations increased from 78 per cent up to practically 83 per cent, and the number of small enterprises – by 25 per cent up to 1.14 million. The incomes from entrepreneurial activities and property incomes contributed 17–20 per cent to the total money incomes of the population. The structure of

Table 7.1 Main socio-economic indicators of the Russian Federation (2004–8)

	2004	2005	2006	2007	2008	2009 (estimation)*
GDP total:						
nominal, $ bn.	591.9	766.0	988.2	1289.6	1668.3	1237
in constant prices, as % to the previous year	107.2	106.4	107.4	108.1	105.6	91.5
GDP per capita –nominal, $	4115	5353	6936	9082	11741	8717
Industrial output as % of the previous year	108.0	105.1	106.3	106.3	102.1	88.5
Agricultural output as % of the previous year	103.0	102.3	103.6	103.3	110.8	100.5
Investments in fixed capital:						
nominal, $ bn.	103.2	125.5	179.6	262.4	353.4	...
in constant prices, as % of the previous year	113.7	110.9	116.7	122.7	109.8	82.4
Foreign investments accumulated in the economy, $ bn.	...	111.8	142.9	220.6	264.6	262.4**
Foreign investments received in the economy, $ bn., of which:	40.5	53.7	55.1	120.9	103.8	54.7**
direct investments	...	13.1	13.7	27.8	27.1	9.9**
portfolio investments	...	0.45	3.2	4.2	1.4	1.0**
other investments	...	40.1	38.2	88.9	75.3	43.8**
Foreign trade turnover:						
$ bn.	280.6	369.2	468.6	578.2	763.7	500.1
as % of the previous year	132.4	131.6	126.9	123.4	132.2	65.5
Export:						
$ bn.	183.2	243.8	303.9	355.2	471.8	305.0
as % of the previous year	134.8	133.1	124.7	116.9	133.1	64.6
Import						
$ bn.	97.4	125.4	164.7	223.1	292.0	195.1
as % of the previous year	128.0	128.8	131.3	135.4	130.6	66.8

Retail trade turnover, as % of the previous year	113.3	112.8	114.1	115.9	113.0	94.3
Consumer prices index, as % December to December of the previous year	111.7	110.9	109.0	111.9	113.3	108.8 – 109.0
Index of industrial producers prices, as % December to December of the previous year	128.8	113.4	110.4	125.1	93.0	106.5
Number of unemployed:						
mn. persons	5.8	5.2	5.0	4.2	5.3	6.5
as % of the previous year	101.6	90.2	95.9	84.9	124.5	122.6
as % of the economically active population	7.9	7.1	6.7	5.7	7.0	8.9
Nominal average monthly accrued wages of an employee, $	234	303	391	529	694	605
Real disposable money incomes, as % of the previous year	110.4	112.4	113.5	110.7	102.9	100.7
Average monthly per capita money incomes of population, $	223	287	380	491	610	…
International reserved assets, at the end of the year, $ bn.	124.5	182.2	303.7	478.8	427.1	439.0
Stabilization Fund of the RF, at the beginning of the year, $ bn.	3.7	18.5	44.7	91.7	155.2	222***
RTS stock exchange index at the end of the year	614.11	1125.6	1921.92	2290.51	631.89	1444.61

* Ministry of Economic Development of the RF; ** January – September 2009; *** Reserve Fund and National Welfare Fund – total in January 2009. All indicators nominated in rubles are converted in USD according to the official annual average exchange rate of Central Bank of Russia.

Source: Russia in Figures 2009: Statistical Handbook. Rosstat. Moscow, 2009; Federal Service on State Statistics (www. gks.ru); www.rts.ru.; Osnovniye parametry utochnyonnogo prognoza socialno – ekonomicheskogo razvitiya na 2010 god i planoviy period 2011 i 2012 godov. Ministry of Economic Development of the RF,31.12.2009 (www.economy.gov.ru/resources/8848)

consumer expenditures came nearer to that of developed countries: the share of non-food purchases has grown from 37 per cent to 42 per cent; 50 million Russians (35 per cent of the population) have regular access to the Internet.

Russia also improved its international trade and investment position. The average coverage imports by exports was almost 180 per cent. Even in 2008, Russian foreign exchange reserves were enough to cover the value of 16 months of imports. During 2004–7, the official exchange rate of roubles to the US dollar improved by almost 20 per cent. Foreign investments accumulated in the Russian economy have increased 1.4 times up to $264 bn. Russian government foreign debts have declined sharply up to $41 bn. (only 3 per cent of GDP at the end of 2008), increasing the state sovereign credit rating and permitting Russian banks and companies to push their total foreign indebtedness to the huge amount of $452 bn. However, even the total state and commercial foreign debts of Russia were, at the end of 2008, about 30 per cent of GDP – the usual rate for developed countries.

This 'economic miracle' formed among the Russian decision-makers (both in state administration and business) the habits and skills of 'welfare growth management', not one of crisis, and a mood of well-being among the population. Continuing economic boom, mostly fuelled by high oil and gas prices and relatively cheap international credits strengthened social confidence and gave additional 'economic legitimization' to Putin's and subsequently to Medvedev's political leadership and strategy. These trends were also supported by international experts who produced the so-called 'BRIC theory'. It had put down to the Russian (as well as to the Brazilian, Indian and Chinese) economy certain built-in development factors supposedly guaranteeing their durable growth and further transformation to 'global economy locomotives'. Warnings from some Russian experts against such an euphoria (although well-grounded) were in fact ignored: '... the economic boom may last for a relatively long time but not endlessly', wrote at the very beginning of 2008 the Rector of the Academy on National Economy of the Russian government, professor V. Mau. 'The Russian successes based on such unstable factors as high prices of energy resources and availability of cheap money at the world's financial markets. Moreover, the Russian institutes (economic and political) are in the meantime hardly capable to foresee easy and correct consequences of both a worsening of the economic conjuncture and a possible economic crisis' (Voprosy ekonomiki 2008/2: 27).

These predictions had come true already in the second half of 2008. The drop in the world's demand for energy products and metals as well as the global credit crunch caused, in September–October, a shock slump in prices of the main Russian export goods and a fall at the Russian stock market of roughly 70 per cent (primarily due to margin calls). Russia's banking system suddenly faced a sharp liquidity problem. The last quarter of 2008 reversed practically all positive trends. Between September 2008 and February 2009, industrial output decreased by 21 per cent. The ruble depreciated to the US dollar by 35.4 per cent, the capital outflow from Russia from October 2008 to January 2009 was almost $200 bn.

(more than 70 per cent of the inflow in 2004–7). The unemployment rate increased to 8.5 per cent, the share of citizens with incomes below subsistence minimum – up to 14.5 per cent, the number of 'well-off' citizens declined by 15 per cent and their total assets by 16 per cent. The number of Russian billionaires (according to 'Forbes' magazine) fell three times to 32 and their total assets by 73 per cent. The economic recession came to undermine the social support and political base of Putin–Medvedev's 'tandemocratic' system of national governance.

The vast resources and the strong inertia of the huge Russian economy prevented it from the 'hard landing' similar to the one of the small and resource-slender economies of the Baltic States the previous year. It proved to be less vulnerable to a global deficit of borrowed resources. In the common estimation of HSBC bank and *The Economist*, at the beginning of 2009, it occupied the ninth position at the 'vulnerability list' of the world's seventeen major emerging economies – six points better than Poland, three better than Mexico, one point better than Brazil and Turkey, but five points worse than India and nine points worse than China (BIKI 2009/32: 4). Nevertheless, the crisis revealed the systemic problems and contradictions of the Russian economy and political governance camouflaged earlier by economic and welfare growth.

The first problem was a predominantly quantitative character of the economic growth. Some Russian experts correctly called it 'growth without development', especially as it was insufficient even to restore some of the main socio-economic achievements of the former 'Soviet Russia'. In 2008 industrial output was only 83 per cent of 1990 year level, the agricultural output 86 per cent, investments in fixed capital 67 per cent, and real disposable money incomes 82 per cent. The average life expectancy in 2008 had in fact returned only to its 1992 level. Moreover, despite the 'growth at any price'-oriented economic policy ('the doubling GDP in 10 years' strategy, etc.) Russia's growth rates were left behind by all neighbours within the BRIC (Brazil, Russia, India and China) group, rising predominantly due to a mass scale export industry and services with new technologies and high value added.

Thus, unsuccessful 'growth rates competition' was determined by *the second contradiction of the Russian economy – a continuation of development according to the logic of the traditional industrial model.* Meanwhile, Russia from the 'transformation' crisis of 1998 mostly lost its competitive advantages in the main factors of this model: labour resources, fixed assets, natural–climatic parameters. During 2001–7, the demographic base was compressed and the quality of labour resources declined. The share of the population at a working age out of the total population increased only by 3.5 per cent points up to 63.4 per cent (90.1 million), while the average yearly manpower loss due to retirements and deaths is about 1.2 million. The economically active population is progressively ageing. The quota of people between the ages from 55 to 72 years increased 1.4 times up to 11 per cent of the number of employed. A substantial part of the former employees lost their professional skills, transferring to a non-production sphere after the 1998 crisis. In addition, a considerable share of newcomers do not have the necessary qualifications. According to opinion polls among Russian industrialists, in 2007 the deficit of qualified workers maintained second place (62 per cent) among the 13 most important limitations of business development. The growing inflow of foreign labourers (predominantly from CIS countries) – officially 7.8 times up to 1.2 million, unofficially estimated at 8–10 million – cannot solve this problem either.

The fixed assets in the economy are progressively ageing and depreciating. The total depreciation rate increased 1.1 times and exceeded 46 per cent, including about 54 per cent in the most export earning mining and quarrying branches. The average age of machinery and equipment in industry was 20 years; the share of relatively new (less than 5 years) equipment was only 7 per cent. In 2007, 30 per cent of the Russian enterprises treated deficit of proper equipment as the main barrier of business activity.

As for natural resources, Russia is among the top three countries as regards gas reserves and among the first ten in oil reserves. However, the old relatively conveniently situated deposits of hydrocarbon and metal fossils are intensively working out and the exploitation of new ones and the transportation of these products has become more expensive and less competitive. Russia holds about 10 per cent of the world's arable land and 12 per cent of freshwater resources, but in the European part of the country, where population and production are mostly concentrated, there are only 8 per cent of national water reserves of rapidly worsening quality. Almost 60 per cent of the Russian agricultural land suffers from degradation processes, climatic conditions are not favourable enough, volatility of agriculture production is high (in 2004–8, the minimum grain harvest was 40 per cent less than the maximum one) (Voprosy ekonomiki 2008/97: 18). Moreover, agricultural, mining and even manufacturing branches with low value added are incapable of ensuring sustainable growth and dangerously dependent on international volatility in the state of the market. The course of developing on their base 'a world superpower' (energy, metal, grain, etc.) without using export revenues to a considerable extent for the formation of new high-technological sectors and a radical technological renewal of traditional branches (including development of renewable energy sources, biotechnologisation of agriculture, etc.) may only hamper modernization and diversification of the Russian economy when addressing the globalization challenges.

The third problem consists precisely of an insufficiency of innovative investment efforts for the diversification and qualitative renewal of the Russian economy and its reorientation to the post-industrial development model. Despite priorities of structural modernization of the Russian economy declared by the state, there was neither a strategy for developing 'growth clusters' nor an effective direct (through government financing) and indirect (through promotion of innovativeness of business as a whole and especially high-tech sectors) financial credit policy. As a result, total expenditure on research and development did not exceed 1.1 per cent of GDP and the respective federal expenditures 0.4 per cent of GDP or less than 2.3 per cent of the total budget expenditures. The number of personnel involved in research and development decreased in 2001–7 by 1.1 times up to 800,000 people (about 53 per cent of the level in the first post-Soviet year of 1992). During 2004–7, the share of basically new technologies out of a total number of advanced production technologies created in Russia increased only from 7.7 per cent up to 9.6 per cent and the share of innovative and active organizations decreased from 10.5 per cent down to 9.4 per cent, part of the innovatory products out of the total of production shipped from 5.4 per cent down to 5.2 per cent. The existing high-tech industries (aerospace, electrical, electronic and optical, biotechnological and defence) are reproducing innovations on their own accord, while the usage of their potential for supporting innovative orientation for the whole economy had been only 15–20 per cent, so the Russian industrial share of the world's high-tech intensive markets did not exceed one per cent (Economist 2009/2: 31).As a result, labour productivity in the Russian manufacturing industry, according to the World Bank, was only 26 per cent of the productivity of the United States, about 50 per cent of that of Poland and 70 per cent of that of Brazil.

Nevertheless, Russia still has a serious potential for innovation- based economic growth. Within the majority of industries, there already exist internationally competitive enterprises: in the manufacture of transport equipment the productivity at 20 per cent of the best firms is 11 times higher than at 20 per cent of the worst ones, in the manufacture of food products it is 24 times, respectively. As for the total level of innovativeness of economy, according to an estimation of the independent organization Information Technology and Innovation Foundation – ITIF, in 2009 Russia occupied 35th position among the 40 world leaders being behind of EU25 (20th position), but leaving behind a number of 'potential locomotives' of the world economy (Brazil, Mexico and India – 38th, 39th and 40th in ranking). Moreover, as for the level of higher education Russia ranked first and for research 15th (Voprosy ekonomiki 2008/11: 111).

The fourth problem – insufficient development of the modern system of political, economic and social institutions ensuring economic growth. (For details, see Voprosy ekonomiki 2008/2: 14–18). An essential prerequisite of the economic modernization is the modernization of the state at all levels, securing real protection of citizen and property rights, the independence of the judiciary and the effectiveness of the law and order system. Presently, 'law nihilism' (as President Medvedev calls it), the low quality of state services and state ineffectiveness seriously impede economic development. In 2007, the Russian industrialists ranked the lack of state support and corruption of state officials as 7th and 9th among 13 limitations for industrial development. The authorities not only badly realized the 'night watchman' functions, but also often behaved as 'daylight robbers', oppressing businesses (especially small and medium-sized enterprises) by laying on excessive administrative burdens and using corruption extortions. An unfavourable influence on the socio-economic situation causes ineffectiveness and excessive expensiveness of *institutes ensuring human capital development (education, public health and pension systems).* They are not adapted to new demographic and social conditions and requirements of 'post-industrial' (based on knowledge and comprehensive human development) social and market models. They also do not easily neutralize the negative consequences of excessive income differentiations and social inequality; in 2004–7, the share of the 20 per cent group with the least incomes out of the total of monetary income of the population decreased from 5.4 per cent to 5.1 per cent and the share of the 20 per cent group with the highest incomes increased from 46.7 per cent to 47.8 per cent, the Gini coefficient increased from 0.396 to 0.422. Not using Russian but Western criteria of relative poverty (incomes below a 50–60 per cent average per capita income of the population) we can include in the category of 'poor' not 13 per cent as in official Russian statistics but more than 40 per cent of the Russian citizens.

The weakness of law enforcement mechanisms determines insufficient effectiveness of *economic institutes*. First of all, this concerns economic legislation and national regulators' system, providing the formation of a competitive environment and a demonopolization of economy (including state regulation and participation in the development of the priority sectors), stimulating entry into the market for new companies (to promote the innovation process), the creation of the land and real estate market (as a base of a private property system), the development of stock and financial markets (as a source for capital for growth). The Russian industrialists in 2007 ranked the suppression from 'natural monopolies' (energy, railway transport and communications), frequent changes in legislation and the low availability of credit in the 3rd, 5th and 10th place, respectively, among the 13 main limitations of business development. In 2008, enterprises had to finance a 50 per cent investment in fixed capital through their own resources (Rubchenko and Talyskaya 2009: 58). The system of *special development institutes* aiming to ensure the growth of concrete sectors of economy, regions and enterprises (through direct state financing or state–private partnership) in fact emerged in Russia only in 2005. That is the reason why its financial (Vneshekonombank, the Russian venture company, the Mortgage credit agency, the Russian Nanotechnology Corporation, etc.), administrative (special economic zones, technoparks and business incubators) did not have sizeable effects before and during the crisis, which may create the risk of replacing private business responsibility by state subsidizing of loss-making firms by non-economic reasons, etc. A certain effect (including institutional reforming) was had through comprehensive programmes (two-year national priority projects continuing with five-year state programmes (for the development of education, public health, housing and agriculture). Nevertheless, the cutting of their budget financing due to the crisis may aggravate macroeconomic destabilization.

A number of experts and business persons suggested that the negative impact of the crisis on the Russian economy may be weaker unless the current government financial–credit policy (especially the formation of a Stabilization Fund and the permanent increase of the international reserve assets), which in fact removes from the economy huge financial resources (about $570 bn. in 2007), would make its modernization and diversification necessary thus enforcing business to obtain loans from abroad.

The aforementioned system problems in many respects determined the specific reasons of the rapid expansion of the crisis in Russia (Voprosy ekonomiki

2008/11: 32–33). First, there was the *credit squeeze*. The Russian banks' direct and indirect losses due to a depreciation of debt and other securities of the domestic companies through a shocking (by four times) fall in the stock market (during August–October 2008) were, by some experts, estimated at 10–12 per cent of the total bank system capital. The risk of credit non-repayment rose sharply. The threat of a collapse of the banking system and of the economy as a whole became quite real. Second, there was *the Russian economy 'cooling-off'* as a consequence of its lingering 'Dutch disease'. In effect, an eight-year period of high oil and gas export prices stimulated the outflow of production resources from the manufacturing industry to mining and quarrying and the service sector, while the connected strengthening of the ruble caused an inflow of foreign capital and a 'warming up' of consumer demand, mostly covered by imports (in 2007 the share of import acquisitions in retail trade structure was 47 per cent). During the first half of 2008, almost two thirds of the Russian GDP was provided for sectors directly connected with the favourable foreign economic state of the market (mostly mineral products) or based on massive foreign borrowing (financial activity, construction and real estate activity). During the second half of 2008, both these factors abruptly ceased to function and a real danger of a 'hard landing' of the Russian economy emerged. The fourth quarter of 2008, the Russian companies ended up with losses of 442 bn. rubles ($18 bn.), and, due to this, the whole year's losses increased 4.7 times and profits decreased by 10 per cent, the balance by 30 per cent. Third, there was the *crisis of foreign debts* connected to the fast growth of corporative debt (almost five times during 2005–8). As a result, the total (state and commercial) debt load of the Russian economy began to grow, aggravating the nation's dependence on the world capital market. Closure of these markets in fact put an end to Russian private borrowing from mid-2008 and further provoked a crisis expansion, which required state assistance for covering debts, restructuring or buying out at least for strategically important banks and companies.

The crisis development in Russia was diversified in regional and sectoral dimensions. As a rule, the negative impact was stronger in previously more dynamic and economically successful branches and regions. In the first quarter of 2009, the fall of all economic indicators was fixed in 22 out of 83 subjects of the Russian Federation. Compared to the first quarter of 2008, the average fall in Russia was 10.4 per cent in investments, 14.3 per cent in industry, 19.3 per cent in the construction industry, and 3.3 per cent in incomes of the population while the unemployment rate was 9.5 per cent. Among the macroregional federal districts, the crisis first struck badly at the Ural region with its high concentration of heavy industry, as the demands for its goods dropped due to the investment squeeze, although it had the potential for a quicker recovery (−7.2 per cent, −8.7 per cent, −17.4 per cent and +0.1 per cent and 10.3 per cent, respectively). A more favourable situation (excluding industry) was to be seen in the south of Russia with a small number of large industrial enterprises but with an advanced agriculture disposing of a large and stable consumer demand (−4.5 per cent; −31.5 per cent; −10.3 per cent; but +3.4 per cent and 12 per cent, respectively).

A rather difficult situation was to be seen in Siberia with its domination of mining, quarrying, manufacturing of basic metals and its orientation towards decreasing and cheaper exports (-13.8 per cent; -17.9 per cent; -36.2 per cent; -8.9 per cent and 11.8 per cent). The Volga region suffered the most from the crisis, dominated by logistic enterprises, which depended on declining trade activities of other Russian regions and industry (mainly the automobile industry), and orienting itself solely to a sharply decreased domestic demand (-23.6 per cent; -22.4 per cent; -33.3 per cent; -0.6 per cent and 10.5 per cent). There was a less difficult but unclear situation in the northwestern region (including Saint Petersburg and Kaliningrad) and in the Far East region. The economy of the former highly depended on decreasing Baltic foreign trade transits and assembly plants (automobiles, TV and electric home appliances) based on more expensive and recently imported components (-7.9 per cent; -19.5 per cent; -27.9 per cent, -12.6 per cent and 7.5 per cent). The latter region is located far away from the European part of Russia and had not developed oil and gas pipelines from Siberia and is therefore very dependent on a growing railway network, aviation, gas and electricity tariffs, with a number of large industrial enterprises mostly engaged in state orders, strongly dependent on current Pacific foreign trade transits but with continuing investment in energy and transport infrastructure (2.4 per cent; -6.6 per cent; -22.3 per cent; -4.4 per cent and 10.3 per cent). In fact, only regions and branches obtaining state or foreign investments had a more or less stable socioeconomic situation (Vedomosti 4 May 2009; Putin 6 April 2009).

The Russian government had no forward plan for a crisis, but after recognizing its inevitability the authorities responded quickly. In October–November 2008 a set of emergency measures was adopted, including: support for the financial sector (for restoring liquidity of lending organizations, preserving people's savings in banks, continuing settlements between enterprises, helping firms to refinance debts); tax decrease as a safety net for industry; a smooth devaluation of the ruble (allowing economy and people to adjust to changed realities); curtailment in the growth of planned tariff rates (in energy and rail transportation); producing a list of 295 enterprises crucial for the national economy (for guaranteed state assistance), and, last but not least, a reduction of quotas for foreign workers.

This rescue plan had cost about \$200 bn., but cushioned the initial most painful effects of the crisis (lending to the real sector of the economy even increased by more than 15 per cent over the five-month crisis). The impact of the emergency measures had also shown itself in the first quarter of 2009.

For the realization of growth, the renewal variants in April 2009 were designed as anticrisis budgets and as an additional government programme to ensure, as premier Putin stated, 'an optimal combination of anticrisis measures and long-term projects, not only to ward off attacks but also to undertake offensives, to build a new and more effective economy' (Putin 6 April 2009). Thus, the new budget remains a development budget. For the first time since 2000, it has a deficit of 3 tn. rubles (about \$85 bn.) or 7.4 per cent of GDP due to an expenditure increase of 7.4 per cent, and revenues decreased by 38.6 per cent. Taking into account quasi-fiscal measures (mostly realized by the Central Bank), the real budget deficit will be 8 per cent GDP, which is close to the EU states' parameters (12 per cent in the United Kingdom, for example). About 47 per cent of the planned deficit is connected to the implementation of direct anticrisis measures and will be allocated for the support of the social sphere (about $15 bn.), real sector branches (about $6 bn.), regional budgets and banks ($8.5 bn. each). Total state financing of the anticrisis relief and economic rehabilitation (from reduction of taxes, state orders, reserves of the Central Bank, the National Welfare Fund and other sources) is estimated at 3 tn. rubles (about $85 bn.). The government plan was also to extend domestic borrowing and renew external borrowing. Some Russian experts estimate the aggregate state aid to the economy as being 4.8 tn. rubles ($136) or 12 per cent of GDP in 2009 (relatively the same rate as in China, although 4.3 times less in absolute figures). In fact, the state became the main buyer, main creditor and main borrower in the Russian economy. This should guarantee the realization of seven key priorities of the government programme: social security and employment of the population; preservation and development of the accrued industrial and technological potential; boosting internal demands; stimulation of innovations and the restructuring of the manufacturing sector; creation of favourable conditions for economic revival by developing the most important market institutes and removing barriers for commercial activity; construction of a powerful and reliable national financial system; responsible macroeconomic policy ensuring macroeconomic stability and securing trust in the country's budget system among Russian and foreign investors.

The Russian government estimated the positive effects of the anticrisis programme to compensate 1.1–3.6 percentage points of the possible decline in the dynamics of GDP, industrial and agricultural output, investments in fixed assets and real disposable incomes of the population thus creating the expectation of the revival of economic growth in the second half of 2010. Therefore its forecast for 2009 was more optimistic while for 2010 corresponded with main international estimations (see Table 7.2).

In the first quarter of 2009, Russian GDP declined by 9.5 per cent (against the officially expected 7 per cent) and the yearly figure estimation is 8.5 per cent, so the OECD (Organization for Economic Cooperation and Development) forecast

Table 7.2 Russian GDP development short-term forecast, %

	2009	*2010*
Russian Ministry of Economic Development	−2.2	1.3–3.5
OECD	−5.6	4.9
World Bank	−4.5	3.2
Goldman Sachs	−5.5	4.5

Source: Vedomosti 2 March 2009, 13 April 2009, 15 December 2009; Argumenty i fakty 25 November 2009

was more realistic. Thus, the decline in Russia will be close to the one expected in Germany (5 per cent) and deeper than in the Eurozone (4 per cent) and the United States (2.8 per cent). The main factors behind this trend were the bank credit 'bottlenecks', hampering the allocation of financial resources in the economy and aggravating recession and a sharper fall of investments and consumer demands together with a rapid growth in unemployment. The situation was worsening due to an excessive concentration of the Russian banking system's capital in a few large banks. The share of the five biggest banks (out of approximately 1200 banks in Russia) in this capital grew in October 2008 – February 2009 from 43.3 per cent to 46.9 per cent (Vedomosti 30 March 2009).

Meanwhile, from January to March 2009 the decline in the dynamics of the main socioeconomic parameters had not worsened. In practice, the Russian economy came to the turning point: further stagnation or renewal of growth. About 62 per cent of Russian companies remain profitable. Due to receiving state financial assistance and some banks' commercial credits, a GDP growth of 1.1 and 1.9 per cent was realized already in the two last quarters of 2009, although the beginning of a general economic recovery is not expected earlier than in the second half of 2010. Nevertheless, a recovery of the Russian economy beginning from 2010 is expected to be slow, and future growth is expected to be very modest compared with the previous five-year period. According to the Russian Ministry of Economic Development, the forecast for the GDP in 2009–12 will increase by 6.4 per cent only (in 2005–8 by 30.8 per cent), investments in fixed capital by 1.3 per cent (in 2005–8 by 72.1 per cent), productivity of labour by 9.8 per cent (in 2005–8 by 23.9 per cent), real disposable money incomes by 4.9 per cent (in 2005–8 by 47.2 per cent). A rapid growth will only show an inflation by 43 per cent in 2009–12 compared with 53 per cent in 2005–8. Industrial output will even decrease by 1 per cent (in 2005–8 there was an increase of 21.5 per cent) (Vedomosti 30 April 2009).

This rather pessimistic estimation corresponds relatively well with the results of the anticrisis competitiveness stress test of 57 countries published in May 2009 by the Swiss IMD international business school. The test puts Russia on the 51st place of country ranking: worse than some other big economies (the 18th place of China, the 24 of Germany and 28th of the United States) but better than a few emerging economies (Hungary, Romania, Ukraine, Argentina and Venezuela) (Vedomosti 20 May 2009). The realization of an anticrisis programme as the economic revival and prospective modernization of the Russian economy crucially depends on economic relations with the EU. Both face very difficult problems and need to find a common approach to solving them.

Russia and the EU: partnership both in the boom and in the crisis?

The world's financial–economic crisis sharply broke off the 'boom period' in trade and investment development both in the EU and Russia and in their relations. This phase began in 2000 and was strengthened from 2004 due to an exceptionally favourable foreign trade conjuncture for Russia and the stimulation of an 'enlarged' EU economy, mainly due to the rapid growth in new member states, predominantly in Central–Eastern Europe (CEE). Until the middle of 2008, it substantially improved the world positions of the EU and Russia. Even 2009 – the second year of the global crisis, the EU began maintaining its first place in the world by its volume of GDP, its export of goods and investment, whereas Russia held the eighth position of its GDP and seventh as regards exports in the world and third place in received foreign investments among emerging market economies.

In fact, during 2004–8 the EU and Russia became 'boom partners' reciprocally supporting their economic growth. The stable economy of 15 old EU member states and the growing economy of 12 new member states was, to a considerable extent, based on energy and raw materials from Russia and its constantly widening market while diversifying and modernizing (though very slowly) the Russian economy actively used the technological and financial resources of the EU. The trade–investment interaction had been developing dynamically and mutually profitably, despite the obvious disparity of the economic power of the partners. In 2008, the EU exceeded Russia ten times as regards GDP, four times in export values and 3–3.5 times as regards an economically active population. Meanwhile, the EU and Russia became dominants of the pan-European development (accumulating in the middle of the 2000s 85 per cent and 9 per cent, respectively, in a 'Wider Europe' aggregate of GDP and 62 per cent and 18 per cent, respectively, of the population) and of the sub-European or regional development (for example, dividing the Baltic Sea).

From 2002, when Russia declared a 'European accent' of its international relations, up to 2007, the average annual rate of EU exports to Russia was 21 per cent – three times higher than of the total extra-EU exports and 1.4 times higher than export to China. Analogous indicators for extra-EU imports were 20.8 per cent, 2.4 times and 1.2 times, respectively. In 2007, Russia for the first time ranked second (together with Switzerland) in extra-EU exports and third in extra-EU imports. The Russian share in extra-EU trade was equal to the aggregate share of Norway and Japan, while in exports it was ahead of China (6 per cent) and in imports just behind the United States (13 per cent). This evolution continued in the first half of 2008 with a 'peak' Russian share in extra-EU exports of 8 per cent and in imports of 11 per cent. On the other hand, the enlargement of 10 new (predominantly CEE) countries in 2004 sharply raised the EU share in Russian exports (from 35.2 per cent in 2003 up to 50.3 per cent in 2004) and imports (from 38.7 up to 43.0 per cent). The accession of Bulgaria and Romania in 2007 increased these shares up to 53 per cent and 43.5 per cent, respectively, and definitively fixed the EU domination in the Russian foreign trade. The financial-economic crisis broke the dynamics (in the first quarter of 2009, the turnover with the EU was 49.6 per cent of the total Russian foreign trade compared to 53.7 per cent in the first quarter of 2008) but without the essence of that trend (see Table 7.3). Moreover, in 2008 the EU–Russia trade in goods grew by a 'peak' of 25 per cent to Euro 280 bn.

Table 7.3 Evolution of the EU27 trade in goods with Russia, 2004–2008 (EUR bn.)

	2004	2005	2006	2007	2008
The EU:					
Exports to Russia	46.0	56.7	72.3	89.1	105.1
Imports from Russia	84.0	112.6	140.9	143.9	173.4
Balance	−37.9	−55.9	−68.6	−54.8	−68.3
Share in Extra –EU exports, %	4.8	5.4	6.2	7.2	8.0
Share in Extra –EU imports, %	8.2	9.5	10.4	10.1	11.2
Russia:					
EU share in exports	50.3	55.2	56.6	53.0	53.3
EU share in imports	43.0	44.2	44.0	43.5	46.7

Source: '*Vneshneekonomicheskiy kompleks Rossii: sovremennyoe sostoyanie i perspektivy*' *VNIKI* No. 2 2007: 19, 26; *EU-27 trade with China and Russia in 2007. Eurostat.Statistics in focus* No. 9 2009; *BIKI* No. 26 2009; www.delrus.ec.europa.eu/ru/news_1036.html.

Table 7.4 Evolution of EU27 trade in goods with Russia by main product groups, 2004–2007, %

Product groups	Energy	Machinery and vehicles	Chemicals	Food products
Exports:				
2004	0.5	46.5	13.8	10.0
2007	1.0	48.0	14.0	8.0
Imports:				
2004	60.4	1.5	3.7	1.7
2007	66.0	1.0	3.0	1.0

Source: EU-27 trade with China and Russia in 2007. Eurostat. Statistics in focus *No 9, 2009.*

Within the whole period, the EU had a trade deficit with Russia. In 2007, only four old EU states (Austria, Denmark, Germany and Ireland) and two new ones (Malta and Slovenia) recorded small surpluses in their trade with Russia. Nevertheless, this evolution was not regular: from 2007, Russian exports to the EU, fuelled predominantly by energy prices, grew slower than imports determined by high value added products and the same applied to the trade surplus. The 'Partnership in boom' was, in fact, based on an export-raw material model ineffective for Russia in the long-term perspective (see Table 7.4).

The crisis only aggravated the problems accumulated during the boom period and was mainly connected with *a double asymmetry of the EU–Russia interdependence. This asymmetry was stronger and more sensitive for Russia as regards both the total trade and main export goods.* In 2007, the EU's share in the Russian export of goods was almost eight times higher, and in imports – four times higher, than Russia's respective shares in extra-EU trade. With Russia's share in EU imports for coal being 17 per cent, enriched uranium 27 per cent, oil 33 per cent and natural gas 42 per cent, the EU's share in the Russian exports of these goods was higher (for gas – about 70 per cent). Besides, in some cases the trade was mutually connected: in 2008 'Gazprom' sold gas to the EU at a value of about $70 bn. and bought gas transport equipment there at a value of almost $30 bn.

The foreign trade asymmetry was exacerbated by a growing negative balance for Russia in services trade with the EU. In the period between 2005 and 2007, it doubled up to Euro 6.6 bn. (which was equal to 12 per cent of the Russian positive trade balance in goods).

In the investment sphere, the interdependence asymmetry was even stronger for Russia. The dominating part of foreign capital Russia obtained from the EU (more than 75 per cent of the total cumulative foreign investment and 77 per cent of the direct one in 2008). Simultaneously, Russia concentrated in the EU more than 60 per cent of its cumulative investment abroad. Moreover, the EU's cumulative investments in Russia (about 200 bn. Euro in 2008) were six times higher than the respective Russian investments.

As for current capital flows, the EU also dominated both in total and in the main sectors of foreign investment in Russia. In 2004–7, the EU27's net foreign direct investments to Russia increased by 2.8 times up to Euro 16.1 bn. In 2008, the EU's share in total foreign investments inward to Russia was 65 per cent for mining and the quarrying of energy producing materials, 69 per cent – for the manufacture of food products, 87 – for the manufacture of basic metals and fabricated metal products and 91 per cent – for electricity, gas and water supply (Eurostat 2008; Eurostat 2009; BIKI 2009/38). Last but not least, Russia became one of the largest world holders of Euro-denominated assets and the crisis had even strengthened that trend. At the beginning of May, 2009, the share of Euros in the total Russian currency assets was 41.5 per cent (about Euro 120 bn.) – almost 1.6 more than average for the world's currency assets (Vedomosti 19 May 2009).

Against this background, laments on an 'overexpansion' of Russian capital in the EU sounded strange, although the flow of Russian foreign direct investments in the EU grew five times during 2004–7. It was an objective process caused, on the one hand, by demands of globalization of the world and European economy, and, on the other hand, by specific conditions of business activity in Russia. These investments were aimed at ensuring the transformation of technological–production cycles of the large Russian companies in the complete upstream and downstream transnational corporations, coming close and easing the accession to the EU's internal markets, assistance for the modernization of the main sectors of the Russian economy through a transfer of modern assets and technologies, and, last but not least, the additional protection from a corrupt Russian bureaucracy. Lately, these processes have practically resulted in the formation on EU territory of some sort of 'external sector' of the Russian economy. The advantages of such a development were rather obvious in the boom period. The disadvantages were brutally revealed by the current crisis, when the Russian government had to allocate about $200 bn. to restore the liquidity of the big banks and to refund the debts of corporations, which for years used large — scaled cheap and long money from the EU.

Table 7.5 EU27 trade with Russia by Member State Groups, 2007 (%)

	Share in extra-EU exports	Share in extra-EU imports	Share in EU–Russia exports	Share in EU–Russia imports
EU27	7.2	10.1	100.0	100.0
EU15	6.8	8.3	83.0	76.0
EU12	18.0	28.5	17.0	24.0

Source: EU-27 trade with China and Russia in 2007. Eurostat. Statistics in focus *No 9, 2009.*

Table 7.6 EU27 dependence on Russian gas supplies by Member State Groups, 2007 (%)

	Share of Russian supplies in gas consumption	Share in Russian gas supplies to the EU
EU27	27.7	100.0
EU15	22.2	67.9
EU12	57.2	32.1

Source: BP Statistical Review of World Energy, *London, June 2008.*

The certain asymmetry concerning trade – investment interdependence with Russia was also observed within the EU. Old member states, while prevailing in the total EU–Russia trade were substantially less dependent on it than the new member countries (see Table 7.5). In 2007, the EU15 that were providing about 80 per cent of trade in goods with Russia depended on this less than on 9 per cent of export–import ties. Conversely, extra-EU trade of the EU12, at an average, was almost three times stronger depending on the turnover with Russia while providing its four times lesser share.

Trade asymmetry was substantial, in particular for the most sensitive of the EU's energy producing materials, including gas (see Table 7.6). New member states having an average share of the total EU gas purchases from Russia two times less than the old ones had an almost three times stronger dependence on the Russian deliveries.

Intra-EU asymmetry was weaker in the services trade with Russia, where the share from the new members was higher (more than 50 per cent of the EU's positive balance in 2006), though mainly due to revenues from the Russian transits to the old member states.

The EU15 also dominated in the balance of received and cumulative EU investments in Russia. Meanwhile, the Russian investment activity to a great extent moved to the CEE countries of the EU (in 2007, more than 50 per cent).

This division between the EU15 and EU12 obviously oversimplified the real distinctions in the approach of old and new member states to interdependence and cooperation with Russia. EU experts regularly tried to identify groups of member states by the extent of 'loyalty' or 'friendliness' towards Russia, according to a wider range of indicators, including the strength of trade and investment flows, energy dependence, considerations for regional security, the existence of economic and political disputes, historical and even religious backgrounds. The European Council on Foreign Relations (ECFR) study of 2007 defined five distinct policy approaches to Russia shared by old and new members alike: the *'Trojan Horses'* (Cyprus and Greece) who often defend the position of Russia within the EU system up to vetoing unfavourably for it in common EU decisions; the *'Strategic Partners'* (France, Germany, Italy and Spain) who enjoy a 'special relationship' with Russia, which occasionally undermines common EU policies; the *'Friendly Pragmatists'* (Austria, Belgium, Bulgaria, Finland, Hungary, Luxembourg, Malta, Portugal, Slovakia and Slovenia) who maintain a close relationship with Russia and often put their business interests above political considerations; the *'Frosty Pragmatists'* (the Czech Republic, Denmark, Estonia, Ireland, Latvia, the Netherlands, Romania, Sweden and the United Kingdom) who also focus on business interests but are less afraid than others to speak out against Russian behaviour on human rights or other issues; and the *'New Cold Warriors'* (Lithuania and Poland) who have an overtly hostile relationship with Moscow up to using the veto to block EU negotiations with Russia (Leonard and Popescu 2007). Research conducted at the University of Siena in 2008 (Braghiroli and Carta 2008) categorized the EU states according to their attitudes towards Russia in four groups: the *'Eastern divorced'* – former Comecon countries and the Soviet Union republics (Poland, the Czech Republic, Estonia, Latvia and Lithuania), who nowadays have the coldest or even a hostile attitude towards Russia; the *'Loyal Wives'*, who maintain good relations with Russia (Austria, Greece and Italy); the *'Vigilant Critics'* – CEE states with mainly 'defrosted' relations with Russia and Western countries totally energy independent from Russia (Bulgaria, Hungary, Slovakia, Slovenia, Portugal, Sweden and the United Kingdom); and the largest and most heterogeneous group of the *'Acquiescent Partners'* comprising only old EU states – small with a lower stake in relations to Russia or European great powers that maintain stable strategic relations with Moscow (Belgium, Denmark, Finland, France, Germany, Ireland, Luxemburg and Spain).

Despite the distinctions in the composition of groups, both studies show the absence of a direct correlation between the rate of concrete trade of EU member states – investments and energy dependence on Russia and their positioning within the system of EU–Russia relations. The researchers revealed a rather similar main paradigm of EU policy towards Russia, precisely described in study.

> At one end of the spectrum are those who view Russia as a potential partner that can be drawn into the EU's orbit through a process of "creeping integration." They favour involving Russia in as many institutions as possible and encouraging Russian investment in the EU's energy sector, even if Russia sometimes breaks the rules. At the other end are member states who see and treat Russia as a threat. According to them, Russian expansionism and contempt for democracy must be rolled back through a policy of

"soft containment" that involves excluding Russia from the G8, expanding NATO, supporting anti-Russian regimes in the neighbourhood, ... excluding Russian investment from the European energy sector. ... The first approach would give Russia access to all the benefits of cooperation with the EU without demanding that it abides by stable rules. The other approach – of open hostility – would make it hard for the EU to draw on Russia's help to tackle a host of common problems in the European neighbourhood and beyond'.

(Leonard and Popescu 2007)

This policy was paradigm reflected *at three main levels of EU–Russia economic cooperation: most strongly at the national level with bilateral relations between Russia and each member state; moderately at the EU level with relations between Russia and the EU as a collective partner; weakly at the all-European level with Russia–EU relations in the context of the positioning of a 'Wider Europe' within a globalizing world.* These levels are different from the point of view of character, development, and balance of economic and political factors. However, their mutual priority aim is to ensure sustainable growth and economic and political security for partners. During the boom period, the EU–Russia partnership can, from time to time, fluctuate closer to the 'soft containment' variant. During the crisis, positions of 'friendly pragmatics' or 'acquiescent partners' are preferable (if not inevitable), especially for CEE countries. These latter countries, as it was formulated in the IMF report, practically instantly lost the 'aureole effect' of EU membership (Vedomosti 14 May 2009) and had to seek any possibility for economic recovery. *Therefore, the optimal model of the EU–Russia crisis partnerships can be called 'fleximatism' – from 'flexibility' and 'pragmatism'.* This model is based on an optimal combination of using all national and intra and extra EU trade-investment resources with the depolitization of economic decisions, including reasonable compromises with foreign partners. It presupposes a mutual rejection of the idea of determined rivalry between Russia and the EU and a vital necessity of a formation of their common economic space as the best way for pan-European economic security, global competitiveness and getting closer in non-economic values.

At the first level, 'fleximatism' is currently vitally important and really possible. Only restraining from unilateral protectionist measures and maintaining open trade–investment regimes can soften negative crisis effects and prepare the base for future growth. This is especially true for CEE countries with a 20–50 per cent Russian share in extra-EU trade. Manoeuvrability of the EU member states is limited, as their trade policy is in the competence sphere of Brussels. Nevertheless, they have a considerably more free hand in cooperation with Russia in transports, energy, trans-European networks, ecology, industrial policy and tourism. National governments and business enterprises have already tried to realize these possibilities with a minimum deterioration of trade balances and internal markets of partners. In 2009, despite a temporary decline in demands, German, French and Italian companies did not stop their production of automobile and electric home appliances in Russia and Russian companies did not stop steel works in Italy and aluminium production in Ireland in order to maintain market positions and personnel.

Moreover, some firms tend to use the crisis for a further expansion of their partners' markets, sometimes on a multilateral basis and with their governments' support. According to the December 2008 poll, 30 per cent of German firms in Russia planned to widen their activities. On the one hand, British–Dutch 'Unilever' bought the control stock (40–50 m. Euro) of the largest Russian ketchup firm for controlling 50 per cent of the Russian market, and the French 'Alstom' – the same stock ($200–300 m.) from the dominating Russian producer of locomotives and wagons, and the Czech financial group PPF bought one of the biggest Russian retailers of TV and electric home appliances 'Eldorado' for $500 m. On the other hand, the biggest Russian private bank 'Alfa bank' seized control of the second world producer of auto brake components – the German 'TMD Friction', preventing its bankruptcy and ensuring work for 4,000 employees in Germany and Italy. The state 'Czech Export Bank' finances ($156 m.) delivery of German-made equipment and its assembling at the Russian Magnitogorsk steelworks by the Czech engineering firm 'PSG' in order to support its expansion in Russia. In May 2009, a Canadian–Austrian world-scale auto component company 'Magna' and the largest state-controlled Russian bank 'Sberbank' won the tender for buying from 'General Motors' the control stake of one of the leading EU automobile companies, 'Opel', (35 per cent for 'Sberbank') promising to keep all 25,000 'Opel' work places in Germany and a main part of its employees in other EU states. Unfortunately in December 2009 the American corporation broke that deal. Meanwhile, the crisis in cutting off component deliveries from the Dutch 'Philips' and the Lithuanian 'Snaige' blocked the production of some types of TV sets and refrigerators in the Kaliningrad region. In May 2009, Russia, Belarus and Lithuania stopped a long-lasting dispute and operatively agreed on railway tariffs being equal for directions to the ports of Kaliningrad (Russia) and Klaipeda (Lithuania) in order to support declining international transits.

Also, Russia did not show fleximatism when it increased export taxes for timber and import tariffs for automobiles, decreasing meat import quotas and criticizing the fact that EU fruit and vegetables have higher limits of pesticide residue. Nevertheless, it is now seeking a reciprocity decision: improving conditions for the Baltic, Finnish and Swedish investments in timber processing in Russia, developing exports of semiproducts for final processing in these states; giving some preferences to EU companies assembling cars in Russia; negotiating bilateral agreements on the safety of horticultural products with a majority of the member states concerned, etc.

The Baltic States are a good case. The large Russian business enterprises operating there are generally more cosmopolitan than their national enterprises. In fact, it rather defends the interests of the Baltic states in Russia, softening the consequences of political disagreements. The state-controlled 'Gazprom' and the private transnational 'Lukoil' are reinvesting most of their Baltics' profits in these countries and are not breaking off the supply. Nevertheless, permanent political frictions decrease, for Russian business enterprises, the predictability and attractiveness of economic relations with the Baltics. Economic patriotism becomes more effective as well as business expansion in the politically more friendly Baltic Sea states (Finland and Germany). This also applies to the enterprises of the Baltic States in Russia, where they from time to time camouflage themselves for Scandinavian companies. Lately, the Baltic States have begun to soften their routine rhetoric concerning their Russian minorities and their common history with Russia, as, in changed conditions, a non-flexible and unfriendly policy negatively influences Baltic business competitiveness on the Russian market. Meanwhile, local authorities and business enterprises in the Baltic States (small and medium-sized business enterprises in regions near the Russian border, in particular) behave more fleximatically than their governments do. A full depolitization of bilateral Russian economic relations with the Baltic States is rather improbable. Nonetheless, the fleximatism becomes much more useful both in the mid-term (the next 2–4 years with a difficult internal and particularly external economic situation) and long-term (10–15 years with an active shaping towards new globalization challenges – cooperation in innovation development, including the organization of joint innovation clusters in the neighbouring regions). This is true for the rest of the EU states, independent of their pre-crisis positions.

At the second level, where Russia is playing with the EU member states as representatives of the combined team, fleximatism is even more necessary. There, the weight of the political factor is much heavier. The member states that are responsible for a relatively small part of the EU's total economy and trade – investment turnover with Russia can much more seriously have an influence on the Union's economic policy towards Russia. It is particularly important for CEE countries from *'New Cold Warriors'* or *'Eastern divorced'* groups. For instance, the Baltic States have about a four per cent membership in the Euro parliament and the Council (twice more than Finland or Denmark and only twice less than France or Germany). Pooling with Poland and one more medium-sized EU state (Romania, for example), they can in fact block any decision that is unsatisfactory for them in the Council. Any member state can veto the preparation and ratification of any EU agreement with Russia, even the new basic Partnership agreement and the agreements on the formation of the four common spaces. In fact, Lithuania did so up until May 2008, replacing Poland who had done this up until the end of 2007. Member states from different abovementioned groups try to 'speak in a single voice', while participating in the development of EU common policies and 'unified rules of the game' in partnership with Russia. In practice, this periodically leads to attempts of putting their group or even national interests higher than common EU interests. All seven member states pressing for sanctions against Russia after the Caucasus conflict in August 2008 (in which the EU later became the successive intermediary) belonged to groups usually unfriendly to it (five new states – Poland, Estonia, Latvia, Lithuania, the Czech Republic and only two old ones – Sweden and the United Kingdom).

Nevertheless, global economic crises have substantially changed the socioeco-nomic situation within the 'enlarged EU' and in the 'wider Europe' and created prerequisites for breaking with the former model of the EU–Russia relations: 'whereas old member states consider the strength of economic ties with Russia as a *relative* asset, new member states consider it as an *absolute* form of dependence, which threatens their overall stability' (Braghiroli and Carta 2008).

The necessity of coordination and joining efforts in overcoming the crisis and a renewal of economic growth creates better conditions for more operative achievements of a new EU–Russia strategic partnership agreement providing for a 'comprehensive legally binding framework to cover all main areas of the relationship based on ... shared interests and the international commitments' of the EU and Russia, including the free trade area (EU Commission 2008b). In an economic context, the strategic goal of a new basic agreement is the formation in the historically limited period (15–20 years) of the 'integrated market' of the EU and Russia, analogous to the EU's 'common market' just before its transformation into an 'economic and monetary union'.

A fleximatic approach can bring with it a mutual benefit compromise even on such old and difficult problems as the EU–Russia energy interdependence and regional security. Russia made serious steps on this path, particularly in the energy sphere. The Russian energy companies (including 'Gazprom') are preparing to act on the EU's gas and electricity markets as on the consolidated continental sized markets and to adapt to regulations of the 'third energy package'. Russia had opened its electricity markets for EU companies and constantly widened the mutual profitable practice of change of assets in upstream and downstream branches of the oil and gas sectors. Russia would neither try to prevent the construction of the 'Nabucco' gas pipeline bypassing its territory from the Caspian region to the EU, nor object to the participation of member states (Bulgaria, Hungary and Austria) simultaneously in this project and the alternative Russian 'South Stream' project. It prefers open competition in these projects, bearing in mind that with similar costs (about 9 bn. Euro), the Russian project has guaranteed gas and financial resources, practically with twice as large a capacity and has not any non-EU transit country ('Nabucco' will have three – Azerbaijan, Georgia and Turkey). Simultaneously, Russia is activating the construction of the 'Nord Stream' pipeline, which may improve stability and cost-effectiveness of EU's gas supply. The project has been delayed and increased in costs due to the counteractions of Poland and the Baltic states and the cautious positions of Sweden and Finland. 'Nord Stream' already had priority status in the EU, and, apart from Russian, German and Dutch participants, would probably include France. These Russian actions correspond to the EU's strategic energy policy. In this context, Russia expects reciprocal acts from the EU. In particular, it awaits a positive EU reaction on its proposal for a new international agreement on energy security, changing or replacing the Energy Charter Treaty that is unfavourable for Russia and other main gas and oil suppliers (Norway, the Persian Gulf and Caspian Sea region states). Russia also insists on returning to the idea of a transnational consortium for the modernization and exploitation of a

Ukrainian pipeline system for transits of Russian (and partly Caspian) gas to the EU with Ukrainian, Russian and European participation.

The relations at this level will obviously not be limited to traditional energy spheres. A wider perspective is cooperation in new 'postindustrial' fields, such as alternative energy (nuclear, biological, wind, waves, etc.), biotechnology and nanotechnology and other directions corresponding to a 'knowledge based economy'.

Regional security is another prospect sphere for a fleximatic approach, especially concerning the EU–Russia mutual notion of a 'close abroad'. In this context, launched in May 2009, 'The Eastern Partnership' aiming at deep bilateral relations and built on a new multilateral framework for cooperation, the EU and six CIS countries (Azerbaijan, Armenia, Belarus, Georgia, Moldova and Ukraine) need a more detailed explanation to prevent suspicions that its real goal is decoupling these countries from Russia and the formation the EU 'zone of influence' on the perimeter of the Russian borders in Europe. The Eastern partnership is oriented towards convergence of the economic policy of the participants with EU policies and includes five profile initiatives ('flagship initiatives'): a border management programme; the integration of electricity markets, energy efficiency and renewables policies; small and medium enterprise facility; 'southern energy corridor', and a common response to disasters. The initial financing of these measures from the EU budget is 600 m. Euro. As Commissioner B. Ferrero-Waldner has formulated: 'We are asking a lot and we have to be willing to give in return. This is not philanthropy. It is 21st century European foreign policy' (EU Commission 2008b). All these initiatives affect the vital interests of Russia and its plans of subregional integration within the CIS region, in particular border management and the 'southern energy corridor' covering three gas pipelines bypassing Russia. Russian concerns that 'there are those who may wish to present the invited participants with the choice – either you are with Russia, or with the European Union' were revealed by Russian Foreign Minister Lavrov (EurActive 2008a).

At the third level, Russia and the EU in the long-term perspective should become members of one 'Wider European team'. Here, the intensification of a Russia–EU strategic partnership becomes necessary for the realization of a common interest of the EU as a part of a 'Wider Europe' in addressing new current globalization challenges (overcoming world financial–economic crises) and strategic (the common fight against climate changes, world food and energy problems). More attention should attract cross-border cooperation, Baltic Sea ecological fishing issues as well as a transcontinental transportation system and other common macroregional problems, particularly within the framework of a 'Northern Dimension' currently being joined by Iceland and Norway. Only with an approach of fleximatism can a Wider Europe effectively influence the reforming of the global financing architecture, participate in the creation of a world food security system, launching of the post-Kyoto process, use connected with global warming opportunities for the opening of new transcontinental trade–transport routes and the access to fossil resources of the Arctic Sea shelf. The first practical results have already been achieved. In March 2009, Russia and the EU adopted a long-term agreement on cooperation in the fisheries sector and the preservation of live marine resources in the Baltic Sea and coordinated their positions before the G20 meeting in the London summit on world financial market reform, including a global system of supervision.

The EU–Russia partnership in crisis creates a rare and real chance to be transformed into a prospective new relationship based on trust and a tighter integration, and fleximatism can be the 'reset knob' for such a development.

8 Germany–Russia relations, 1992–2009[1]

Angela Stent

Introduction – the legacies of history and geography

Russia's relations with Germany have undergone a transformation since the end of the Cold War and, as Russia has recovered from the Soviet collapse, Berlin has become increasingly significant for Moscow, making Germany Russia's most important Western economic and political partner today. Three moments since 1992 capture this remarkable evolution: in August 1994, President Boris Yeltsin seized the baton from the conductor of the military band in Berlin at the parade as the last Soviet troops left Germany and tried to conduct the farewell march for the soldiers who had occupied the GDR for more than forty years. At that point, Russia was economically weak and in the throes of a difficult transition. In September 2005, a confident President Vladimir Putin presiding over a dynamic economy and his friend Chancellor Gerhard Schroeder signed a deal to build an undersea gas pipeline that will increase gas supplies to Europe and deepen Germany's energy dependence on Russia. In June 2008, newly elected President Dimitry Medvedev, leading a country which at that point had the third largest currency reserves in the world, chose Berlin for his first major foreign policy speech, proposing a new Euro–Atlantic security architecture that could eventually replace NATO and the OSCE (Medvedev 5 June 2008). Even though, unlike Putin, Medvedev has no German background and does not speak German, Germany is viewed as Russia's central partner in many different fields and as the Western country that best understands Russia's complex situation.

History and geography have ensured that Russia and Germany would enjoy a close relationship. Russians have always admired Germans for their technological prowess and their cultural achievements. Peter the Great, Russia's first modernizing Tsar (1689–1725), brought Germans to Russia to assist in developing his country that lagged behind Europe. His successor, Empress Catherine the Great (1761–96), herself a German princess, created a large German immigrant colony along the Volga river to develop Russia's backward agricultural sector. Moreover, the Baltic German nobility played a disproportionately important role in the administration of the Russian empire. In the nineteenth century, about one-third of high government officials in Russia were of German origin at a time when

Germans constituted about 1 per cent of the total population (Laqueur 1993: 53). Historically, the lack of natural frontiers for both Germany and Russia and the complementarity of their economies – Russia exported raw materials to Germany and imported manufactured goods from Germany – created a pattern of relations between the two powers where periods of cooperation alternated – and sometimes coexisted – with periods of confrontation.

The interaction between Russia and Germany in many ways defined the history of the twentieth century. Germany played a major role in both the birth and death of the Soviet Union. The Rapallo Treaty of 1922 established diplomatic relations between the USSR and Germany at a time when most major powers shunned Bolshevik Russia, facilitating the Soviet Union's entry onto the international stage. The clandestine military cooperation between the USSR and Germany from 1921–33 enabled Germany to evade the restrictions on its rearmament imposed by the Versailles treaty of 1918 and facilitated the Soviet Union's military modernization. World War II – in which more than 25 million Soviet citizens perished – 1 million alone during the Wehrmacht's siege of Leningrad – left an indelible legacy on Germany–Russia relations. Until the collapse of the USSR, the driving force behind Moscow's foreign policy was the need to ensure that Germany could never again attack the USSR; hence the creation of the Soviet-dominated GDR and the belief that it was necessary to control the governments of the Eastern European countries that lay between the GDR and the Soviet Union to prevent them allying with Germany against Russia (Stent 1999: Chapter 1). Germany also played a key part in the demise of the Soviet bloc. The fall of the Berlin Wall on 9 November 1989, set in motion a process that ended with the collapse of both the GDR and the USSR. Since then, Russia has looked to Germany to be its chief advocate in Europe. Germany appreciates Russia's role in allowing Germany to reunify and continues to feel a special responsibility toward it.

There are, therefore, three major legacies that continue to influence Moscow's relations with Berlin. The first is the centrality of Germany for Russia's domestic economic and political evolution and for its rise as a major European power. The second legacy is of the historical cooperation between the two countries, which has often had a positive impact on Russia and its neighbours. Given their close economic ties for more than two centuries, Germany played a key role in Russia's modernization and in the attempt to reinforce Russia's evolution as a European power, despite Russia's ambivalence toward Europe (Stent 2007: 393–442). Russia has been a key energy supplier to Germany for decades, and Russian gas has heated German homes for over forty years (Stent 1981: Chapter 7).There is, however, also a legacy of less benign cooperation in the twentieth century – the interwar secret military collaboration, the Molotov–Ribbentrop Pact (1939) that enabled the USSR to occupy Eastern Poland, the Baltic states and part of Romania, and Soviet–GDR collaboration that repressed the populations of Eastern Europe. The third legacy is of Russia–Germany conflict which produced two World Wars in the twentieth century and made Berlin the flashpoint for East–West tensions during the Cold War. With the rise of post-communist Russia, and Germany's commitment to bind a reborn Russia closely to the European

Union, the positive cooperative legacy will no doubt prevail in the twenty-first century.

The process by which Germany was reunified between 1989 and 1990 has also had an impact on Russia–Germany relations. Key decisions were made by Soviet president Mikhail Gorbachev and a few close advisors over the opposition of the traditional Communist party German specialists, known by Gorbachev's advisors as the 'Berlin Wall', because they were against German unification and against the entry of a united Germany into NATO (Gorbachev 1995: 712–14, Cherniaiev 1993: 305–9). After the Soviet collapse, these Soviet Germanists accused Gorbachev of betraying Russia. They argued that Gorbachev had promised that German unification would be the last act in a gradual process of the overcoming of Europe's division during which both the Warsaw Pact and NATO would be dissolved. The Warsaw Pact did end, but NATO, of course, survived, leaving the issue of the future architecture of European security and of Russia's place in Europe unresolved. Many Russians believe that the United States and its allies violated promises made to Gorbachev in 1990 by enlarging NATO to Central Europe and the Baltic states (Karaganov 2009). Medvedev's call for a new European security pact during his first visit to Berlin as president in June 2008 was a response both to this conviction and to the prospect of further NATO enlargement to the post-Soviet space, a reminder that Russia's belief that its interests were ignored during the unification negotiations continue to influence its policies toward both Germany and Europe.

Key issues under Boris Yeltsin

In the aftermath of the Soviet collapse, Germany was the major source of economic support as Russia embarked on its difficult post-communist transition. Moscow realized that German gratitude for unification and concern that Russia's weakness had the potential to disrupt European security were two key sources of Russian leverage. Moreover, Germany believed that it understood Russia's situation better than other countries because of its own experiences after its defeat in 1945. During Yeltsin's tenure as president (1992–99), Germany (despite having to deal with the daunting economic and social challenges of its own unification) was the stronger partner, supporting the Yeltsin administration politically and economically, and acting as Russia's advocate in European structures. It was an asymmetrical interdependence which Russian leaders recognized but sometimes resented. Four major bilateral issues dominated the relationship in the 1990s: troop withdrawals, ethnic Germans, economic ties and Germany's support for Russia's domestic evolution.

In 1990, there were 546,000 Soviet troops and their dependents in the GDR. As a result of the negotiations that ended Germany's division, Russia had agreed to withdraw its troops in four years, but the process was a major logistical challenge for both Russia and Germany. To which country would these 'Soviet' soldiers return? Now that the USSR had been replaced by fifteen independent countries, how would these military personnel (for instance, 30 per cent of the officer corps was ethnically Ukrainian) determine where they belonged? Who would provide housing for them? Withdrawing such a huge military machine held many potential pitfalls.

Despite these challenges, the soldiers had left by 1994. Given all the obstacles to completing the troop withdrawal, it is remarkable that the operation proceeded as smoothly as it did (Brandenburg 1993: 76–88). The reasons for the success were a combination of efficient German organization and Russian calculation that it was ultimately better to cooperate than complicate the process. Nevertheless, perceptions of mistreatment and humiliation on the part of officers and enlisted men fuelled domestic opposition to the policies of the Yeltsin government.

A second bilateral issue in the 1990s and another Cold War legacy was the situation of ethnic Germans in post-Soviet Russia. Many of these 2 million descendants of settlers brought over by Catherine the Great and deported to Kazakhstan and Siberia from the Volga region after the German invasion of the USSR in 1941 had sought to emigrate to the Federal Republic during the Soviet period. After unification, Germany tried to encourage these ethnic Germans to remain in Russia, but was largely unsuccessful. Many ethnic Germans have left Russia and resettled in Germany since 1992.

After assenting to German unification, Russia anticipated that Germany would be a major source of both economic assistance and trade and investment for its emerging market economy. The complementary character of the economic relationship continued, with Russia providing oil, gas and other raw materials to Germany and importing German finished goods. The German private sector remained actively involved in the Russian economy, but was cautious about investing during the Yeltsin era, given the absence of the rule of law and the paucity of enforceable legal structures to protect their investments. Indeed, the most dynamic period of Russia–Germany economic ties only began after the Russian financial crash of 1998, when the economy had begun to recover. In the 1990s, Germany also contributed to and supplemented American programs designed to help secure and dismantle Russian nuclear weapons and materials and reduce the dangers of proliferation by providing safeguards for nuclear facilities. Germany offered training to unemployed nuclear scientists in the Russian Federation so that they could find alternative jobs instead of selling their skills to countries or non-state actors seeking to acquire nuclear weapons.

During Yeltsin's last year, the Germany–Russia relationship experienced strains when Gerhard Schroeder became Chancellor and vowed to reassess the 'sauna diplomacy' of the Kohl–Yeltsin era. The Russian economic crisis, the succession of five prime ministers between March 1998 and August 1999, and Yeltsin's failing health and erratic behaviour adversely affected ties. Moreover, other European developments – the 1999 enlargement of NATO to include Poland, Hungary and the Czech Republic as well as the Kosovo war – both of which Germany supported and Russia opposed – caused further strains in bilateral relations. By the end of 1999, therefore, with Yeltsin on his way out, a fragile economy and growing Russian alienation from the West, the Germany–Russia relationship appeared increasingly challenged.

The Putin presidency, 2000–2008

The Schroeder years – 2000–2005

When Vladimir Putin became President, his official biography highlighted the fact that he had spent five years serving the KGB in the GDR, spoke German and was, as one German writer noted 'der "Deutsche" Im Kreml (The German in the Kremlin)' (Rahr 2000). It became clear that he would focus on Germany as a key interlocutor. Yet, his experiences in East Germany had been contradictory. His 2000 autobiographical interview *First Person* emphasizes how much he enjoyed his time in the GDR, where people lived much better than they did in the USSR. His work in Dresden consisted of 'the usual intelligence activities: recruiting sources of information, obtaining information, analysing it and sending it to Moscow'. Even though he understood that the GDR could not last, 'I only really regretted that the Soviet Union has lost its position in Europe, although intellectually I understood that a position based on walls and dividers cannot last. But I wanted something different to rise in its place. And nothing different was proposed. That's what hurt' (Putin 2000b: 69, 80) Putin's experience of the collapse of East Germany and the hasty Soviet exit from the GDR influenced his foreign policy views and made him determined to restore Russia's position in Europe.

Early on in his tenure, Putin singled out Germany as a key interlocutor. At the beginning, Chancellor Schroeder, like his other European counterparts, was cautious about Putin, with his dual biographies of KGB officer and assistant to the reformist mayor of St. Petersburg, Anatoly Sobchak. Although early on he had emphasized that Russia was a European country and had committed himself to pursuing economic reforms and further modernization of Russia, he had also launched the second Chechen War and created a political system termed 'managed democracy', which eventually included a compliant parliament and new political parties created and largely controlled by the Kremlin. Nevertheless, his initial commitment to greater economic integration with the West, to more effective governance and to battling corruption found a sympathetic ear in most of Europe. While he stressed the need for better ties with the EU as a whole, his focus was on key bilateral relations, with Germany as the most important interlocutor. While relations with the United States were tense in 2000, he concentrated on cultivating ties with Germany as the first step to restoring Russia's position as a great power.

The German government soon responded to Russian overtures and the relationship recovered from the difficulties in 1999. Both sides spoke of their 'strategic partnership' designed to integrate Russia into Europe and strengthen the rule of law. Contacts between representatives of civil society in both countries were also strengthened, student exchanges grew and the Petersburg Dialogue, a Putin–Schroeder initiative, involved regular meetings aimed at bringing a wide range of Germans and Russians together. Although Schroeder himself had criticized Kohl for the over-personalized nature of the Germany–Russia relationship during Yeltsin's tenure, he now admitted that 'without President Putin little gets done in Russia' ('ohne Präsident Putin geht in Russland wenig') (Schroeder 2001).

Germany–Russia relations had, therefore, improved after Putin came to power, but they were given an additional impetus by the terrorist attacks on the World Trade Center in New York on September 11, 2001. Putin's support for the United States reflected what appeared to be a strategic choice in siding firmly with the West over the issues of international terrorism. Shortly after the attacks, he made a landmark visit to Germany. He made a historic speech in the newly restored Berlin Reichstag, which is full of historical symbolism – including the graffiti on the walls left by triumphant Soviet soldiers as they completed their conquest of Berlin. Speaking in German, he regretted that he had not warned the West more directly about the possibility of such a catastrophic attack, made a direct connection between Al Qaeda and Chechen separatists, and pledged support in the international fight against terrorism. Referring repeatedly to the historic ties between Russian and Germany, he highlighted their shared destinies:

> Russia has always had special sentiments for Germany, and regarded your country as one of the major centers of European and world culture (-) Between Russia and America lie oceans,-while between Russia and Germany lies a great history (-) Today's Germany is Russia's leading economic partner, our most important creditor, one of the principal investors and a key interlocutor in discussing international politics.
>
> (Putin 25 September 2001)

Putin was largely successful in ensuring that Germany developed a strong stake in its bilateral economic and political relationship with Russia and viewed itself as Russia's major advocate within the European Union. After the initial US–Russia rapprochement soured by the end of 2002, Russian and German opposition to American plans to attack Iraq crystallized into a joint front (with France) against the war. This 'coalition of the unwilling' did not develop into the alliance that Putin might have wished for, but it ensured that Russia was not isolated in its opposition to the war. Thereafter, until the end of Schroeder's tenure in office, the bilateral relationship flourished. Roughly 200,000 Russians had come to live and work in Germany since 1992, moving between the two countries and building a network of personal and business ties. Venues for civil society interaction grew, as did the stakeholders in the relationship. Despite criticism in the German media about Putin's moves towards a more centralized and less competitive political system and his muzzling of independent electronic media, the Chancellor himself, when asked 'Is Putin a pure democrat?' answered 'Yes, I am convinced that he is' (Schröder 2002). Putin rewarded the Chancellor for his support by naming him the Chairman of the consortium of Nord Stream, the Russia–Germany pipeline project that will transport Russian gas to Germany under the Baltic sea, a deal that Schroeder and Putin negotiated while Schroeder was still in office and from which he profited once he left office.

Indeed, energy is a key factor in the Russia–Germany relationship. Under Putin, Russia emerged for the first time ever as a major global economic player, largely as the result of a sharp rise in energy prices. The most significant aspect of Putin's tenure as President was the rise of Russia as an energy superpower, combining traditional geopolitics with instruments from the world of globalization to implement them. Russia is the world's number one producer and exporter of oil and has the world's largest natural gas reserves. More than 90 per cent of its energy exports go to Europe. During the Cold War, the USSR was known as a reliable supplier of gas to Germany. However, under Putin, energy became a problematic issue in relations between Berlin and Moscow when Gazprom (closely linked to the Kremlin), without prior warning, interrupted its gas supplies to Europe on several occasions, affecting many EU countries, with no resolution to the issue apparent. The first cut-off came on New Year's Day in 2006, after a pricing dispute with Ukraine, which transports 80 per cent of Russian gas to Europe. A longer cut-off occurred in January 2009, again because of a pricing dispute with Kyiv (Kiev). The global financial crisis, which affected Ukraine significantly more than Russia, exacerbated these gas disputes and made it more difficult for Ukraine's to pay Russia, heightening Berlin's concerns about its own energy security. Germany imports 35 per cent of its oil and 40 per cent of its natural gas and from Russia, a figure that will rise once Nord Stream is complete, but some parts of Germany – for instance, Bavaria – are more dependent on Russian gas than others.

The Ukraine cut-off introduced an element of uncertainty within Germany about the implications of further dependence on Russian gas, and led to calls for better bilateral communication over energy issues. It had counterproductive consequences for Russia as Germans questioned whether Russia might in the future use its energy supplies for political leverage. The Russia–Germany energy relationship is one of interdependence, albeit with asymmetrical vulnerability. Some argue that Germany is more vulnerable to Russian supply disruptions than Russia is to the loss of German energy revenues. In their view, the German government could not credibly threaten to stop importing Russian gas if that would leave a resident of Munich freezing and in the dark, whereas Russia can and has threatened to take measures that could interfere with supplies of gas, because it could more easily tolerate the loss of revenues. Others argue that this is not the case and that the relationship is one of mutual interdependence. The impact the global financial crisis has strengthened the view that Russia needs the revenues from gas sales, and Russian leaders have emphasized that they define energy security as security of demand, while Europe defines energy security and security of supply (Putin 9 September 2006). The paradox of the Russia–Germany energy relationship is that Germany will become more dependent on Russian gas, while at the same time expressing concerns about the future consequences of that dependence. Since the EU lacks a unified external energy policy, Russia is able to take advantage of competitive energy bilateralism in its ties with Germany and other European states (see Engelbrekt and Vassilev elsewhere in this volume).

The Merkel years – 2005–2008

When Angela Merkel became Chancellor, there was some concern in Russia that the Russia–Germany relationship might change. Since Mrs. Merkel had grown up in the GDR and experienced decades of Soviet occupation there, and since she is from the CDU (Christian Democratic Union of Germany), it was expected that she would have a more sceptical view of Russia than her SPD (Social Democratic Party of Germany) predecessor. Although she was more cautious than Schroeder in her public praise of Putin, and the male bonding relationship (*Männerfreundschaft*) of the Schroeder–Putin years was gone, there was great continuity in Russia–Germany relations after 2005. Merkel's SPD Foreign Minister, Frank-Walter Steinmeier, was partly responsible for that. His concept of 'rappochement through integration' (*Annäherung durch Verflechtung*) committed Germany to intensified engagement with Russia and both he and the chancellor described the relationship with Russia as a strategic partnership (Singhofen 2007). Nevertheless, it was also a result of lobbying by Germany's business community, which has a strong stake in the Russian market. Germany is Russia's most important trading partner, and bilateral trade grew from US$18 billion in 2003 to US$52.8 billion in 2007, representing 9.6 per cent of total Russian trade. Eighty per cent of Russian exports to Germany are oil and gas, whereas the majority of imports from Germany are machinery and finished goods. In 2007, German investments in Russia totalled US$12.6 billion and, by 2008, Russia, with its growing sovereign wealth funds, was poised to become a significant investor in Germany (Kuznetsov 2008). Thus, the economic dimension of Russia's ties to Germany will ensure that the relationship remains close.

Russia continues to view Germany as its chief advocate for integration into Euro–Atlantic multilateral structures. Berlin has often led EU efforts to improve ties to Russia, both with the initial Partnership and Cooperation agreement in 1994 and in 2007 and 2008 trying to resolve the problems after the Georgia war that prevented the start of negotiations for a renewed agreement. It also played a crucial role in the difficult negotiations over the status of Kaliningrad prior to the enlargement of the EU to Poland and the Baltic states in 2004. It has encouraged Russia's inclusion in ESDP (European Security and Defence Policy). Germany has promoted the NATO–Russia Council and favoured intensifying Russia's contacts with NATO, just as it has opposed what it considers premature NATO enlargement to the former Soviet space, partly because of Russia's opposition. Indeed, Germany was the key player in ensuring that Ukraine and Georgia were not offered a NATO Membership Action Plan at the Bucharest summit in April 2008. Its views have thus been opposed to those of Poland and the Baltic states (and the administration of George W. Bush), all of whom supported NATO enlargement to Russia's neighbourhood. Germany also encouraged Russia's fuller participation in the G-8, postponing its chairmanship by a year to give Russia the chair in 2006. The G-8 summit in St. Petersburg that year was a major milestone for Russia, demonstrating that it had recovered from the economic collapse and political weakness of the 1990s and was once again a world power.

Despite Germany's importance as Russia's closest Western partner, Russia has also criticized Germany for backing US policies that it opposes and for Germany's support for the new EU members when they have had disputes with Russia. Germany–Russia solidarity in opposing the Iraq War was an exception rather than the rule. Germany has supported the United States in its war in Afghanistan and has contributed troops to the NATO forces there, albeit with increasing reservations. It supported NATO enlargement to Central Europe and the Baltic States, as well as the independence of Kosovo and the stationing of US missile defence components in Central Europe, all of which Russia opposed. Some Russian officials still view Germany's Atlanticist policies with suspicion (Mezhdunarodnaia Zhizn 2006). Moreover, Germany was a key mover in the EU's new neighbourhood policy of reaching out to Ukraine, Belarus, Moldova and the states of the South Caucasus. It also supports the 2009 EU Eastern partnership initiative, which further develops this engagement. Russia considers the post-Soviet space to be within its legitimate 'sphere of privileged interests', according to Dimitry Medvedev, and it views the EU as a rival in its backyard. It has accused the EU of trying, through the Eastern Partnership, to establish its own sphere of influence in its neighbourhood. Germany, because its political identity is so bound up with the EU, has worked hard to maintain EU solidarity when there have been problems between Russia and individual EU members, particularly the new members. For instance, when Russia took economic and political measures against Estonia after it had moved a Soviet war memorial from the centre of Tallinn to its outskirts in 2007, Germany worked to ensure that the EU backed Estonia in its dispute. However, Russia also understands that Germany often tries to moderate the policies of some new EU members who favour a tougher policy towards Russia than Germany itself pursues.

Germany–Russia relations under the Medvedev presidency

Germany, like its European partners, greeted the accession of 42-year-old Dimitri Medvedev to the Russian presidency with cautious optimism, hoping that a younger post-Soviet leader not connected to the intelligence services might eventually liberalize domestically and pursue a less assertive foreign policy. Foreign Minister Steinmeier was Medvedev's first official visitor after his election. Before leaving, Stenmeier had underscored the importance of the relationship, saying that: 'Russia is and will remain an indispensable strategic partner if we want to create an all-European peace order' (*Russland ist und bleibt ein unverzichtbarer strategischer Partner, wenn wir eine gesampteuropaeische Friedensordnung verwircklichen wollen*) (Steinmeier 2008). The new concept was that Germany would engage Russia in a 'partnership for modernization', encouraged by Medvedev's verbal commitment to strengthening the rule of law and civil society and to fighting Russia's pervasive corruption. Medvedev's first foreign visitor after his inauguration in May 2008 was Chancellor Merkel. Although Medvedev, unlike Putin, had no German background, he, too, singled Germany out as a key partner for Russia. Moreover, since Putin was now Prime Minister, but still very influential, Russian foreign policy under Medvedev was very similar to that under Putin.

President Medvedev's first foreign trip to a Western country was to Germany, and he used the occasion both to court German business and to make a speech proposing a new Euro–Atlantic security architecture. Sergei Prikhodko, his national security advisor, emphasized the symbolic significance of Medvedev's choice of Germany, highlighting the priority that Russia attaches to Germany. He claimed that there were no serious problems in the bilateral relationship (Moslakin 2008). The focus of Medvedev's speech was to call for a new security system based on a legally binding treaty covering 'the whole Euro–Atlantic area from Vancouver to Vladivostok', adding that 'Atlanticism as a sole historical principle has already had its day' (Medvedev 5 June 2008). The new president, invoking the argument that the West reneged on its promise to include Russia in a post-Cold War European security structure, made a case that resonated across much of the German political class. Russia looked to Germany to play a leading role in the design of this new architecture.

Two months after Medvedev's speech, the outbreak of armed conflict between Georgia and Russia highlighted the fact that one major aspect of the post-Cold War agenda remained unfinished, namely how to regulate relations between Russia and its neighbours. Because the USSR, unlike Yugoslavia, had collapsed relatively peacefully, the West had paid little attention to the emerging relationships within the post-Soviet space. After Russia and Georgia went to war, some argued that NATO's failure to offer a Membership Action Plan to Georgia, combined with remarks attributed to Chancellor Merkel to the effect that Georgia could not have a NATO membership perspective unless it enjoyed full territorial integrity, had intensified both President Saakashvili's determination to reintegrate South Ossetia and Abkhazia and Russia's determination to ensure that this did not happen. Germany was supported by France, Italy and other NATO members in 2008 in its opposition to a MAP for Georgia but, given its role as Russia's most important Western interlocutor, its policies on this question carry disproportionate weight. Although it was France that took the initiative in conducting the negotiations after the end of hostilities, Germany has played an active role in seeking to assist Georgia's recovery. Nevertheless, it remains opposed to NATO membership for either Georgia or Ukraine in the foreseeable future.[2]

The advent of the Obama administration in Washington will also affect Russia–Germany ties. Prior to President Obama's election, German officials were concerned that the United States would not make Russia a sufficiently high priority on its new agenda. In fact, Obama early on emphasized the importance of improving ties. As Vice President Joe Biden put it at the February Munich Security Conference, 'It is time to press the reset button', with Moscow – although he warned that Washington 'will not recognize a sphere of influence' (Biden 2009). Although Moscow and Washington may well have different definitions of what 'reset' means, Berlin welcomed Obama's pledge to re-engage Russia and his emphasis on the need for a new strategic arms reduction agreement. German officials presently support any moves that reduce US–Russia tensions and that could encourage a more productive relationship between Russia and the West. The new US administration also acknowledged the desirability of closer US–Europe consultations on relations with Russia, but the greatest challenge both for Germany and for the United States will be the issue of Russia's neighbourhood and how far

Berlin and Washington will pursue policies that encourage the integration of such neighbours into Euro–Atlantic structures.

German society remains divided over how far Berlin should push the issue of common values as its relationship with Russia evolves, but most parts of the political spectrum favour continued and intensified engagement with Russia. Thus, Putin's policy towards Germany – which his successor continued – was largely successful, and the outlook for the years ahead was one of continued close economic, political and societal ties. While some in the Russian leadership might hope that a reunified Germany no longer dependent on the United States for its security as it was during the Cold War might revert to a *Schaukelpolitik* (see-saw policy) balancing between the West and Russia or even tilting toward Russia, this view overlooks the centrality of the European Union in Germany's foreign policy (von Klaeden 2008). Nevertheless, Germany will remain Russia's most important Western partner for the foreseeable future, seeking to moderate the more hard-line views of some of its allies. It is unlikely that either France or the United Kingdom could supplant Germany in this role, although Britain remains the favourite educational and business destination in Europe for the Russian elite. The end of the Cold War and the inclusion of Central Europe in the European Union have removed a major source of competition between Russia and Germany and have ushered in an era of closer cooperation that will last for decades.

Notes

1 An earlier, German-language version of this chapter is available in Schroeder (2010).
2 Author's conversations with senior German government officials.

9 Russia and the European great powers

France

Isabelle Facon

friendship is about telling things as they are, about facing disagreements where there are disagreements and about trying to find solutions.

(Nicolas Sarkozy, 10 October 2007, Press conference with President Putin, Moscow)

The 2008 Georgia war prompted, among other things, many analyses about the specific nature of France–Russia relations. Because the Sarkozy–Medvedev six-point plan for resolving the acute crisis was quite light and did not contain any reference to Georgia's territorial integrity, many observers in Europe argued that France, which by that time held the presidency of the European Union (EU), was yielding to Russian pressure. A few months before, France's opposition to NATO's granting a Membership Action Plan (MAP) to Georgia and Ukraine had confirmed, in the eyes of many analysts, Paris' traditional bias towards Russia. France, one of the countries which adopted an open-minded stance on President Medvedev's proposal on a new Euro–Atlantic security architecture (which many European states view with suspicion) has even been accused of privileging relationship with Russia over partnership with the United States.

It is true that during the Georgia crisis, on August 12, 2008, President Sarkozy took many by surprise by saying that Moscow has a right to defend Russian-speakers abroad.[1] The fact that Foreign Minister Bernard Kouchner opted for a much more alarmed position – wondering about possible Russian 'other objectives … , in particular the Crimea, Ukraine and Moldova' (quoted in Harding 2008), was not enough to dissipate the impression in many EU member states, especially among the new ones, that France is not vigilant enough with Moscow, if not excessively complacent.

France's behaviour towards Russia is the object of careful observation by other European countries because of the tradition of good relations between the two countries. These countries have long seen each other as useful allies in countering their rivals on the old Continent (Romer 2000: 439). In De Gaulle's times, France relied on its connection with Moscow as an instrument among others to balance the US influence in Europe, a rationale which has kept its strength until now. Jacques Chirac, in particular, maintained this approach, even adding a more 'affective' tint to it due to his friendship with Boris Yeltsin and, above all, to his own personal interest in Russian culture and literature.[2] In February 2003, President Chirac's remarks on the attitude of some of the future new members of the European Union, amid France's diplomatic efforts aimed at preventing US military intervention on a unilateral basis in Iraq,[3] was interpreted as an additional symptom of France's inevitable practice of disregarding the visions of 'small' European states, which many view as a direct consequence of the priority attached by Paris to its partnership with the Russian great power. Some of Chirac's other initiatives, such as granting the *Légion d'honneur* to Vladimir Putin in September 2006, at a time when many in Europe were disturbed by the trend towards increasing pressure on political pluralism and on human rights in Russia, have created the image of staunch stability in France's Russia policy and in the bilateral relationship.

Nowadays, this tradition of France–Russia relations is apparently maintained. French officials keep repeating they 'reckon and understand the Russian specificities ...' (Sarkozy and Putin 10 October 2007). They stay committed to encouraging Russia to take its full-fledged place and role in Europe. After a presidential campaign where he took critical positions on Moscow's deeds in Chechnya and its poor record on human rights, President Sarkozy seemed to come back to a more conventional stance on Russia, saying, a few months after his election, that Paris and Moscow were 'natural partners', as a result of which France 'seeks to be a privileged partner' for Russia – a Russia that is 'a major player. ... a strong country ... a strong country with international responsibilities' (Sarkozy and Putin 10 October 2007).

However, taking a closer look, one can see that beyond the rhetorical dimension, the factors of France's policy towards Russia have become more diverse, and its nature – more flexible. This reflects changes in Russia – which in recent years has become a great economic opportunity but, at the same time, a tough political and security partner – but also in Europe (EU enlargement, 'constitutional crisis', declining dynamism of the 'French–German' engine). This is also the result of a new background in France – with a president who has shown determination to define a 'rupture' in foreign policy. As regards Russia, that means integrating the treatment of the bilateral relationship into the broader framework of France's EU ambitions as well as a new parameter: the desire to develop a renovated relationship with the United States. This has produced different impulses in France's approaches to Russia, and partly explains Moscow's veiled scepticism and doubts about the orientations of France's new foreign policy.

Still convergent interests and visions

France, Russia and the world architecture

At the official level, the two countries continue to assert that they are close to each other. The official website of the French Ministry of Foreign and European Affairs describes the ties between the two countries as 'particularly strong' (French Foreign Ministry 2009). The Russian MFA (Ministry of Foreign Affairs) portrays relations with France as 'one of the priority orientations of the foreign policy of Russia' (Russian Foreign Ministry 2007). In 2007, President Putin described the bilateral relationship as a 'centuries-old friendship' and as being made of 'mutual attraction'[4] and of 'glorious pages' (Sarkozy and Putin 10 October 2007).

Many Russian experts see a natural proximity between the self and world visions of the two countries, if only because they are former imperial powers, which is supposed to create mutual understanding. There is also a feeling in Russia that the economic models of the two countries are quite similar, which creates a basis for a 'natural convergence' of interests: 'Etatist French business culture is much closer to the Russian business culture than the liberal Anglo–Saxon or German', says Yuri Rubinskiy, Head of the French Studies Center at the Russian Academy of Sciences Institute of Europe (Rubinskiy 2008: 26–27). And it is probable that France is not the European country that experiences the greatest problem in understanding the Kremlin's rationale in reinforcing its control over the energy sector since for Paris, this sector is a strategic one and necessitates a more or less strong degree of state control. Other Russian experts consider that the two countries are faced with the same kind of problems in the field of immigration and that this could constitute an axis of cooperation (both bilaterally and at the EU–Russia level) (Streltsova 2007: 54).[5]

In the new Foreign Policy Concept that Moscow adopted in July 2008, France comes second on the list of the European countries which Russia singles out as states with which it seeks 'the development of mutually advantageous bilateral relationships' (after Germany and before Italy, Spain, Finland, Greece, the Netherlands, Norway and 'some other West European States') (Russian Foreign Ministry 2008c). France is presented as 'an important resource for promoting Russia's national interests in the European and world affairs'. And indeed, France and Russia continue to share a number of common visions on the international architecture. As two nuclear powers and permanent members of the UN Security council, they have the same inclination to see themselves as having special international responsibilities. They share the same commitment to maintaining a central role for the United Nations in solving international security problems – which was a key factor in their joint opposition to the US–UK war in Iraq (with Germany and China).[6] When they insist on the primacy, in international relations, of the UN Charter's legal principles and on the role of multilateral institutions, France and Russia are motivated, among other things, by the desire not to let the 'unique hyperpower' downgrade their status on the world stage. The common aspiration to see the world become (or remain) really multipolar has certainly encouraged France to support the transformation of the G7 into a G8 including Russia as well as this country's integration into the WTO. Since 2002, a bilateral 'Security cooperation council' regularly meets at the level of Foreign and Defence Ministers. Viewed in Moscow as an incarnation of the effort of the two countries to 'coordinate their steps on the international scene' (Kamynine 7 March 2008), its agenda has included: Russia–EU relations (external security issues),

UN reforms, Russia–NATO cooperation, the CFE treaty, OSCE issues, non-proliferation of weapons of mass destruction, Global Partnership, as well as regional issues (Iraq, the Middle East and Afghanistan) (Russian Foreign Ministry 2007).

It is not clear, though, whether these orientations can be modified as a result of France's changing attitude about its relationship with NATO and the United States. Much will depend on whether president Obama will show a greater availability than his predecessor to take into account the visions of others in dealing with certain sensitive international issues. This is true, in particular, of Middle Eastern issues, where Paris and Moscow have long shared a number of similar opinions.[7]

France, Russia and Europe

Moscow has found in Paris a positive partner in defending its 'rightful place' in Europe. France has always been convinced that if Russia is not firmly anchored to the Old Continent, European security cannot be complete. This is a long-term trend in French foreign policy thinking and policy making, transcending all political currents (Romer 2000: 444). This was, again, obvious in the aftermath of the Georgia war. At the World Policy Conference, President Sarkozy was clear about this vision: 'The Russia–European Union relationship has just lived through a serious trial. From this, I draw the conclusion that the European Union and Russia should not drift apart, triggering fears about a new division of Europe, even reviving fantasies about a "new Cold war". Such a new Cold War would be a historical mistake' (Sarkozy 8 October 2008).

Therefore, Paris has always been at the forefront of the effort to demonstrate to Russia that it should not feel isolated and that its interests are taken into account by its European partners. In particular, France has been trying to limit the negative political consequences of the parallel enlargement of NATO and the European Union on relations with Moscow. Paris was one of the promoters of the Russia–NATO council and of the idea of having a formal strategic agreement with Russia before starting to enlarge the Atlantic alliance (the Founding Act on Mutual Relations, Cooperation and Security). It is thus not by chance that the 'Permanent Joint Council' was inaugurated in Paris in May 1997. Later on, in the post-September 11 context, Paris supported the British plan to establish a 'new quality' of relations between Russia and NATO – which led to the creation of a new NATO–Russia Council, where Russia's voice is equal to that of all NATO members on a number of subjects deemed as being of common interest.

France is perceived in Russia as having tried, more than other EU members, to give flesh to the Union's Common strategy on Russia (adopted in June 1999) when it held the rotating presidency of the Union in 2000 (Zueva 2008: 56). Paris, together with Berlin, was also an active proponent of the EU's 2003 decision to create the 'four spaces' of cooperation with Russia, aimed, among other things, at stimulating the EU–Russia partnership, which was stagnating, and at stemming Moscow's frustration with the EU's proposal to include it in the

European Neighbourhood Policy (ENP), something it found incompatible with its great power status. France was the first European country to sign with Russia an agreement on visa facilitation (France/Russia 2008) – a very sensitive issue for Moscow from a political, symbolic and economic point of view. This accelerated developments on this issue at the EU–Russia level.[8]

For all these reasons, some Russian experts believe that France holds an 'intermediary role in Russia–EU relations' and that, starting in the early 2000s, it 'was the initiator of further rapprochement between Russia and the EU and NATO' (Streltsova 2007: 59; Zueva 2008: 57). This may be an exaggeration but it shows how important France is in Russian representations of how to deal with the new European political landscape. Russia has always relied on its relationship with France (as well as with Germany and Italy) in order to have its visions and interests taken into account within the European Union. Russian officials and foreign policy experts are unambiguous about the necessity for Russia to use to its own benefit the variety of foreign policy postures within the EU. This has become even more important after the enlargement of the Union, the Russian Federation being extremely concerned that the integration of countries that have mixed or bad feelings towards Moscow may make the EU's policies increasingly contrary to Russian interests – especially in the economic field and in the common neighbourhood. Paris (like Berlin) is frequently trying to balance those initiatives by some of the new members which they find can have a negative impact on relations with Russia. One illustration is their opposition to NATO's establishing a Membership Action Plan with Ukraine and Georgia, while Warsaw or Vilnius have militated in favour. It is not clear, however, whether president Sarkozy's absence from the Prague Summit in Spring 2009 launching the Eastern partnership (initiated by Sweden and Poland) has to do more with his perception that the project is a competitor to the 'Union for the Mediterranean Sea', or with the fact that it can be interpreted as a sign of resistance of the EU to Russian exclusivist impulses in the common neighbourhood.[9]

As concerns Russia's place in Europe, the 'French–German' engine has indeed played an important role. In the late 1990s, France, Russia and Germany established a 'triangle' in Ekaterinburg (the last meeting in this trilateral format took place in 2006). Originally, for France and Germany, this triangle was supposed to be a contribution to giving more weight and substance to a European pole in a multipolar world order. The strength of this France–Germany–Russia political 'axis', which served to discuss many international issues,[10] has often been exaggerated by commentators, especially when Paris and Berlin opposed the Iraq war together with Russia (Iraq Mission Activities 2003), and, more recently, when Germany and France jointly rejected MAPs to Georgia and Ukraine. However, this triangle is a reflection of a common belief in Paris and Berlin that it is worth supporting Russia's transformation (in the sense of its Europeanization) in order to make sure that this country would remain anchored to Europe. Recently, in a joint tribune in *Le Monde*, Angela Merkel and Nicolas Sarkozy again asserted the need to have Russia closely tied to the Old Continent, by 'restoring and developing trustful and fruitful relations' with this country (Merkel and Sarkozy

3 February 2009).[11] The two countries have been eager to ensure that the parallel EU and NATO enlargements would not damage European security by antagonizing Russia. Considering that the first two rounds of NATO expansion have badly damaged relations between the West and Russia, they are convinced that a third round to bring in Georgia and Ukraine would only make matters worse. In addition, they share the vision that the United States, since the end of the Cold War, has more or less consciously tried to separate Russia from the rest of Europe and that this should be prevented. They also agree that the post-Soviet space does not have the same status in US and EU strategic perspectives. For the United States, the post-Soviet space is a theatre, while for the European Union, it is a neighbourhood. This, in their view, does not call for the same visions, instruments and solutions (Facon 2008a: 8–10). As a result, both sometimes feel annoyed by the rigidity of the positions of some of the new EU/NATO members towards Russia, and by their assertive support of US initiatives in the post-Soviet space.

Concerning this problem of the 'common neighbourhood', France is a political ally also indirectly for Russia in the sense that its focus is clearly on the Southern neighbourhood (as shown, among others, by the Union for the Mediterranean Sea launched in summer 2008). As viewed from Moscow, this is an important factor in preventing the EU from spending too much political energy in the eastern neighbourhood. In addition, France is clearly not a supporter of further EU enlargement, at least for the near future, which for Russia is tantamount to indirect assistance to its own effort to prevent the other former Soviet republics from getting too close to the Union.

For Paris, before the election of president Sarkozy and the changes in foreign policy he has started bringing about, France's willingness to balance the influence of the inputs of new members into the EU's Russia policy was in part dictated by the fact that the new members' attitudes could be perceived as deriving from their strong pro-Atlantist predisposition. This was even more salient under the Bush administration, when many people in the French foreign policy decision-making circles proved quite critical of Washington's democracy promotion agenda in the neighbourhood common to Russia and the EU (an agenda supported by, among others, Poland and the Baltic States), which, in the aftermath of the coloured revolutions, became a crucial issue for the Kremlin. These dimensions led some Russian academics to applaud 'the independence of the French position, stronger than Germany's, in relations with Poland, the Czech republic, and the Baltic states' (Streltsova 2007: 60).

Since then, Nicolas Sarkozy's personal support to George W. Bush, his sympathy for 'the universalist, pro-democratic rhetoric of the United States and for its effort to expand democratic values in the world ... including with military force', as well as his choice of Socialist Bernard Kouchner to head the Foreign Ministry created apprehension in Moscow that Sarkozy's France may be willing to opt for a French style of transformational diplomacy. Bernard Kouchner is notorious in Moscow for being one of the most active advocates of a policy of intervention (including militarily) when human rights are deemed to be in danger. The Russians are angered by his role on the issue of Kosovo, and many of them expected him 'to plunge into public rhetoric about violations of democratic norms and human rights in Russia'.[12]

So far, these apprehensions have not really materialized. However, some Russian intellectuals consider these trends seriously, and view them as not compatible with 'Russian representations about sovereign democracy' and as 'leaving few chances for partnership in the sphere of values' between Paris and Moscow (Streltsova 2007: 58). Many other factors have come to tarnish France–Russia relations over the past years.

Reciprocal fatigue? Economic and political disillusionment

President Sarkozy has often described his dialogue with Russian leaders as 'frank' and 'free', which is a disguised way to assert that France does not hesitate to criticize Russian policies. Moscow observers – both officials and analysts – have not forgotten the critical tone of candidate Sarkozy on Moscow's human rights violations in Chechnya. During his first visit to Moscow as head of the French state, Sarkozy met with Memorial – an NGO that has, among other things, vividly denounced war crimes in Chechnya. The French president also castigated Russia's 'brutal' use of its energy assets (Sarkozy 27 August 2007). However, a few months later, in December 2007, Nicolas Sarkozy telephoned President Putin to congratulate him for the success of his party in a parliamentary election that Germany called 'neither free, fair nor democratic' by Western standards (Reuters 2007). In addition, he congratulated Dmitry Medvedev on his victory in the widely criticized Russian presidential election in 2008.

France–Russia partnership: a relatively thin economic foundation

This ambivalence in France's attitudes could be explained by the hope, which was widespread in Europe in 2008, that President Medvedev could take on a more liberal line both domestically and internationally. However, it may also reflect France's 'economic pragmatism'. President Sarkozy's team has put economic interests high on its agenda, which has consolidated a realistic approach in dealing with Russia as a dynamic market where French interests should be developed in a resolute manner.

The experts of the European Council on Foreign Relations rightly note that 'the political relationship does not have much of an economic foundation' (Leonard and Popescu 2007: 31). France is certainly satisfied that its exchanges with Russia have increased in recent years,[13] but its viewpoint on bilateral economic cooperation with this country is that it has a strong potential to develop (French Foreign Ministry 2009). France stands behind many European countries in terms of tackling the expanding Russian market,[14] and so far it is primarily big industrial companies that have been able to penetrate this market, while medium-sized and small enterprises are absent from the picture (Mariton 2009). In Moscow, in October 2007, President Sarkozy underlined the wish of French economic players to enter the capital of big Russian companies such as Gazprom, and to invest more substantially in the Russian economy, and his hope that Russian investors would reciprocate (not only in energy but also in the nuclear field, space, the car and aviation industries, transportation, informatics ...). Just a few months later, Total and Gazprom signed an agreement granting Total 25 per cent of the capital of the future company that will operate the Shtokman gas deposit in the Barents Sea. Gaz de France, for its part, has started consultations on its potential participation in the Nord Stream gas pipeline project.

Moscow has long hoped that 'uniting our efforts on world markets will become an important competitive advantage for Russian and French producers' (Putin 10 October 2007). Furthermore, in its Foreign Policy Concept, Russia still places France among the countries which have a potential to 'contribute to putting the Russian economy on an innovative track of development'. Besides, the French Ministry of Foreign Affairs indicates a desire to build with Russia 'a partnership that can guarantee this country's sustainable development' (French Foreign Ministry 2009). A number of prestigious projects have indeed taken shape that have a strong symbolic value – for example, the creation of a launch pad for Soyuz launchers in French Guyana (however the project has suffered many delays), or joint work on the Russian Superjet 100 regional aircraft (involving Safran and Thales). In addition, Russia has sold to foreign clients various military platforms with French subsystems (Cowan 2008).

However, mutual investments remain rather low (ITAR-TASS 2008). France's economic image in Russia has suffered from having concentrated on speculative investments more than on productive ones, which is not Russia's preference (Romer 2000: 445). Russia is looking for industrial partnerships that could help it compensate for its backwardness in various technological fields. However, in many of these, Russia and France appear more as competitors than as real partners, which has contributed to hampering the development of full-fledged industrial and technological partnerships. With France, as with many other countries, Russia has the feeling that foreigners take more than they give. The Russian state bank VEB's entering the capital of EADS (European Aeronautic Defense and Space Company) (five per cent) on behalf of the Russian Aircraft Corporation (OAK) in 2007 was partly an answer to its disillusionment on this front (Lahille 2007: 4).

France, for its part, is interested in setting up industrial partnerships (not just commercial agreements) with Russia in the energy field (Gomart 2007: 10–11) – precisely one of the domains where the Russian state has, in recent years, limited the possibilities for such joint undertakings. While visiting Moscow in October 2007, the new French president stressed that 'a good agreement is an agreement that is good for the two sides' with 'no taboo, no limit', but with 'transparency and reciprocity'. This was a direct reference to some suspicions in economic relations, including those tied to Russia's marking a clear tendency towards ousting foreign investors from the energy sector. France also views the monopolist status of Gazprom as risky, since it bears the potential danger of unstable gas prices. President Sarkozy said that while France finds it normal for Moscow to defend its interests in the energy sector, he called for transparency in this connection. On top of that, Russia and France are competing for energy partnerships with

Box 9.1

Trade relations as presented by the French Foreign and European Affairs Ministry

French exports to Russia: 5.6 bn. Euros / French imports from Russia: 11 bn euros (2007).

Russia is the second gas and oil provider of France. In December 2006, GDF and Gazprom (which have been partners for several decades) signed a long-term contract (2012–2030) for Russian gas exports to France.

France occupies the 9th rank on the list of countries exporting to Russia behind Germany, China, Ukraine, Japan, the United States, Belarus, South Korea and Italy.

France is the 7th investor in Russia (with about 500 subsidiaries) behind Cyprus, the Netherlands, Luxemburg, the United Kingdom, Germany and the United States.

A few success stories: sales of Airbus to Aeroflot, preparations for facilities to launch Soyuz launchers from Guyana (Kourou), Total's participation to the exploitation of the Shtokman gas field, investments by Alstom (contract in September 2008 on the equipment of rolling stocks for the high-speed line Helsinki–Saint Petersburg; partnership with Transmasholding for the production of high-speed trains in Russia; contract with Rosatom on the creation of a JV for the production of turbines for Russian nuclear plants), Renault* and Peugeot PSA, expanding positions of Société générale (the first foreign bank in Russia) and Axa.

* Renault controls 25 per cent of the capital of Avtovaz.

North African countries. Finally, France is concerned, like many other countries, that Russia's financial *force de frappe* can be used in the future for hostile raids on European assets (Gomart 2007: 19).

It is not clear how France will react to the trend for Russia to politicize economic relations and to grant 'privileges' to the countries that appear open enough to Russian political interests. Many people interpreted Moscow's choice of Total for developing the Shtokman gas field precisely in this light (Kommersant 2007a). What is clear is that any important deal with Russia in energy or in other prominent sectors will trigger criticism by some other European countries.[Box 9.1]

France's Russia fatigue

'This strong engagement in Georgia and the important efforts in favour of strengthening relations between the EU and its Eastern neighbours have not prevented the French presidency from keeping open the path of dialogue and cooperation with Russia, accompanied, however, by enhanced precautions about the respect of its commitments by Moscow. The Nice Summit on November 14th allowed to continue the assessment of [EU–Russia] relations, to obtain a few concrete signs of Russia's reengagement on economic and trade issues as well as in foreign policy ...' (French Presidency of the Council of the European Union, Balance Sheet and Prospects).

Yuri Rubinskiy explains that there is an opinion 'in the Russian expert community that the traditional Russia–France dialogue in the sphere of security has shrunk. This allegedly produced the fairly modest results of the working visits of the French President and the Foreign Minister to Moscow where they discussed the most sensitive issues (the Kosovo status, sanctions against Iran, elements of the American ABM system in Poland and the Czech Republic, the prospects ratification by the West of the modified CFE Treaty) on which France supported the positions agreed within NATO' (Rubinskiy 2007: 28). This opinion is not erroneous. Moscow, in order to have its interests taken into account on some of these issues (in particular NATO enlargement to Georgia and Ukraine and US missile defense plans in Poland and the Czech republic), has put strong pressure on the Europeans (threatening to target missiles, suspending CFE application and building up energy pressure) in the hope that they would pass on such pressure to Washington. France, like most European states, has reacted to such pressure by taking on a low profile on these issues and by receiving and commenting Russian proposals more cautiously than usual. This is an important element, since many people in French official and academic circles consider that on these two issues, the Russian position is worth considering at least in some of its aspects.

In analyzing this evolution in the posture of the French, Moscow has deliberately underestimated the 'Russia factor' and inflated the 'US' one (Facon 2008b). While French officials recognize Russia's stature as a 'great power' that 'is back on the world scene' (Sarkozy 9 October 2007), growing Russia fatigue can be felt in Paris, which, like other Western capitals, is faced with the difficulty of achieving real partnership with a country which rejects any constraints on its action and any criticism. In addition, this, of course, has contributed to making the atmosphere of bilateral relations less cordial. In other words, Russia fatigue (more than France's desire to improve relations with the United States) has made Paris very guarded on supporting or at least responding to Russian positions on certain security issues. Russian commentators, however, in their effort to explain this evolution, focused on what they view as the more Atlantist outlook of France's foreign policy. Many Russian observers have been concerned to see Nicolas Sarkozy engage in a strong rapprochement with the United States and try to end the situation in which anti-Americanism is 'one of the invisible pillars of French foreign policy' (Gomart 2007: 16). It is in this perspective that Moscow has warily observed Paris's growing interest in 'other' post-Soviet states (Rubinskiy 2008: 24), visible, under the EU French presidency, with the decision of the Union to develop a specific partnership with Ukraine.

The Russians had been especially amazed by the effort paid by the new French president to get closer to the United States while George W. Bush was still in power, which came as a surprise even in France and in many European capitals given the poor foreign policy record of the Bush administration and its odious image in the public opinion of most countries (Gomart 2007: 7). However, the effort was continued after the French presidential election with the announcement of a greater military effort in Afghanistan and the return of France into the integrated military structure of NATO.[15] All these developments have challenged Russia's traditional vision of France as a staunch political ally. This is closely tied to the two countries' increasingly divergent approaches to how to deal with Washington. While France seems to seek normalization of its relations with the United States, Russia is still trying to impose itself as a great power by resisting and opposing the unique superpower. This major factor of the traditional solidarity of Russia and France on international affairs – fighting US ambitions and unilateralism – is now of lesser importance in the definition of French foreign policy, and may become even lesser should Barack Obama keep his promise of giving the United States a more multilateralist outlook. The fact that Russia often seems to seek realization of its great power status by seeking a 'partnership of equals' with the United States, instead of really trying to do so by building stronger ties to the European Continent, certainly generates frustration in various circles in France.

Some Russian experts rightfully note that president Sarkozy, while working on the *rapprochement* with the United States, has reminded his American friends that alliance relations do not mean that the ally should be obedient.[16] They seem to conclude that this leaves Russia room for manoeuvre. However, such hopes seem to be 'neutralized' by the fact that the French leadership has put forward the same kind of reservations concerning Russian policy and international behaviour. Even the most traditionally minded French diplomats have tended to change their minds about Russia as they have understood the limits of Moscow's commitment to multipolarity, which has been increasingly restricted to Russia's focusing primarily on its own interests. The effort by Russia to present itself as an alternative player in a world which is no longer dominated by Western models and values has been received with unease in Paris, like in all Western capitals. Paris also understands that Russia, which keeps repeating that it should be free to pursue its own choices both politically, economically and diplomatically, has ceased to systematically proclaim that it is Europe-focused in defining its destiny. This does not fit France's traditional vision that a strong Russia–Europe partnership is positive for the building up of a multipolar world order. Besides, for Nicolas Sarkozy, multipolarity already exists, and, in his view, it is not only a positive phenomenon since it can also take the form of clashes of power policies (Sarkozy 27 August 2007). The aggressive way in which Russia has been pursuing its interests in recent years clearly corresponds to this vision.

The fact that 'Russia is not a strategic priority for the new team' in Paris (Gomart 2007: 6) has not gone unnoticed in Moscow. This was obvious from Sarkozy's first declarations after his election, which did not mention Russia. Neither was the relationship with Russia a priority in the agenda of the French presidency of the EU in the second semester of 2008[17] – even though the Georgia war came to change this. In addition, foreign policy decision-making in France is growing diverse in its sources and determinants, including on Russia. Moscow understands that this means it may have, over the longer term, less leverage on French policy makers. A new generation of diplomats is entering into the administrations that are less influenced by the Gaullist visions and more familiar with the image of Russia as an irritant power. However, their views are often overshadowed by strong words about Russia from top level officials. In November 2007, François Fillon, the Prime Minister, whose personal interest in Russian affairs is well known, said that Moscow has certain legitimate interests with regard to the countries in its immediate vicinity, which was immediately interpreted in many European countries as his recognition of the post-Soviet space as Russia's sphere of influence. In November 2008, the Quai d'Orsay apparently tried to obtain a postponing of the visit of the nuclear cruiser Peter the Great to Toulon, while the French General staff wanted to maintain it. The Elysée decided in favour of the latter option (Sarkozy 21 November 2008). It should be noted, however, that the appearance of a lack of coordination in foreign policy, which was also illustrated by the variety of the declarations by French high-level officials during the Georgia crisis, is not reserved for Russia.

The diminishing importance of Russia on Paris's foreign policy agenda is also a consequence of France's coming to the conclusion that the traditional tonality of the Russia–France dialogue has not served its image and positions within the EU in recent years, on the contrary. The bilateral relationship is increasingly entangled in the broader framework of EU–France relations – a priority for the French leadership.[18] Therefore, such is the vision of French diplomatic circles, the bilateral partnership with Moscow should not be viewed by other members as being detrimental to EU interests.[19] The realization of the lack of real concrete concessions on Moscow's part which Paris could boast as a positive input for the EU, plus the integration in its calculus of the political weight of the new members, have altered the political background of Russia–France relations. Russian experts have noted that being the son of a Hungarian immigrant, Nicolas Sarkozy is likelier to yield to the pressure of former Warsaw Pact members, thus leaving them a larger margin for manoeuvre and influence within the EU and NATO to promote anti-Russian measures. Indeed, French diplomats now seem more inclined to heed the potential impact of their initiatives towards Russia on the perceptions of Central and Eastern European members. This was the case during the discussions on how to respond to the Russian proposal on a new European security treaty. Another example is president Sarkozy's asking his Russian counterpart to demonstrate 'openness' towards the Baltic States in the energy field, especially for Lithuania, obliged by the EU to decommission its Ignalina nuclear power plant in late 2009 (Sarkozy 14 November 2008). This reflects not so much the president's family background as the new reality of a more diverse EU, which France, though not without difficulty, is taking into account; and the fact that Russia is viewed as complicating the European game. For France, the first priority is asserting itself as a leader within the EU. This has been made a difficult task by the 2005 referendum on the European constitution, and it seems clear that Paris does not wish to be hampered in its European effort by a Russia that seems to proclaim itself less European.

Questions about the Germany–Russia partnership

In dealing with Russia, France may also be motivated by its concern about the strong rapprochement between Germany and Russia. Things have changed substantially since 1998, when the first meeting of the Russia–France–Germany triangle took place. The France–Germany 'engine' of the European Union is still a reality, as demonstrated by the frequency and quality of bilateral meetings and initiatives. However, it has been challenged in recent years not only by the new EU members but also by the hectic personal relationship between Angela Merkel and Nicolas Sarkozy. While Berlin was irritated by the French initiative on the Mediterranean Sea, Paris is intrigued by the growing depth of the Russia–Germany partnership.[20]

A major factor of this Germany–Russia rapprochement has been Vladimir Putin's career profile – with his several-year sojourn in Eastern Germany as a KGB agent. From a political point of view, the fact that Russia has started doubting the ability of France to play the role of an engine in the European construction since the Constitution referendum and consequently its capacity to have enough political strength and will to counterbalance the new members, it has tended to focus even more on Germany to do so – and Berlin is presumed to have more concrete reasons to deliver, given the importance of its economic interests in Russia. On the economic front, France is clearly distanced by Germany, which is by far Russia's largest economic partner, both in terms of trade ties and investments (in 2007, France held 3.9 per cent of the Russian market, while Germany held 12 per cent) (French Foreign Ministry 2009). It is quite possible that some Russian economic choices have increased the political preoccupation that some French experts express regarding the depth of the French–German partnership. For example, Rosatom recently picked Siemens as a partner for creating a joint venture in the nuclear field which claims to become the world's leader in its field and will certainly challenge Areva's leading positions in the sector while depriving it of its German partner (L'Expansion 2009). This strongly disappointed the French, especially as president Sarkozy, while in Moscow, had told his Russian counterpart that he was ready to engage in bilateral cooperation in this field (Putin 10 October 2007).

These circumstances may produce two different orientations in French 'Eastern policy'. Paris may resolve that developing closer ties with new Central and Eastern European members, some of which are concerned by the intensity of the political and economic partnership between Moscow and Berlin, may allow France to try and play its part also in the Eastern policy of the European Union – and thus avoid being relegated to the status of 'specialist' on relations with the Southern neighbourhood. A second option could be for France to keep strong channels of communication and cooperation with Russia in order not to 'lose contact' with the evolution of the new realities of the Germany–Russia partnership.

Conclusions

Many things are going smoothly in Russia–France relations. The two countries are preparing to organize, in 2010, a year of France in Russia and a year of Russia in France. Cooperation on Afghanistan is going on, with the agreement allowing transit military supplies to the ISAF (International Security Assistance Force). French and Russian officials say they expect the economic relationship to grow.

The Georgia war has been analyzed amid emotion, triggering comments about the unconditional support that France, by virtue of its traditional diplomatic line, is willing to offer to Russia. This has obscured the silent adjustments that have changed the nature and atmosphere of the partnership between the two countries in recent years (adjustments that have not been easy to discern also because of, among others, the impression of instability in the French foreign policy in the last few years). Nuances in the France–Russia relationship are not coming from nowhere – already in the 1990s, there were hurdles and discrepancies. Russia took a critical stance on President Chirac's decision to resume nuclear testing in 1995, on Paris' participation in the Allied Force operation for Kosovo and Bernard Kouchner's subsequent handling of the issue – with the ensuing divergence of views on the subject of Kosovo's independence in 2007–8. Moscow has also been offended that public opinion in France could be as negative as it is on Russian issues. Moreover, there have always been ambivalent feelings in France about Russia, which are certainly due to specific political circumstances in France. On the one hand, there is an enduring sentiment of solidarity with Russia which derives from the years of resistance during the war (with the mythology surrounding the Normandie–Niemen squadron's epic). On the other hand, the singular weight that the Communist Party held in French political life in the postwar years and the difficulty of its leadership to recognize the criminal nature of the Soviet regime has certainly had its impact, and paradoxically, continues to generate, depending on who thinks about it, positive or negative emotions about Russia.

Until recently, official France had found that Russia was worth supporting anyway for the sake of European security and multipolarity. Nowadays, the two players are often pursuing different interests and they are no longer a priority for each other, which has contributed to 'normalizing' their partnership. In fact, Paris and Moscow are living through a new phase of their relationship. Russian officials and experts feel that France, like many European countries, is in the middle of a crisis of its political, social, economic and diplomatic model.[21] This happens at a time when Russia is getting more self-confident, if not over-self-confident. The resulting perception of a gap in power between the two countries, which does not reflect the reality of a Russia that is much weaker economically than France, creates an atmosphere that is not conducive to positive interaction. This is all the more the case that, as in most European countries, Russia's domestic policy, on Chechnya in particular,[22] and Moscow's reaction to the coloured revolutions, to which the Kremlin has responded in claiming more explicitly than ever privileged interests in the post-Soviet space, have 'forced Paris further to consider the Russia–Europe disparity of values' (Gomart 2007: 151). This was visible, in substance, in one of the first major foreign policy speeches by President Sarkozy: 'Russia is imposing its comeback on the world stage by using its assets, in particular oil and gas ones, with a certain dose of brutality, while the world, Europe above all, expect from it a substantial and positive contribution to solving the problems of our times, which its restored stature justifies. A great power should ignore brutality ...' (Sarkozy 27 August 2007).

The new or different visions brought about by the slow generational change in French political and diplomatic élites combine with the influence of the French traditional vision of the need to promote universal humanitarian values (democracy and human rights) to make France uncomfortable with the evolution of the domestic political regime in Russia, the way it dealt with the Chechnya problem and the hardening of its foreign policy. When Nicolas Sarkozy said in Evian that he was prepared to discuss Russian proposals for a new pact of Euro–Atlantic security, he added that 'the balance of power is not a guarantee of sustainable security, democracy is also needed, human rights also are needed, they are essential stability factors' (Sarkozy 8 October 2008).

The new leadership in Elysée is probably not as fond of the tradition of the bilateral partnership as its predecessors used to be. It has less reason to cultivate this tradition, with a new US administration that may prove more open to its partners' opinions, at a time when Russia seems, for its part, less interested in real integration with its European partners. After the failure of the 2005 referendum on the European Constitution, Paris has become sensitive to anything that may complicate 'France's return to Europe' which Nicolas Sarkozy promised to achieve during his presidential campaign. Russia is one of the factors that have made it impossible for Sarkozy's France to 'put its friendship with Russia at the service of all Europe' (Sarkozy and Putin 10 October 2007). France has grown increasingly annoyed with Russia's temptation to use Europe in its illusory strategic competition with the United States and would like Russia's 'European desire' to be more pronounced. In such a context, where the political foundation of the relationship gets weaker, the failure to strengthen the economic partnership (with economic links weaker than those between Russia and the other big EU states) makes the general fabric of Paris–Moscow relations lighter. Besides, this situation seems more coherent with the mood of the French public opinion, which has grown increasingly negative about Russia (the French press being highly critical of developments in Russia, sometimes not without excess, and a large part of the research community being at best skeptical about the chances for Russia to become a democracy and a more constructive player in world affairs).[23]

France remains convinced that isolating Russia or criticizing it systematically as a player that is 'unavoidably' foreign to European practices and life is wrong and detrimental to European security. Paris also believes that one should live with geopolitical realities, which means, among other things, accepting the 'inevitability' of partnership between Russia and Europe – 'Europe's and Russia's fates are tied together. They are tied by geography, countries seldom change addresses ... they are tied by the growing interdependence of our economies ... To build confrontation between us would be foolish', Nicolas Sarkozy said (Sarkozy 8 October 2008). This explains the conviction of the French government that it is necessary to answer Russia on its proposal for a new European security architecture, as well as Nicolas Sarkozy's suggestion that Russia and the European Union should build a common economic and human space that will 'bring about a decisive contribution to the stability of the international order that we have to create' (Sarkozy 8 October 2008). However, the French now seem to be more inclined to tell Russia that while it is very welcome in Europe, it should demonstrate more clearly the sincerity of its commitment to the Old Continent.

Notes

1 'It is perfectly normal that Russia wants to defend its own interests and the interests of Russians inside Russia, as well as Russian-speakers outside Russia', quoted in RFE/RL 2008. He added that 'it is equally normal that the international community wants to guarantee the integrity, the sovereignty and the independence of Georgia'.

2 Jacques Chirac, when he was a student, worked on a translation of *Eugene Onegin* from Russian into French, which is symbolically very important as viewed from Russia, given the prominence of Pushkin in Russian cultural life (Romer 2000: 441).

3 Reacting to their alignment with the position of the United States in the Iraq crisis, expressed in a letter of support to Washington, he said that they had 'missed a good opportunity to shut up' and accused them of being 'not very well brought up ...'.

4 This word was used also by President Chirac in 1997, in a speech at MGIMO (Moscow State Institute of International Relations) (26 September 1997).

5 There is a trend, in the Russian debate, to find some similarities between Russia's problems with Chechnya's separatism and France's Corsica issue, although it is hard to assess whether this is a sincere analysis or pure rhetoric stemming from Russian irritation due to French criticism of the Chechnya war (see section that follows).

6 Later on, these countries have held quite similar positions about the necessity to restore as quickly as possible Iraq's sovereignty over its territory and state structures.

7 The more open attitude of Paris towards Syria under Sarkozy has contributed to making their positions closer on an issue which is quite sensitive – since France has always tried to limit Syrian influence over Lebanon while Damascus is one Russia's major partners in the region.

8 Germany and Italy also signed separate bilateral visa facilitation deals with Russia. The European Commission decided to suspend them, because it found they breached the Schengen rules. This forced the European Union to negotiate a visa facilitation deal with Russia (Leonard and Popescu 2007: 16).

9 However, it should be noted that it was under the French presidency that it was decided to reactivate the Eastern partnership initiative.

10 Iraq, Iran, reform of the OSCE, Kaliningrad, Kyoto protocol, Georgia, energy partnership, economic cooperation, EU–Russia relations ... (Zueva 2008: 58).

11 In this joint article, the two leaders, while criticizing Russia's behaviour on Georgia and calling on Moscow to observe the principles of the Helsinki Act and the Paris Charter, also suggested it was necessary to 'stretch out a hand to Russia and to restart our cooperation in the framework of the NATO–Russia Council and between the EU and Russia, if the latter wishes so. We want a tighter political and security dialogue between the EU and Russia which enables to engage it further in the Euro–Atlantic security space'.

12 Bernard Kouchner is widely perceived in Russia as a supporter of what the Russians see as 'NATO's aggression in Serbia and of granting Kosovo its independence' (Streltsova 2007: 55). 'His previous experience as the UN Administrator for Kosovo had done nothing to bring the Albanians and the Serbian minority (an object of Russia's concern) closer together', Russian experts believe. For this reason, he is considered in Russia as having contributed de facto to 'the realization of the plans of Albanese separatists on the creation of an independent state of Kosovo' (Rubinskiy 2008: 24; Zueva 2008: 57).

13 The economic partnership is supervised by the Economic, Financial, Industrial and Trade Council, the last session of which took place in June 2009.

14 Among the objective factors that deter French businessmen from exploring the Russian market: the fiscal burden, legal complexities, administrative rackets and corruption, the poor protection of property rights.

15 In October 2007, just a few days before going to Moscow for his first presidential visit there, Sarkozy welcomed his Ukrainian counterpart Yushchenko at the Elysee Palace. In October 2008, at the World Policy Conference, he stressed that 'Russia's "near abroad" is also … the European Union's. … It should be a field for cooperation, not a space of rivalry'. Sarkozy asserted, during a press conference with president Obama, that he indicated to his Russian counterpart Medvedev that 'the USSR was over, that the Berlin wall had fallen, and that around Russia, there was no satellite country, and that he had to respect this' (Sarkozy 3 April 2009).

16 This, however, and paradoxically, has been no big deal in Moscow, probably because Russia has long ago grasped that Paris, sooner or later, would be willing to take its full seat at NATO's table.

17 France, for example, has already indicated its disagreement with Washington concerning Turkey's integration into the EU.

18 This was also a result of the analysis that the French leaders had made of Germany's own EU presidency where it had proved impossible to move forward on the partnership with Russia because of the blocking power of some of the members.

19 Asked by the Russian daily *Kommersant* whether France could be an 'intermediary' between Russia and the European Union, former French Ambassador to Russia, Stanislas de Laboulaye, replied he did not like this term, and that France wished to 'develop relations with Russia above all within the framework of the European Union – and here, European solidarity is a priority' (Kommersant 2007b).

20 At the World Policy Conference in Evian, Nicolas Sarkozy said that he hoped Russia to make the choice of partnership with Europe 'which is not to be resumed to the particular relations that Russia has with every [EU] member'. Besides, a Russian expert understands that by constantly triggering conflicts with some of the new members, Russia works against its own interests: by doing so, 'Russia undermines the historical and often mutually profitable ties it has built with the "big"' EU members (Streltsova 2007: 60).

21 Of course, times have changed since then. However, it is worth recalling here that Mikhail Gorbachev's accepting the German reunification triggered a lot of concern in Paris, which feared that this could produce a major change in European balances of power.

22 See, for example, Streltsova's analysis on the crisis of the French model (Streltsova 2007: 54).

23 For a few years, in various international fora, France was one of the most vocal critics of Russia's deeds in the North Caucasus (Zueva 2008: 55–56).

24 For a Russian reading of this, see Zueva (2008).

Part 3
'In-Between-Europe'

10 European energy policy meets Russian bilateralism

The case of Southeastern Europe

Kjell Engelbrekt and Ilian Vassilev

For more than a decade the European Union (EU) has been working towards creating a common and cohesive policy to energy issues. An EU directive concerning common rules for the internal market in electricity was adopted in 1996, and two years later, the Union's energy regime was extended to the natural gas market. These initial steps arose from a consensus among member states that the European electricity and gas markets were to be restructured and modernized by way of breaking up formerly state-owned energy monopolies. With regard to natural gas, EU directives adopted in 1998 and 2003 sought to 'unbundle' transmission facilities from marketing business and to create independent regulators. Recently, however, EU governments have come to realize that this approach was too narrow given the challenges that the Union faces in the future in terms of shrinking supply and ecological repercussions of fossil fuel consumption.

In publishing a Green Paper entitled 'Secure, Competitive and Sustainable Energy for Europe' in March 2006, the EU Commission launched a broader approach than hitherto. The 2006 Green Paper singled out three overarching objectives, namely competitiveness and reasonable price levels, the security of supply and environmental sustainability (EU Commission 2006: 17–18). While energy problems affect EU member countries to varying degrees, governments agreed that competitiveness, energy security and sustainability are all generic to the Union and its common market. From the vantage point of an integrated European market, the functioning of member state economies at present rely on 50 per cent imports for gas and oil deliveries, a share which is expected to grow over the next 10–20 years. Some estimates project a 70 per cent dependence on outside suppliers by the year 2030 (Monaghan and Montanaro-Jankovski 2006: 7). Historically, strategic oil reserves exist in part as a result of the arrangements made by the International Energy Agency (IEA). (The IEA was established during the oil crisis in 1973–74 and is tasked with managing the excesses of oil supply dependency at the global level. It has helped alleviate the consequences of four major oil supply emergencies.) However, so far the EU lacks strategic gas reserves as well as a set of common rules and regulatory framework governing gas transit tariffs and price policies.

As an external supplier and political actor, Russia features disproportionately when it comes to overall EU vulnerabilities on energy issues. At present Russia supplies 11–12 per cent of the crude oil and over 30 per cent of the natural gas consumed in the EU, with seven member states receiving at least 90 per cent from that source (Baran 2007: 132). Of the natural gas used by Union members, Russia's state-dominated giant Gazprom alone supplies 29 per cent, the equivalent of some 130 billion cubic metres. By comparison, Norwegian Statoil appears as a distant second supplier, with 17 per cent and Algerian Sonatrach as number three with 13 per cent (Röller, Delgado and Friederiszick 2007: 7). In terms of total volumes Germany, Finland, France, Poland and Italy count among the biggest consumers of Russian natural gas. With regard to supplier dependence, Slovakia, Finland and the Baltic countries lack alternative supply routes and are therefore in a comparatively weak bargaining position vis-à-vis Gazprom. Only some 10 per cent of EU imports are in the form of liquefied natural gas (LNG), as opposed to that carried in pipelines.

By the same token, though, Russia and its energy companies are currently highly dependent on European consumption (Monaghan and Montanaro-Jankovski 2006: 16). First of all, the EU looms much larger in Russian overall economic relations than Russia in those of the Union (Proedrou 2007: 334). EU countries are the recipients of some 56 per cent of Russian exports and account for 70 per cent of natural gas sales whereas Russia imports merely 6 per cent of Union sales and (as aforementioned) provides a little more than a fourth of its gas consumption. However, perhaps more importantly, given the relative absence of Russian modes of transportation to alternative markets in Asia and beyond, Moscow presently has little option but to go on selling gas to EU countries (Leonard and Popescu 2007: 8). The structure of market relationships thus causes Russia to pursue, in part, a strategy that economists refer to as rent seeking, rather than profit earning (Tullock 1993).

This will not change in the short to medium term. In 2008, Gazprom launched the Yamal 'megaproject' and announced an ambitious investment program worth $37 billion that reflects the company's desire to shift emphasis towards the Far East and Siberia in terms of extraction as well as delivery of hydrocarbons (Gazprom 2009; Interfax 2008). In addition, in October of that year, the Russian oil pipeline monopoly Transneft signed an agreement with the China National Petroleum Corporation on building an extension of an existing east Siberian oil pipeline to China. According to the details agreed with Rosneft in February 2009, the link would carry some 15 million tons of crude per annum (China Daily 2008; Fischer 2009). While barely any fields are located so that they can furnish either European or Asian markets, these developments indicate that Russian energy companies hope to become less reliant on the hard currency provided by EU economies in ten to fifteen years from now. At the same time, the severe financial crisis of 2008–9 will no doubt slow down the process of shifting from European to Asian markets, as it negatively affects most investment schemes in the energy sector.

A series of political and economic events in the past few years have made EU–Russia relations jittery, with diplomatic and energy matters blending into each other and a tendency toward 'securitization' of related policies (Wæver 1995). The official EU–Russia Energy Dialogue stagnated due to conflicting perceptions of its utility, and occasional disruptions in supply to EU member states have shaken the previously stable reputation of Russia as a supplier of hydrocarbons. Most consequential in this respect was the early 2009 'gas war' with Ukraine, through which four fifths of Russian exports to EU consumers are transited. Neo-authoritarian political processes in Russia, the military confrontation in Georgia in the early autumn of 2008, further fanned European fears that Moscow would be prepared to go far in protecting its already strong position as provider of crude oil and natural gas to EU countries (Oil and Energy Trends 2008). Together with the financial crisis, the 'gas war' led to a dramatic 56 per cent drop in Russia's gas exports in the first quarter of 2009 (Handelsblatt 2009).

Almost equally uneasy with the presently high levels of interdependence in the field of energy, both the EU and Russia have in recent years tried to reduce vulnerabilities and enhance their respective bargaining positions (Hadfield 2008; Proedrou 2007). The Union's strength lies, as in other areas, in its systemic power to establish and uphold *constitutional rules* of markets and states with implications for all players, enhancing the potential influence of individual member states at the same time as it reshapes national preferences in the long term (Beck and Grande 2004; Engelbrekt and Hallenberg 2008). Conversely, Russia's influence is the strongest when it strikes bilateral agreements with separate governments that affects *collective choices* more broadly, as well as allows national energy companies greater freedom to maneuver in terms of adjusting *operational rules* to their own advantage (Ostrom 1990: 50–55).

An examination of the EU–Russia energy relationship could be broken down into a fine-grained examination of the respective challenges in the spheres of resource depletion of particular energy resources, the mode and distance of transportation and distribution, political uncertainties, cost fluctuation on the part of producer as well as consumer, etc. The following analysis, however, will not enter into a detailed discussion of the different dimensions and sources of risks associated with various energy supplies; it will remain focused on the development of EU–Russia relations over the mid to long term and thus to 'external energy policy' (EU Commission 2006: 14–17). A second delimitation is that we pay no attention to energy saving measures, emissions trading and renewable resources, as our analytical emphasis is on the security of energy supply (Skinner 2006). Without entering a wider conceptual discussion on security of supply, we accept that the notion of energy supply security can be broken down into 'operating reliability' and 'resource adequacy' (Röller, Delgado and Friederiszick *et al.* 2007: 12–13).

In particular, our analysis will focus on Russia's evolving, bilateral approach to energy issues in Southeastern Europe, as part of the overall strategy towards the EU and as a tool of enhancing Moscow's political influence and extract large revenues from European markets. We will first describe the rise of Russia's novel approach in this field, roughly dating from 2003, and then turn to the specific conditions of energy demand and supply on the Balkan Peninsula. We go on to analyse Russia's evolving activities towards that region and its repercussions on

EU ambitions to forge a unified policy. At the end, we discuss the prospects that the Union will be able to either thwart Russian bilateralism in that important region, or at least embed Moscow's approach within a wider policy that would render it compatible with the EU's strategic needs and preferences.

Moscow changes tack

Russian oil and gas have been imported to Western Europe since the 1960s and 1970s, a trade pioneered by the Italian energy consortium ENI and German Chancellor Willy Brandt. When the Soviet Union collapsed and Moscow embarked on free market reforms, US and West European companies began to develop an interest in the exploitation of Russian gas and oil fields. For most of the 1990s, however, the political and economic situation was quite unsettled and the majority of Western investors remained on the sidelines. Meanwhile, Gazprom was successfully reconsolidated under the leadership of Viktor Chernomyrdin, a process greatly facilitated by a general tax exemption instituted by the government (Goldman 2008, Chapters 4–5; Margolina 2008).

Also worth noting is the role of Germany in establishing its internal gas price formula as standard for CEE and the Commonwealth of Independent States (CIS) region, and thus indirectly forming the basis of the Russia–Germany strategic gas relationship. The German option of gas price setting was by no means the only alternative available. Until that moment, gas prices between Russian and its former allies in CEE had been based on a cost plus profit base, which was substantially below the internal German market reference based price model. The German government thus played an instrumental role in designing the current price model that determines gas prices, that in turn formed the backbone of Russia's 'market price' justification for aggressive gas policies against most CEE countries. The spread between the internal German market and Russian internal prices (which reflect the old cost–profit model) therefore generated much of the new wealth that has fuelled Russia's new assertive foreign policy.

As the Russian economy and its hydrocarbon industry started to make a handsome profit, West European energy companies and its Russian counterparts, often with the direct involvement of their respective governments, engaged in serious trade negotiations. Several major Western energy corporations expressed an interest in buying stakes in Russian oil and gas fields, and offered their assistance in developing them. For their part, Gazprom and the Kremlin began playing out such companies against each other (Baran 2007: 133). Most poignant is the Shtockman interplay pertaining to a major Barents Sea gas field, in which French Total and Norwegian Statoil eventually were allotted minority stakes (EurActiv 2008b). West European governments were also repeatedly asked to exchange exploiting rights in Siberia or other oil-rich areas against downstream resources in terms of stakes in refineries and national distribution companies in Europe. Later Russian resistance against foreign participation in the exploitation of the island of Sakhalin led to the unravelling of such deliberations. In late 2006, the Kovykta and Sakhalin II deposits were restored to Gazprom and, accordingly, contracts with major Western companies such as British Petroleum and Royal Dutch Shell were revised and their operating licenses revoked (Leonard and Popescu 2007: 20).

Russia's so-called 'gas war' against Ukraine in 2006 and 2009, a country that earlier had benefited from trading a preferential gas price rate for lower transit rates, changed the tone of the diplomatic conversation as well. The dispute was in part triggered by longstanding disagreement regarding how much 'leakage' was acceptable during transit on Ukrainian soil, with Kiev (Kyiv) claiming that 15 per cent was a reasonable level and Moscow demanding 5 per cent (Avioutskii 2008: 8). Another component was the failure to strike a balance between transit prices charged from Russia and Ukrainian expenses for gas destined for domestic consumption (Hadfield 2008). The complete interruption in deliveries in 2006 only lasted three days but the instant politicization of the dispute, which was handled directly by Vladimir Putin and Viktor Yushchenko rather than by Gazprom and Naftogaz Ukrainy, caused widespread concern in Russia's neighbourhood and in Western European countries that rely on imports.

The January 2009 interruption lasted over two weeks and affected gas distribution in eighteen European countries – in some quite severely, before deliveries were resumed. EU attempts to diplomatically intervene in the acute phase of the crisis, led by the Czech presidency and supported by France, were only partially successful. A bilateral Ukraine–Russia agreement signed on 18 January cut an intermediary corporation, RosUkrEnergo, out of the loop, a demand long raised by Ukrainian Prime Minister Yulia Tymoshenko while opposed by President Viktor Yushchenko. As a result, the domestic political rivalry between Tymoshenko and Yushchenko actually intensified at the same time as the conflict with Russia was scaled back.

Nevertheless, the EU then engaged in serious talks with the Ukrainians, and soon adopted a structural view of the problem. In mid-March, Gazprom finally consented to a gas price for 2009 that was twenty per cent lower than that offered to EU members states (Kramer 2009). A week later, a memorandum was signed between Kiev and the Union on a comprehensive overhaul of Ukraine's 13,300-kilometre-long pipeline network, with the European Investment Bank and the World Bank providing $3.4 billion of funding (Lungescu 2009). The premise of these plans is that the principles of EU energy directives are extended to Ukraine, which would seriously undermine Moscow's hopes that Russian companies could further penetrate and control the transit business in this part of the European continent (Krutihin 2009).

Quite understandably, normal market principles are difficult to apply to the Ukrainian situation (whose Soviet-style transit fee formula never covered replacement and renovation costs) and the stakes are extremely high on each side. Whereas Ukraine relies heavily on Russian deliveries, some 120 billion cubic metres of Russian gas annually passes Ukrainian territory on its way to Western Europe via its pipeline system. In fact, this transit gas constitutes as much as 80 per cent of Russia's overall export to the EU market. Needless to say, this is a major bargaining chip in Kiev's favour (whose actions, in turn, are far from flawless). It is also an explicit, long-term objective of Transneft, charged with creating and maintaining the oil export infrastructure, to eliminate or at least mitigate dependence on transit countries such as Ukraine (Fredholm 2005: 13–16).

In terms of direct effects on EU member states, the temporary disruptions of deliveries to Lithuania in July 2006 and to the Czech Republic in mid-2008 also sensitized EU governments to the potential problems of high levels of dependence on Russian energy supply. The latter disruption occurred three days after Prague had signed a deal with the United States to station a tracking radar system for antiballistic missile that would be located in Poland, against the will of the Kremlin, and commentators were quick to link the two events (Kramer 2008). More than anything the Czech government objected to the lack of a convincing explanation for the supply cut (Dempsey 2008). As a result of these irregularities directly affecting a member state, EU governments grew even more inclined to regard Russia as a 'risky supplier', meaning an energy supplier that might default on its contract-based commitments (Le Coq and Paltseva 2008: 46).

Russia, meanwhile, already the world's number one producer of natural gas (21.5 per cent), in 2007 closed in significantly on Saudi Arabia in terms of the biggest output of crude oil (12.4 per cent) (IEA 2008). Moscow's more assertive political stance and its growing profit from hydrocarbon exports thus began to transform the relationship with the EU. As two commentators put it, Russia has lately found itself in a position to start 'picking off individual EU member states and signing long-term deals which undermine the core principles of the Union's common strategy' (Leonard and Popescu 2007: 7). However, more troubling, at least to the international business community, is that Gazprom and other Russian energy companies like Rosneft and Transneft seemingly do not always make judgments based on projections of economic dividends, but rather on political affinities, interests and perceived loyalties of foreign governments (Gorst 2004).

The same has been said about Russian energy policies in the Caucasus and Central Asia, where pressures also have been more overt and explicit (Nygren 2008b). For instance, at one point Russia was selling vast amounts of natural gas to Georgia for $230 per thousand cubic metres while paying merely $100 per thousand cubic metres for gas purchased from Turkmenistan. It is apparently Georgia's pro-Western stance that precipitates Russia to ask a higher gas price from Tbilisi than it does from most of its neighbours in the Caucasus and Central Asia (Baran 2007: 137). Indeed, there can be little doubt that the Russian political leadership and the major energy companies in the past few years deliberately employed both the 'tap weapon' (using the on and off switch) and the 'transit weapon' (in a capacity as pipeline 'gatekeepers') to control energy resources, bolster revenues and pull countries in the post-Soviet space back into an economic region dominated by Moscow (Nygren 2008b).

Without much attention paid in the West, moreover, Russia successfully disciplined Moldova and Belarus into economic submission even after the 2006 'gas war' with Ukraine (Proedrou 2007: 339). In this case, Moscow could exploit the full force of its structural advantages to set constitutional rules as well as to exert its direct influence over ongoing operations. When Kishinev refused to agree to a doubling of the gas price from $80 to $160 in 2005, Gazprom simply shut the tap in January 2006. Moldova could not hold out but eventually agreed to upward annual price adjustments for 2007–8 as well as to handing over a major stake in the MoldovGaz state energy company. The Belarusian leadership, which believed it had a stronger position due to the hosting of a significant transit pipeline to Western Europe, also gave up resistance to Gazprom's demands for control over the transit route and a major price hike for 2008. With debts to the Russian energy giant accumulating each month and fresh threats to shut down deliveries, Lukashenko agreed to pay gas at 67 per cent of West European prices and to a $1.5 billion stabilization loan that allows Belarus to resume payment of debt instalments (Nygren 2008b: 6–8).

Whereas Belarus and Moldova have been successfully 'disciplined' to accept higher prices for Russian gas and oil, major transit countries like Ukraine and Poland are in a stronger bargaining position and so are some of the Central Asian oil-producing states. The spread between purchasing prices and those charged to end consumers in Europe has been substantially narrowed due the fall in oil prices and the accompanying drop in gas prices in the second half of 2008. Consequently, the profit margins for Gazprom have dropped substantially and further down the road Russia will probably have greater difficulties justifying and obtaining reduction in primary gas supply contract from Central Asian countries. The pledge to offer 'European prices' made to Central Asian governments in the summer of 2008 at high-level talks by President Medvedev and Prime Minister Putin has yet to be reconciled between negotiating parties, but affirms the general tendency that Asian markets are becoming more important.

In all of these contexts, it is clear that Moscow has sought to integrate elements of economic statecraft with foreign policy in the realm of energy; it is more difficult to assess whether one is subordinated to the other. The tendency in Western Europe and the United States has been to view Russia's energy policy as part and parcel of a 'great game' that induces rivalry over natural resources, transport routes and economic dominance (Hadfield 2008; Stern 2006: 4–7). Against the backdrop of the major economic crisis of 1998, though, an alternative interpretation is that Russian leaders are primarily bent on extracting maximum revenues from oil and gas in order to stabilize the country politically and socially. A nuanced and perhaps more realistic reading of Russian 'gasplomacy' in recent years is that elite constituencies have emphasized geopolitical and geo-economic strategies, respectively, depending on whether they belong to the defence establishment or the business community (for more on these 'super-groups', see Fedorov's chapter in this volume).

The pivotal location of Southeastern Europe

As European decision-makers debate energy security and the possible repercussions of enhanced reliance on Russian fossil fuels in the future, it is becoming increasingly clear that Southeastern Europe is integral to any long-term solution of the continent's energy challenges. One reason is that for the EU it makes a lot of sense to encourage Central Asian, Middle East and Gulf producers to provide gas and oil directly to the European market, in part to balance the growing significance of Russian supplies. Especially when it comes to the supply of natural gas, there are simple geographical reasons to expand the existing delivery system of pipelines by way of the Balkan Peninsula. Even in the case of oil and liquefied natural gas, access to ports in the Aegean and Black Seas, along with the Adriatic, can shorten transport routes and thus lower costs (IEA 2000: 25–40).

The promise of new infrastructure projects further lies in the opportunities to reinforce the security of external energy supply. That is, by building new pipelines and adding alternative routes, energy delivery can be diversified and the number of independent suppliers increased (Le Coq and Paltseva 2008: 27). Turkey, for instance, is heavily dependent on imported oil and gas for the overwhelming majority of its consumption. Lacking significant energy resources besides domestic lignite deposits – which accounts for the lion's share of Turkey's supply in coal – Ankara can primarily enhance the security of its supply through diversification. More recently, Turkmenistan has emerged as a feasible alternative, or at least supplementary, source of Turkish gas imports.

Meanwhile the Bosphorus and Dardanelles straits – which are practically inaccessible to LNG tankers – are increasingly congested by oil tankers and need relief in the form of pipeline transportation. The positive precedents of creating new infrastructure that ease Europe's reliance on Russia that US policymakers point to are the US-backed Baku-Tbilisi-Ceyhan (BTC) and the South Caucasus Pipeline (SCP) oil pipelines. Completed in 2000, the BTC is the more significant of the two, with the capacity of carrying a million barrels per day (Baran 2007: 136). A so-called Pan European Oil Pipeline to Constanta, Romania, has long been promoted by the EU, along with an interconnecting system between Turkey, Greece and Italy (ENS 2007).

However, the biggest and potentially most consequential of EU-backed projects is the Nabucco natural gas pipeline that would connect eastern Turkey with southeast Austria via Bulgaria, Romania and Hungary. The Nabucco pipeline would stretch 3,300 kilometres and transport some 31 billion cubic metres of gas per year. The weakness of the concept stems from the lack of major suppliers waiting to be connected. The original idea was that gas would be drawn from resources in Azerbaijan, Kazakhstan, Egypt, Iran or Iraq, neither of which today is fully committed to the scheme nor possess the necessary infrastructure. In a first construction phase, the Baumgarten (Austria)-Ankara (Turkey) portion of the pipeline would be built, and the full stretch completed in a second phase in 2013–14 (Nabucco website). The basic legal framework was put in place in mid-July 2009, with Austria, Hungary, Romania, Bulgaria and Turkey signing up to the project (EurActiv 2009).

A second component in transforming Southeastern Europe into a building block of a sustainable, long-term solution to the European continent's energy problems would be the expansion of its refining and storage capacities. For the time being, Romania possesses ten of the region's eleven refineries and produces almost half of its own consumption, some 115,000 barrels per day. Romania's refining capacity exceeds its own consumption and the country's oil reserves amounted to 956 million barrels in 2005. The Bulgarian city of Burgas also hosts a significant refinery controlled by Russian Lukoil, which makes petroleum products for domestic and regional markets. In former Yugoslavia only Serbia has a limited refining capacity. Like Bulgaria, also Serbia, Romania, Greece and Macedonia operate power plants fuelled by lignite, but are otherwise more dependent on imported crude oil, gas and petroleum products for the operation of their economies (IEA 2000; IEA 2007).

At present none of the Balkan countries hosts a major transit route for Russian hydrocarbons (Russian companies transit a little over sixteen billion cubic metres via Bulgaria to Turkey, Greece and Macedonia through special transit pipelines that are not part of the national gas pipe grid). Based on its own experience of recent years, Moscow is acutely aware that by extending the status of a transit area to a country it simultaneously enhances significant leverage to the host country itself, leverage that can be used against it to renegotiate price levels at a later stage. Indeed, Russian energy companies have had major disputes with each of the three countries hosting natural gas pipelines to Western Europe in the past five–six years, namely Ukraine, Belarus and Poland. And it is in no little measure the Russian leadership's desire to deny Poland and Ukraine such leverage that has precipitated friction with the EU as a whole (Fredholm 2005).

Manipulating the constitutional rules of the hydrocarbons game by creating new realities on the ground is a strategy already tested by Russia, with partial success. The Blue Stream pipeline from Russia (Izobilnoye) to Turkey (Ankara) was projected and built within the space of a few years and completed in late 2002. On the one hand, the Blue Stream pipeline became a useful instrument for Moscow in negotiations with Ukraine and other transit countries in the late 1990s. On the other hand, Blue Stream also revealed the often poor economic record of hard-nosed Russian 'gasplomacy', examplifying that assertive energy policies may eventually precipitate high costs. Until the gas crisis of January 2009, the Blue Stream was used at below 50 per cent capacity. And whereas Gazprom never made good on its related promise to let ENI in on ownership and exploitation of gas reserves in the promising field near Archangelsk, Turkey ended up gaining additional leverage over pricing as the Russian company could not afford the pipe to idle.

Nevertheless, Moscow seems ready to accept certain side effects as long as its overall leverage over Central and Eastern Europe is enhanced. With regard to Ukraine, Russia has tried to employ its 'transit weapon' indirectly so as to offset Kiev's potential influence over the level of Russian revenues from the EU's common market. Notably, one aspect of the 2006 Ukraine–Russia gas war concerned the fact that Turkmenistan sells gas to both countries but relies entirely on Russian pipelines and can therefore not demand much higher prices for the product. In recent years, Ukraine's 'orange government' has explored alternative routes for importing gas from Turkmenistan via countries like Iran, Armenia and Georgia. Western governments, concerned about foreign policy implications resulting from cooperation with Iran and the political uncertainties in the Caucasus, have been reluctant lo lend their support (Avioutskii 2008: 25–26).

With regard to Poland, Russia has successfully worked out an ambitious pipeline project with Germany, the Nord Stream project (mentioned earlier) that is set to place a pipeline on the bottom of the Baltic Sea. Nord Stream is believed to cost some four to five times more than a new land-based pipeline through Lithuania and Poland, and to some observers the project therefore illustrates how political logic trumps economic calculation in the minds of Russian leaders (Baran 2007: 135). In spite of several EU member states becoming more wary of their dependence on Russian gas and oil, Germany has been moving ahead with plans to expand its own import from the eastern neighbour (Proissl, Hecking and Wetzel 2009). The projected \$7.3 billion (€4.97 billion) Nord Stream project would clearly enhance the transport capacity of gas to Germany in particular, but new intra-European transport routes might eventually allow others to benefit from the additional inflow. In Poland, however, the concern is that Russia will want to bypass the existing 'Yamal–Europe' pipeline in order to lower the transit rates that Gazprom pays Warsaw at present. In addition, the Baltic countries, Finland and Sweden have serious misgivings about the latter project (Larsson 2007; Fuster 2008).

Russia's new energy diplomacy: bilateral, brisk and pre-emptive

As far as we know, the Putin administration first developed its paradigmatic approach of deliberately linking foreign policy with its economic influence as a producer and distributor of gas and oil resources during the course of 2003 (Lukyanov 2008: 1108). Since then the methods that Kremlin and the major gas and oil companies – in particular Gazprom – apply to states and markets have evolved and become more elaborate. On the other hand, due to the high stakes that are involved in the political, economic and strategic 'games' played when it comes to energy issues, the awareness of Moscow's growing assertiveness has grown almost as fast among proximate and distant neighbours. In other words, few of Russia's energy customers are naïve enough to fail to pay close attention to the small print written at the back of draft contracts put in front of them.

Crucially, the Russian government and the major gas and oil companies have consistently dealt with its counterparts on a one-by-one bilateral basis. In this way, Moscow has been able to tempt a number of Balkan countries to seriously consider entering long-term agreements with Russia and its energy companies. Another key element of Russia's energy diplomacy lies in its emphasis on inducing swift decision-making at the national level. Moscow does not leave the offer on the table for longer than necessary and uses diplomats as well as business leaders to lobby fiercely for striking a deal as soon as possible.

Notably, this pre-emptive 'gasplomacy' cost Russia and Gazprom next to nothing. In most cases, binding agreements that impact current behaviour of transit governments in return for future potential gains were achieved. An integral part of Russia's gasplomacy is the objective of raising public awareness on energy security at EU and national levels, often exaggerating the real dependence (Vassilev 2008). The public then puts enormous pressure onto public figures in these transit countries, people who in the absence of a clear, consolidated, 'constitutional' response at EU level often are inclined to opt for the lesser evil of bilateral agreements. Especially in Southeastern Europe, where energy companies are small compared to West European 'national champions' in that sector, bilateral bargaining with Russia and Gazprom means accepting the role of the smaller party.

As alluded to in the foregoing, though, the approach also has a pre-emptive component, aimed at creating facts on the ground that will speak louder than lofty multilateral project plans. The Russia–Turkey Blue Stream pipeline, which came online in early 2003, offset at least part of the implications of alternative projects promoted by the United States (Baran 2007: 138). The same tactic has in recent times been applied in efforts to try to thwart the Nabucco project in the Balkans. The latter project is led by Austrian OMV Gas & Power, but the consortium includes German, Hungarian, Turkish, Romanian and Bulgarian partner companies. Very shortly after the Nabucco project had been conceptualized and a consortium formed, in June 2007, Putin proposed the massive South Stream pipeline and approached Hungary as the site of a central hub of that project, thus turning the tables on Austria (Schmid 2008). In an apparently related effort to undermine Nabucco, the (then) Russian president reaffirmed provisional backing from Kazakhstan, Turkmenistan and Uzbekistan to upgrade existing pipelines and build a new one on the eastern shore of the Caspian Sea (Baran 2007: 138–39). In February 2009, President Medvedev invited his Turkish counterpart Abdullah Gül to Moscow to discuss energy issues, and promote South Stream at the expense of Nabucco (Ackeret 2009). Another element in this strategy was the package of bilateral agreements signed between energy companies from Italy (ENI), Greece (DEPA), Serbia (Serbia Gas) and Bulgaria (Bulgarian Energy Holding) in mid-February 2009 (Socor 2009a).

But potentially more important than undermining or neutralizing Nabucco – which some experts believe is unviable anyway – are the inroads that Moscow managed to make in Southeastern Europe during 2007–8 (Gärtner 2008). Within the space of a few months, Russia and its largest energy corporations were able to consolidate the development of delivery systems in Southeastern Europe and secure long-term control of downstream assets like refining and transportation capacities. Pivotal to these ambitions were successful negotiations with governments of Bulgaria and Serbia, resulting in bilateral accords and major stakes in key domestic energy assets (Oil and Energy Trends 2008). A critical element in the South Stream project is that Sofia agreed to fifty per cent ownership of South Stream transit infrastructure on Bulgarian territory, thereby creating a precedent to which Gazprom can refer in future negotiations.

The January 2008 agreement between Russia and Bulgaria means that the South Stream pipeline will start crossing Bulgarian territory at the Black Sea coast and then be subdivided into two separate connections, one directed towards Austria and the other to Italy. The official signing process took place during Putin's last trip as President of Russia. On the same occasion, Russian nuclear power construction company Atomstroyexport committed to building a second nuclear power plant at Belene, worth €4 billion. Belene is planned to start working in 2014 and will consist of two 1000-Megawatt reactors (Tzermias 2008). Speaking at the signing ceremony in Sofia, Putin brushed off European concerns about Russia's increasing energy clout, saying that South Stream would 'seriously increase the energy security of the Balkans, Europe as a whole and, of course, Bulgaria' (Brunwasser and Dempsey 2008).

Mark Leonard and Nicu Popescu (2007) reasonably categorize Bulgaria as a 'friendly pragmatist' country with regard to its relationship to the Russian Federation. Already in 1999, the two countries agreed to build the Burgas–Alexandropoulis pipeline on Bulgarian territory, featuring Greece as the third partner. In this case, Russia gained a fifty-one per cent controlling stake in the project through a careful blend of legitimate concerns on specific oil supply guarantees with an exclusive general oil supplier status that lacked solid guarantees. By implication, Bulgaria and Greece at this point forfeited their independent capacities to attract Central Asia oil producers. In this context, it should be recalled that Bulgaria since 2001 has been ruled by a president and a coalition government dominated by the Bulgarian Socialist Party, with historic and personal ties to the ex-Soviet elite. Moreover, in 2003, the Russian energy corporation Lukoil generated more than 5 per cent of Bulgaria's GDP, and thus around 25 per cent of that country's tax revenues (Leonard and Popescu 2007: 37) While these levels have come down to an estimated 18 per cent, Lukoil helps sustain Russia's position as Bulgaria's second biggest trading partner.

By virtue of the January 2008 agreement, Gazprom and other Russian companies were poised to enter the fray in a big way. Yet, the July 2009 elections ousted the Socialist-led cabinet and brought a centre-right party into power that is less likely to favour Russian business and political interests. In the aftermath of the January energy shortages, where Bulgaria suffered most among EU member states, Sofia also decided to endorse the construction of a 160-kilometre-long Bulgarian–Greek interconnection (Stara Zagora-Komotini) so as to reduce vulnerability on Russian supplies via Ukraine in the future. Further links between Romania and Bulgaria, and between Romania and Hungary, will follow suit and pave the way for a broader energy supply route between Southeastern Europe and Azerbaijan (Athens News Agency 2009).

In Serbia, the process of trying to find common ground between Gazprom officials and Serbian politicians met with greater difficulties, but ended in an ambitious agreement in December 2008. The agreement envisages three projects, namely the construction of South Stream through Serbia, the expansion of an underground gas storage facility in Banatski Dvor and the purchase of 51 per cent of Naftna Industrija Srbija (NIS) – the state oil company – by Russia's Gazpromneft (Vecernje novosti 2008a). Under the South Stream project, 400 kilometres of pipelines are planned to pass through Serbian territory. While some commentators concluded that Serbia had extracted economic advantages from Moscow in return for pro-Russian diplomatic moves, others suspected that

Russia's unyielding support for Belgrade's position on Kosovo was an implicit part of the bargain (Mojsilovic and Vukajlovic 2008). Judging from the accords signed, the Russian side did not offer a big discount anyway. The €400 million price tag was at least two times lower than business valuations reported in the press, and Gazprom's 51 per cent stake in NIS means that the company acquired a controlling share in oil rights and other privileges.

Concerns that the deal ultimately was politically, rather than economically, justified were repeatedly raised throughout the process, even inside the Serbian cabinet. Prime Minster Vojislav Kostunica consistently advocated, and Finance Minister Mladjan Dinkic opposed, the proposed accords. The fragile leadership consensus threatened to unravel when on 26 August 2008 Moscow suddenly reversed course on South Ossetia and Abkhazia and supported the independence of those provinces at the expense of Georgian sovereignty. Some Serbian politicians immediately drew the conclusion that the generic pro-sovereignty argument, bolstered by Russian endorsement of the non-secession of Kosovo, was seriously undercut. A symptom of the political relationship cooling were charges of foreign meddling directed against the Russian ambassador as he urged the Serbian parliament to speedily ratify the Serbian–Russian energy deal in the early autumn of 2008 (Woker 2008). In early December, top officials from both countries slightly revised and reconfirmed the energy deal, insisting that it would not be affected by Russia's new diplomatic stance on the Caucasus (Vecernje novosti 2008b).

The latter developments also illustrate that the much-touted historic dimension of Russia's ties to Belgrade and Sofia is often perceived simplistically, as are facile assumptions of West European capitals as consistently supportive of the interests of Central and East European countries. Going back to the seventeenth and eighteenth centuries, the major West European powers have a well-known tradition of 'leaping over' the CEE region – including Poles, Czechs, Hungarians, Romanian, Serbs and Bulgarians – so as to engage directly with Russian heads of state (beginning with Catherine the Great). On the Russia side, there is also the traditional self-perception of Moscow as 'the Third Rome' that implies a patronizing stance toward Southeastern Europe in particular (see Engelbrekt and Nygren elsewhere in this volume). The latter may go some way towards explaining the substantial and material difference in Russian policies towards key Western countries, based on a reciprocity concept granting them access to upstream energy assets in Russia proper, and the skewed reciprocity vis-à-vis Southeast European countries that are kept at arm's length and not allowed to control energy assets on Russian territory (for more on the direct relations between Moscow and the capitals of the EU 'Big Three', see the second section of this volume).

Policy, patchwork or parameters? The EU and its options

A longstanding approach of the EU Commission has been, as expressed by Commission president José Manuel Barroso in late 2006, that energy should not be allowed 'to divide Europe and Russia as Communism once did' (Bilefsky 2006).

Today some observers say that the EU should assertively respond to Russia's 'gasplomacy' in Southeastern Europe and elsewhere. With regard to diplomatic levers the EU has been criticized for abandoning the demand that Moscow ratifies the Energy Charter Treaty and joins the Transit Protocol that would allow for third-party access to its pipelines (Baran 2007: 141). One view is that the Commission, similar to how it dealt with Microsoft's leading role in computer software markets, could charge Gazprom with abuse of its dominant position in the natural gas market and imposition of unfair trading conditions, in accordance with EU treaties (Proedrou 2007: 342). In the aftermath of the Russia–Georgia crisis in mid-2008, British Prime Minister Gordon Brown expressly urged the Union to increase its support for the Nabucco pipeline, and so try to beat Moscow at its own game by creating infrastructure realities on the ground (Meister 2008).

However, whereas such options are on the table, the main challenge continues to be internal to the Union and its failure to coordinate the actions of its own member states. The general set of recommendations integral to the 2006 Commission directive (mentioned in the beginning of this chapter) is that the EU at the same time pursues competitiveness, security of supply and environmental sustainability, instead of giving top priority to any of the three objectives. More specifically, it has been suggested that the advantage of a European energy policy is to relax the domestic national trade-offs inherent in the three objectives through measures that would strengthen the underlying political and institutional framework, as well as develop the existing energy mix and encourage innovation. This would imply the creation of a pan-European network of regulating agencies and a central EU body of 'last resort', formulating common environmental targets, renewing efforts to enhance synergies and stimulate joint research projects, and providing 'an umbrella of external supply security for all Member States' (Röller, Delgado and Friederiszick *et al.* 2007: 51–52).

So is more coordination the solution that Europe needs? In another study of EU energy security and efficiency, Chloé Le Coq and Elena Paltseva analyse three hypothetical 'union constellations', namely: (i) an autarky, (ii) an uncoordinated union and (iii) a coordinated union. Le Coq and Paltseva describe a union constellation as 'characterized by the degree of solidarity between the member states (i.e., the percentage of losses covered by the non-affected party) and how such a degree is decided inside the union constellation' (Le Coq and Paltseva 2008: 47–48). This would mean that each union constellation develops its own rules for a common energy policy and that an autarchic system lacks mutual insurances in that each country's energy policy is not associated with the risks accepted by other countries. A 'coordinated union', on the other hand, is characterized by 'perfect coordination among the member states and all the policy decisions (including the risky energy consumption and the level of mutual insurance) are chosen to benefit the entire union' (Le Coq and Paltseva 2008: 47–48).

The authors found that a well-coordinated policy would be the optimal outcome, as it reduces the impacts of energy supply disruptions, but that autarchy still has the advantage of inducing member states to 'they take their energy consumption decisions separately, not accounting for the effect on the other union members.

In other words, they do not coordinate on the choice of the risky energy consumption' (Le Coq and Paltseva 2008: 47–48). What on the other hand could be problematic is a situation of insufficiently coordinated action, in which a mutual guarantee is offered but it remains unclear who assumes the burden of risk. In an uncoordinated union no longer characterized by autarchy, a mutual insurance approach may actually strengthen free-riding incentives and therefore increase the consumption of risky energy (Le Coq and Paltseva 2008: 11–12).

According to this logic, there would be distinct advantages to a coherent energy policy among the twenty-seven EU member states, but important weaknesses associated with a patchwork solution that would only go halfway. In Ostrom's terminology introduced at the outset, EU would settle for an energy regime formulated at the level of collective choice, not at the constitutional level. In order to achieve a well-coordinated (constitutional) approach Le Coq and Paltseva therefore recommend the establishment of a powerful regulatory agency at the core of a Union-wide regime. As noted before, the problems associated with a patchwork solution have to do with the lack of clarity concerning the ultimate responsibility and accountability of action, as well as the diffuse burden of costs and risks among the participant countries. Especially a patchwork solution that includes an element of mutual insurance is likely to be suboptimal and create a 'moral hazard' problematic among the member states (Le Coq and Paltseva 2008: 43ff).

However, the assumption that centralized energy policies – also associated with the eventual abolishing of the veto power of individual countries – will result in more efficient policies towards Russia is perhaps flawed in a more fundamental sense. At the end of the day, there is nothing to preclude traditionally friendlier countries to Russia to influence centralized decisions and achieve common denominators of EU foreign policy more in line with what Moscow would prefer to see as the dominant EU energy security policy. This could even accentuate the (legitimate) national security concerns of individual member states and of NATO as an alliance. The core problem here seems to lie in the mismatch of values in relation to energy policy as well as incompatibilities between public–private matrixes in Russian and European energy sectors. More broadly, one also needs to account for the different sets of priorities towards Russia in various corners of the European Union, with countries like Spain and Portugal lacking political sensitivities regarding Russian energy dependence. Even among CEE countries, Slovenia stands out as virtually unconcerned, suggesting that geographic distance and a robust economy lead to relatively balanced relations with Russia.

It is similarly useful to reflect on the role of power groups in EU countries that have vested interests in privileged relations with Russia. These are foremost the large energy companies engaged with the value-added chain of Russian imported gas within the EU zone. Statistical projections made by the International Energy Agency (IEA) and private energy think-tanks may in fact have helped to exaggerate expectations of dependence on Russian gas (Monaghan and Montarano-Jankovski 2006: 18). The IEA projections are based on the idea of a massive build up of gas-powered power plants and a subsequent radical increase in gas demand in the retail segment, predicted by the industry. Neither is proving true as one can glean from the factual history of gas demand. Such scenarios tend to overemphasize the importance of gas and generate virtual vulnerabilities that can be exploited by Russia and EU corporations in the downstream energy business (Vassilev 2008).

In fact, reverting to a low ambition, autarchy-oriented approach that would merely set some common parameters so that member states avoid unnecessary duplication of energy delivery and distribution systems might actually make more sense from a competitiveness standpoint. But then this would threaten to contradict the other two objectives, security of supply and environmental sustainability. From the vantage point of the latter, it is critical to reduce vulnerabilities and integrate the national infrastructures of energy supply, transportation and storage into a broader European framework. Short of 'perfect coordination', which is perhaps not a realistic goal for the foreseeable future, a 'grid code' or some set of common standards and rules for sharing resources – at least in the case of short-term disruptions of energy supply – seems politically astute as well as economically defensible (EU Commission 2006c: 6).

The latest set of proposals from the Commission, issued in November 2008, lists six priorities in the field of energy. Featured among these priorities are the creation of a common regulatory agency in the field of energy and the construction of the Nabucco pipeline (EU Commission 2008d). While either initiative may have beneficial effects on the security of supply, there are less expensive and less complex undertakings that could be equally – if not more – effective. A common agency is not a prerequisite for closer coordination of transit tariffs and gas import contracts. Nor does the building of cross-border pipelines (interconnectors) and storage capacities need the explicit agreement of all twenty-seven EU member states. Indeed, sufficient gas storage facilities might play a greater role in managing gas demand than pipelines buffers provided by 'line pack' gas. From this vantage point, it is curious that European governments were only swayed by Gazprom's rather aggressive practice of upgrading traditional gas contracts with access to gas infrastructure and retail markets. These include recent successful deals in taking over Banatski dvor gas storage in Serbia as part of the NIS deal, arrangements made with OMV concerning joint use of key gas hub Baumgartner and stated intentions to take over gas storage facilities elsewhere in the region.

Pipeline projects still clearly receive most of the attention from decision-makers and pundits. A key claim made by the Russian and German side on the merits of Nord Stream is that it enhances energy security to producers as well as consumers. Of course, the addition of an additional major pipeline would limit the risk of supply disruptions. However, the Nord Stream project does not reflect a genuine shortage of transit capacity, since there is at least 40 billion cubic metres of spare transit capacity through Belorussia and Ukraine, according to Kiev. In Ukraine's case, this capacity could be enhanced from 140 to 200 billion cubic metres with $7 billion, that is, a fraction of the cost of subsea pipeline (Lungescu 2009). In addition, both Nord Stream and the South Stream projects have an open project account in excess of US$40 billion. This means that the financial costs of these projects would add roughly ten per cent to current price levels

($350 per 1000 cubic metres by mid-2009) in gas supply contracts, not only with Germany and Italy, but also with each EU country party to Russian gas supplies, and eventually to the bills of EU consumers. An alternative that will achieve at least three times greater cost/capacity efficiency is to invest in maintenance management, and in an upgrade of transit infrastructure through Ukraine and Belorussia under joint direct EU–Russia supervision (Vassilev 2008).

Another important aspect of a more coordinated stance, diversification of the energy mix, seems to be a minor challenge in some countries though a formidable problem in others. Nuclear fuel and lignite are resources that some countries have resorted to in order to increase the share of domestic energy production. France, Sweden and Finland, for instance, have a sizeable nuclear power industry, while Poland draws heavily on its own brown coal deposits. The actual problems of relying on few or single non-domestic sources for energy supply are compounded by the circumstance that most Russian suppliers are quasi-monopolistic and, at least in part, driven or at least affected by political aspirations. In this respect, the near-total dominance of Gazprom in the export of hydrocarbons, as well as in Russia's domestic energy supply, constitutes a major obstacle to a more diversified market (Proedrou 2007: 338–41).

This should change as independent Russian gas producers gain prominence. So far, however, the steps taken by the Russian government toward breaking up the Russian electricity monopoly have been limited. Back in 2003, President Putin did sign off on a reform plan that involved large-scale privatization, and some twenty new electricity and heating producing companies emerged as a result. Although nuclear energy was excluded from the scheme, foreign energy companies – such as Italian ENEL, Finnish Fortum, German E.ON/Ruhrgas and RWE Power – subsequently made investments and gained control of some ten per cent of Russia's total production capacity (the equivalent of 20.4 Gigawatt). In mid-2008, RAO UES was dissolved in order for new investors to engage in the field. To this day, though, Gazprom still controls a staggering 80 per cent of the domestic natural gas deliveries (Hosp 2008). Moreover, since May 2008, government approval is required for foreign ownership above 25 per cent in 'strategic sectors', including companies providing electricity (Federal Law 2008).

Nevertheless, what eventually may end up economically weakening Russia's hand in Europe are developments in Central Asia over the next couple of years. Reference has already been made to the 'gasplomacy' conducted by President Putin to the countries of Central Asia, and Gazprom CEO Alexei Miller has pursued a largely similar set of priorities. One problem is that there has been a slight but growing discrepancy between the Kremlin and the Gazprom leadership as to what Central Asian countries are actually being offered. A second challenge is that China is becoming increasingly active in the field of energy, and is building bilateral relationships on the basis of its growing regional presence as a manufacturing and trading nation. Third, and crucially, it appears that Moscow has overplayed its cards by going far in wooing Kazakhstan and Uzbekistan and thereby neglecting its key supplier, Turkmenistan. Throughout 2008, there were signs that Ashkhabad was responding positively to overtures from the EU regarding the supply of gas to an east Caspian pipeline and from Beijing when it comes to future deliveries from east Turkmen fields (Bensmann 2008).

A bigger challenge in political terms is the EU's entry – at the invitation of Kiev – as a party that could alter the logic of confrontation between Russia and Ukraine on energy matters and potentially on a wider range of bilateral issues. Besides the apprehension regarding the Union inserting itself into what used to be referred to as Russia's 'near abroad', Russian leaders appear incensed by the mere prospect that the balance of power in Russia–Ukraine relations might be tilted in the latter's favour, and substantially reduce the former's leverage (Krutihin 2009). It is clear that the economic benefits that Moscow had hoped to reap by way of maximizing the number of levers that it could control can only account for part of the salience of this issue. Other dimensions that need to be considered are related to the political and military importance of Russian presence on the Crimean peninsula, as well as the significance of Ukraine as 'the cradle of Russian statehood'. Finally, critical to the prospects of 'Europeanizing' the Russia–Ukraine gas dispute is the stance taken by Germany, the biggest importer of Russian gas. Yet, Berlin's heightened sensitivity to Russian concerns in recent years sometimes makes it ambivalent on whether to promote a multilateral solution or a package of bilateral ones (Bonse 2009; Stent in this volume).

Conclusion

The EU's search for a cohesive approach and the at least partially successful efforts by Moscow to undermine and subvert that ambition begs the question of whether the Union should at all pursue a collective solution to energy problems. Reminiscent of the dilemma faced by participants in a 'stag hunt' parable once described by French philosopher Jean-Jacques Rousseau, there are incentives for each hunter to abandon the common objective in order to capture a passing rabbit, since the collective catch is an uncertain prospect anyway. Lacking a clear perspective that a common energy policy eventually will result from the initiatives of the EU Commission and those member states that strongly favour this notion, Germany, Bulgaria, Serbia and Hungary seem to have yielded to the temptation to settle for small game. (Several other states would quite probably follow suit, if an analogous offer were made.)

The years 2003–8 saw an increasingly active and assertive Russian political leadership, coordinating its measures closely with the country's energy giants, and above all with Gazprom, in order to affect a wide spectrum of rule-making in relations with the EU. Whereas Moscow has been in a position of extraordinary strength towards the majority of former Soviet republics of Central Asia and Caucasus – at least those that lack energy raw materials but are reliant on Russian exports – such high levels of leverage were tempered in the EU context. Even countries with no other supplier of natural gas than Gazprom have been able to pay good money for their energy supply; as 'valued customers' they have for the most part not been subjected to drastic price hikes or political pressures. The 'operational rules' cannot easily be manipulated unilaterally or for political reasons, unless Russia accepts the possibility of being viewed as a 'risky energy supplier' by other EU members.

Nevertheless, Moscow has at the same time been quite successful in at least temporarily tilting the playing field in its own favour. It has done so by skilfully exploiting existing incentive structures and breaches between neighbours to pursue a bilateral form of energy diplomacy that offers 'sweet deals' to individual partner countries. The biggest member state of them all, Germany, entered into a far-reaching agreement with Russia on building Nord Stream that (if realized) will circumvent Poland and the Baltic states as transit countries. Elsewhere, Bulgaria, Serbia and Hungary signed deals that allow for the construction of a South Stream pipeline that will similarly consolidate the hold of Russian energy corporations in the Balkans and beyond.

In fact, until mid-2008, Russia's energy diplomacy was impressively effective and appeared virtually unstoppable. However, at that point, things began to change with the political resistance that Russia's assertive strategy was encountering in Central Europe, the scepticism precipitated by the Russian military intervention in Georgia and the January 2009 gas crisis, as well as the reverberating implications of the worldwide financial crisis. For practical reasons alone, the politically defined horizon of 2015 for completion of South Stream is a non-starter. Given the rapid deterioration of the world economy and the recent decline in the prices of crude oil and natural gas, Russia's possibilities to weigh heavily on its more or less influential neighbours will in all likelihood diminish in the immediate future. Revenues will drop significantly and investments in infrastructure need to be postponed. Clearly, a crisis-stricken Russia will have fewer tools and resources to sustain its confrontation course unless it resorts to sabre (read: missile) rattling to sooth the feelings of domestic public opinion, for emotional wounds that the country's own political and business leaders arguably have inflicted.

In many ways, Russia has thus fallen victim to its own policies. The clumsy state intervention in the energy sector is likely to weaken the competitiveness of most Russian energy companies by undermining the fundamentals of that same assertive Russian foreign policy. Logically speaking, there are two outcomes in terms of a 'liberal' and a 'hard-line' scenario. The liberal scenario rests on the assumption that Russian leadership will soon appreciate that sustaining its confrontational course imposes novel constraints on its activities and revert to a more balanced and cooperative relationship with the EU, even at the expense of loosing the grip over state monopolies and allowing more space for political freedoms and opposition. In the long-term, this would presumably be a win-win solution for EU–Russia relations. Conversely, the confrontation could deepen as the current slide into crisis generates social tension and the ruling elite feels threatened. In that case, a 'fortress Russia policy' might prevail with greater control over public and corporate life. Mass protests and excessive use of force in containing public discontent will make the hard-liner's 'vintage Soviet' mode of government more likely, perhaps with renewed intervention in Georgia under a different pretext.

No doubt, a few years down the road, Russia will be back as a major actor on the European continent, reinvigorating its already strong position as the overwhelmingly important energy provider. By that time, however, it is imaginable that the EU or some of its member states will already have taken action to limit dependence on Russian exports for single providers, be they countries or individual companies, or even on hydrocarbons more generally. Notwithstanding the problems of fairly distributing costs and risks associated with a partially coordinated European energy policy, such an endeavour would of course also provide new opportunities to mitigate lock-in effects and rent-seeking, while encouraging ambitious, large-scale innovation in the field of energy policy. Similar to the climate issue, the EU could then emerge as a leader in forging a credible approach to energy more broadly, one that would also embed Russia in a strategy of sustainable growth and diversification of clients as well as of suppliers.

Such diversification would be healthy for Russia, too. Given that most projections predict that Russian hydrocarbon production will peak in the next few years, Moscow and the major energy companies would do well in promoting policies and programs that save and harness natural resources domestically in order for exports to continue yielding substantial revenues and robust economic ties with neighbours in the West.

11 Ukraine's emerging democracy and the Russian factor

Petro Burkovsky and Olexiy Haran

Introduction

Ukraine gained its independence in 1991 based on the slogans of a democratic political system in the place of Soviet authoritarianism, a social-oriented market economy instead of an administrative one and entering the community of European nations as a sovereign member. However, the former communist nomenklatura remained in power, which resulted in setbacks in the political, economic and foreign affairs of Ukraine. At the beginning of the new millennium, it was commonplace to view Ukraine as a 'no man's land', 'new lands in between' the EU and post-imperial Russia, and a 'new grey zone' of Europe (Legvold and Wallander 2004: 227). The official CIA World Fact Book of 2004 depicted Ukraine as a country where 'true freedom remains elusive, as the legacy of state control has been difficult to throw off. Where state control has dissipated, endemic corruption has filled much of the resulting vacuum, stalling efforts at economic reform, privatization, and civil liberties' (CIA 2004).

Some analysts called the country a 'blackmail state' fuelled by the desire of the former nomenklatura members and new oligarchs to 'maintain division, disorientation and intimidation within the state' (Riabchuk 2002). At the same time, they noticed the 'windows of opportunities':

> The press is indeed controlled, but not completely; the masses are threatened and fragmented but still able to resist; and the oligarchs, although greedy and selfish, are far from being a homogenous group, and more and more "dissidents" defect to the opposition. And finally, there is one more player on the Ukrainian scene who can metaphorically be called international civil society and international public opinion … None of these factors taken alone can bring about radical change; but working together, they could bring about the much-needed transformation of Ukraine's crypto-Soviet political system.
>
> (Riabchuk 2002)

That made the presidential campaign of 2004 crucial for the future development of the country: will the post-Soviet political regime reinvent itself, or will it be reconstructed on a democratic basis. Examples of political developments in neighbouring post-Soviet countries with the same kind of political culture and similar problems with a malfunctioning democracy, with the exception of Georgia in 2003, forced a prediction of high probability of the former.

Therefore, the Orange Revolution appeared to be an event of crucial importance not only for Ukraine, but also for the whole post-Soviet space. On the political level, a peaceful popular uprising and peaceful transition of power from ruling elites to the opposition and the defeat of a Russia-supported candidate showed the alternative to the whole post-Soviet space. On the cognitive level, the events of 2004 proved that the analytical assumptions about a 'vacuum' inside Ukraine and its positioning as the 'grey zone' of Europe, which was popular among a majority of Russian and Western researchers, were to a great extent erroneous.

In the 2006 Freedom House ratings, Ukraine was recognized as the only free country among all other post-Soviet states (with the exception of the Baltic States) (Freedom House 2006). Were all these developments unexpected? Was the Orange Revolution a Western plot as it was often depicted in Russian media or was it primarily a product of internal developments? What are the main transformations and how deep are they? What are the handicaps and pitfalls for further democratic development? How does this influence relations with Russia?

To answer these questions, this chapter begins with an analysis of key political developments in Ukraine on the eve of the Orange revolution, and trends in institution building that determined the political process in post-orange Ukraine. It shows how the emerging democratic rules of the game moved Ukraine closer to European practice, i.e. a balance of power, free elections, cohabitation, coalition building, and, at the same time, made the situation unstable in the post-Soviet environment. Against this background, it then discusses the role of the 'Russian factor', the policies and strategies of accommodation used by Ukrainian leaders, and their results for internal and regional politics.

Pluralism by default? From independence to the crisis of Kuchma's regime

After Ukraine gained independence, there were discussions in the international press whether or not the new country was viable, with predictions about future inter-ethnic conflicts, and Ukraine's turning 'back to Eurasia'. This did not happen. Ukrainian independence transformed the status of the previously provincial elite. The independence of the country became one of the dominant values of the elite. Russian-speaking leaders do not feel excluded from the political struggle in Kyiv, and they find it more realistic to compete for seats and resources in Kyiv than in Moscow (for more, see Haran and Tolstov 2002).

Since the late 1980s and the peaceful transition to independence, the crises in Ukrainian politics resulted in political compromises within the ruling elites. In 1994, for the first time in the post-Soviet space, a peaceful transition of power was achieved as a result of the presidential elections. The 1996 Constitution reflected a reasonable compromise between the President and the parliament and between the left-wing and right-wing forces in the parliament itself. At the same time, as a result of compromises constantly made in Ukrainian politics since 1990, the former Communist nomenklatura remained in power, democratic opposition was weakened and the economic reforms were stalled.

After his re-election in the fall of 1999, Leonid Kuchma promised to become 'a new President'. In December 1999, the pro-market head of the National Bank of Ukraine, Viktor Yushchenko, was approved as Prime Minister. Since 2000, the country has demonstrated impressive positive economic growth rates after a decade of economic decline (for example, in 2000 the GDP increased by six per cent, in 2001 – by nine per cent). This trend is usually attributed to favourable external market conditions and the policy of Prime Minister Yushchenko who started to reduce arbitrary administrative interference in the economy, provided for stable payment schemes in the energy sector and cut inflation down.

Despite this progress, the authoritarian trend in Kuchma's policy was growing. European institutions strongly criticized the 2000 referendum which was held on a very loose legal basis to give the President wider authority. As a result of this criticism as well as opposition within the country, Ukraine's Constitutional Court threw out two of six proposed questions and stipulated that the results of the referendum should be implemented through the proper parliamentary procedure. Because of opposition to his plans, Kuchma lacked the force to implement the results of the referendum through the parliament.

In general, during Kuchma's second term (1999–2004), Ukraine faced a serious decline of civil rights, the rule of law and fair government. The 'tapegate' scandal in 2001 around journalist Gongadze's murder, the intimidation of political opposition and independent media resulted in growing social dissatisfaction with state institutions. In the Western press, Ukraine was often depicted in black colours (criticism of Kuchma and the corrupted state bureaucracy). In addition, the West, after terrorist attacks on 11 September 2001, was concentrated on dealing with Putin, and Kyiv faced the danger of being left in the shade by Russia.

However, in this approach one important factor was missed. In Russia, there was no real opposition to President Putin. In Ukraine, democratic forces still fought for power and the political system remained quite pluralistic, which was explained by: 1) a quite strong democratic opposition in the parliament; 2)competition between oligarchic groups within the ruling elite; 3)disillusionment of the majority of Ukrainians in Kuchma and the ability of his administration to provide economic development of the country.

Non-violent revolution: potential for breakthrough

At the beginning of the 2000s, Ukrainian politics faced the evolution of a new strong non-leftist opposition, which presented an alternative view of the modernization of the Ukrainian state and nation in conformity with democratic values. The main force of this opposition, the Our Ukraine bloc, lead by former Prime Minister Viktor Yushchenko, included not only the traditional national–democratic opposition but also former state executives who protested against Kuchma's crony capitalism and corruption. Despite restrictions from the authorities, Our Ukraine won the first place on party slates in the 2002 parliamentary elections, and Yushchenko emerged as the leading candidate in the 2004 presidential race.

The main 'card' against Yushchenko was to present him as a radical nationalist who was going to oppress the Russian-speaking population, whereas Prime Minister Yanukovych was portrayed as a decent public servant and a great friend of Russia.[1] Russian and Ukrainian consultants of Yanukovych started to promote an idea of a 'schism' in Ukraine between the 'nationalistic' West and the 'industrial' East.[2]

On the other hand, the Yushchenko team declared its desire to get rid of 'state seizure' by clans, to decrease administrative pressure on businesses, abolish the tax police and lighten tax burdens and achieve a breakthrough in Europe. Thus, small and medium-sized business enterprises supported Yushchenko. Moreover, he was supported by the 'second layer' of Ukrainian large business enterprises. The first layer supported Kuchma, but in reality many of them were not happy about Yanukovych either and some of them even showed signs of such disagreement that they were playing on both sides. Yushchenko's claim to fight the 'criminal regime' was justified in the eyes of voters after an alleged attempt at his assassination and numerous examples of administrative pressure during the elections. According to a sociological survey (ISPP 2004) revealed in December 2004, before the first round 59 per cent of Yushchenko's supporters were ready to protest against falsification of voting, while only 20 per cent of Yanukovych's supporters thought they would need to protect results from the counting of votes.

On the other hand, it was not clear whether the opposition was fully prepared to resist falsification and what scale and mechanisms of election fraud would be used by the authorities. Each side expected a weak or hectic response from the other. However, when the protests started, the authorities did not manage to provoke the violent clashes which they had expected. Opposition leaders hoped that gigantic non-stop rallies all over Ukraine, which, in a combination of the celebration of the 'orange' victory (the colour of Yushchenko) with protests against falsifications, would take enough time to persuade Kuchma to give up. It is also important to bear in mind that protests demonstrated to Supreme Court judges who examined Yushchenko's complaint on the decision of the Central Electoral Commission (which named Yanukovych the winner) that a decision in favour of Kuchma and Yanukovych would be followed by a crackdown of the whole court system. People in the streets who behaved in a peaceful and self-organized manner against falsification made an impression on the senior officers of the security services and the Army, which helped opposition leaders to convince them to support the protest and not the authorities.

Mass civic activism spurred a rapid international response that included a very determined EU and US call to Russia not to interfere or endorse a crackdown on the part of the Ukrainian authorities. It also included European mediation in the talks between opposition leaders, the president and the prime minister. Kuchma, Yanukovych and Putin were eager to have a new presidential campaign (from the very beginning). It seems that it was one of the initial scenarios of Kuchma's administration, though they did not expect such strong protests from Ukrainians as well as from the international community. On the contrary, the opposition and the West insisted on repeating the run-off. Thus, the decision of the Supreme Court on 3 December to repeat the run-off on 26 December was a powerful blow to Kuchma and Yanukovych. In fair elections, Yushchenko received 52 per cent of the votes and Yanukovych 44.2 per cent.

To sum up, although supported by Moscow and the powerful business groups, the Ukrainian President Leonid Kuchma did not manage to pass on power to his designated 'successor', prime minister Viktor Yanukovych, and, in that way, to repeat the 'Yeltsin–Putin scenario' of 1999–2000. Neither was Kuchma successful in redrafting the Constitution so that he could retain power after his resignation by being elected with the help of the loyal parliamentary majority as a prime minister, which would have been a prelude of the 'Putin–Medvedev scenario' of 2008 (for more, see Haran, Pavlenko 2003). The main factors leading to Yushchenko's victory and the success of the peaceful protests were:

(1) A weakness of the regime and the relative pluralism of the Ukrainian political system compared with Russia and most post-Soviet states.
(2) Support of small and medium-sized enterprises (middle class), especially in the big cities.
(3) Popular disappointment in state patronage and readiness to protest against blatant manipulations and protect results from the counting of votes.
(4) A split within large business groups dissatisfied with growing authoritarianism and the threat of acquisition from the Donetsk group.
(5) The maturity of civil society when people clearly understood the difference between 'Us' (the nation in the broadest civic term) and 'Them' (the corrupt Kuchma elite and the oligarchs).
(6) Non-interference from the side of the army and intelligence services in politics and the rejection of possible violent crackdowns on protests.
(7) International condemnation of the falsifications and the firm Western position in demanding Kuchma to refrain from the use of force.

Constitutional reform and new rules of the game

After their inauguration, the new leadership of Ukraine enjoyed the highest level of public support since 1991. The absence of effective opposition and high social expectations created a window of opportunity for accelerated reforms. On the other hand, there were at least three main handicaps for implementing reforms:

1) The constitutional reform, which would weaken the role of the presidency, leaving him only one year to implement reforms.
2) Within a year, the country slid from a presidential to a parliamentary campaign, hence the growing populism in Ukrainian politics.
3) Differentiation and internal disagreements within the broad coalition in power.

Therefore, first, the president should have either accommodated to the new constitutional rules of the game or changed them by cancelling constitutional reform with the help of the Constitutional Court. Second, the new government should have chosen a coherent plan of action so that its policy could be compatible with the demands of the people both in the eastern and western parts of the country. Third, a broad and diverse camp of winners should have made an agreement on sharing power instead of dividing spheres of influence and clashing for the control of executive bodies.

One of the main impediments for Yushchenko's course arises from a contradictory compromise reached in the Ukrainian parliament on 8 December 2004 (between the fraudulent second round and the runoff). It stipulated that the constitutional reform would be effective from 1 January 2006. According to the new constitutional provisions, the Prime Minister is to rely on a parliamentary majority and he/she cannot be removed by the President at any time, as was the case before, which was a step in the right direction and had been demanded by democratic forces for many years. On the other hand, the reform appeared to be tangled and inconsistent. For example, the parliament may dismiss any minister by a simple majority, which would make ministers dependent on lobbyist groups in the parliament. Thus, after Yushchenko's victory, there was a debate among Ukrainian politicians and analysts about the question of whether the reform should be implemented, as several constitutional changes had not been approved in advance by the Constitutional Court. During the course of 2005, it was possible to check with the Constitutional Court the legality of the whole procedure of constitutional changes. However, Yushchenko did not use this possibility even in spite of the fact that his broad scope of authority would diminish in a year. It is possible that he did not want to change on his own initiative the compromise reached by the Ukrainian elites. In this situation, to get rid of inconsistencies of the reform might become possible only when the new parliament would be elected. This suggestion, along with the weakened role of the presidency, increased the importance of the March 2006 parliamentary elections.

One of the most serious impediments to reforms was Yushchenko's inability to control his new system of checks and balances. If Kuchma forced different groups to follow his own decision, whether correctly or not, Yushchenko tended to believe that competing actors could achieve a compromise without his direct involvement. In practice, it looked as if Yushchenko supported politicians or groups that were more successful in building personal trustful relations with him and looked more comfortable for him. The most difficult task for Yushchenko was to harmonize competing variants of reforms represented in the leadership of this coalition (which also included left–centre Socialists) into a unified vision.

Thus, a number of important political presidential initiatives did not bring about the expected breakthrough. His promise to conduct transparent and liberal economic policies contradicted the political actions and plans of his allies and companions. Nothing was done in the sphere of reforming the judicial and prosecution system infiltrated with judges and prosecutors nurtured in the atmosphere of corruption and protectionism. He not only failed to stop competition between law enforcement agencies and ministries but also fostered new conflicts. Members of his team, who were empowered to administer these institutions and balance each other's influence, used them for personal gains in the political struggle.

The administrative incapacity of the new team was especially apparent in the sphere of administrative and territorial reforms. Yushchenko's Secretariat wanted to introduce a more service-oriented approach to local government and wanted to change the way local government interacts with people. However, the actual policy planning process was full of misperceptions, due to the lack of open consultations on the topic and bad timing on the eve of the 2006 parliamentary elections.

Growing populism in Ukrainian politics on the eve of the upcoming parliamentary election in March 2006 led to a decline in economic performance and to the split between the liberal Yushchenko and the more populist and state-oriented Prime Minister Tymoshenko. In early September 2005, Yushchenko dismissed the entire cabinet but failed to gain sufficient support on the first vote of approval of his close associate Yuriy Yekhanurov as Prime Minister. In order to collect enough votes Yushchenko had to enter into negotiations not only with oligarchs but also with his rival and leader of Party of Regions, Viktor Yanukovych.

This was the first example after the Orange revolution of a political compromise between institutionally stronger authorities and a weakened opposition. Neither did Yushchenko start to suppress the opposing Timoshenko when she accused him of reconciling with Yanukovych and former pro-Kuchma factions. A placid resolution of the first crisis in the 'orange' power relaxed the political atmosphere in the country and allowed for the implementation of parliamentary elections in March 2006 in a free and fair manner, although very tense in the dimension of political struggle.

Yushchenko tried to compensate his loose style of handling domestic affairs by an active advocacy of Ukrainian interests in the West and in the European Union in particular. He stressed Kyiv's aim to join the WTO, the EU and NATO. Yushchenko tried to stimulate the EU by abolishing visas to EU citizens, but at that time, Brussels had not decided in what form to ease visa requirements to Ukrainian citizens.[3] In general, Brussels being preoccupied with the French and Dutch veto on the EU constitution was not ready to send positive signals that Ukraine would have chances to join the EU in the foreseeable future. Other problems involved Tymoshenko's government being slow to push through the parliamentary legislation necessary for joining the WTO or for becoming closer to the EU. The task of joining NATO improved technically due to the real reforms carried out in the defence sector and in the armed forces. However, politically it was an issue which divided Ukrainian society and, therefore, played into the hands of the opposition on the eve of the 2006 parliamentary elections.

In sum, political freedoms and democratic 'rules of the game' were the most visible achievements of the Orange revolution and the first year of rule of the new authorities. Nevertheless, these freedoms made it possible that all the mistakes of the orange government were easily exploited by the opposition. The Party of Regions led by Yanukovych benefited from the Yushchenko–Timoshenko rivalry, blamed democrats for economic difficulties and in the 2006 parliamentary elections gained the biggest share of the votes (32.1 per cent). Tymoshenko's anti-oligarchic and anti-corruption rhetoric resulted in increased electoral support for her party (22.3 per cent). The pro-presidential party, Our Ukraine, finished only third (13.9 per cent).

Actually, the voting pattern was quite stable when measured against the 2004 presidential elections. It was evident that if the Party of Regions presented itself in a manner that demonstrated that it had learned from its experience at the end of 2004, there would be a certain differentiation and reconfiguration of forces within it, which would enable them to become a more respectable political player. However, Yushchenko's electorate could not understand the alliance with Yanukovych and vice versa.[4] Most of the electorate preferred the restoration of the Orange coalition and the forces who had claimed to 'protect' the revolution again received a majority in 2006, although they then failed to build a coalition.

Coalition-building as a new stage of Ukrainian political development

The 2006 parliamentary elections conducted in a free environment created by the Orange revolution resulted in five political parties winning seats, with no party in the majority. Still, the political process in Ukraine was shaped by shadow decision-making, sharp conflicts between executive and legislative bodies, central and local governments. The outcome was a new political standoff in June–August 2006, determined by hard bargaining in the coalition building process.

Its resolution and the emergence of a self-appointed 'anti-crisis' coalition, established by Yushchenko opponents – the Party of Regions, Communists and his former allies the Socialists – was formally completed by the signing of the Universal for National Unity. It included an abridged program and priorities of the new coalition government and mutual commitment to cooperation. According to the informal part of this agreement, the President was given the right to submit for approval several loyal ministers.

At first, it seemed as if a new system had emerged: the President and the Prime Minister tried to secure separate and sometimes parallel structures of power. For the first two months, this system to a certain extent reminded of the French 'double executive' model in the *cohabitation* periods where the Prime Minister and the President compete with each other but, at the same time, must cooperate to ensure stability and govern the country.

However, the Universal was not a legally binding document. The Prime Minister declared that it was too early for Ukraine to join the NATO Membership Action Plan, which was one of the issues in the Universal, and de facto postponed joining the WTO until 2007. When Yushchenko appealed to the provisions of the Universal and reminded Yanukovych that he had to follow the President's instructions in foreign policy, the Cabinet of Ministers and the anti-crisis coalition struck back. The President responded with numerous suspensions of governmental decrees and even vetoed the state budget for the first time in Ukrainian history.

Then, leaders of the Party of Regions decided to overcome presidential power by creating a situational alliance with the largest opposition faction of Yulia Tymoshenko in order to pass a law on the Cabinet of Ministers that made Yanukovych almost invincible for Yushchenko.

The next step of the anti-crisis coalition was to recruit deputies from opposition factions to forge a constitutional majority (300 votes out of 450) and amend the Constitution to diminish presidential power or to threaten Yushchenko with impeachment. By the end of March, the political confrontation between the coalition and the President and the opposition grew rapidly. In this situation, Yushchenko decided to strike first, and on 2 April, he issued a decree on the dissolution of the Verkhovna Rada. In spite of the fact that the President's step was quite doubtful from a legal aspect, the resourcefulness of his Secretariat and support from two opposition factions helped him force the coalition to submit to his decision.

The compromise between the major political forces was formalized on 27 May in the Joint Statement of the President of Ukraine, Prime Minister and Chairman of the Rada 'On the Immediate Measures for a Resolution of the Political Crisis through the Early Elections of the Verkhovna Rada of Ukraine'. The elections proved that the Party of Regions remained the most powerful group in the country but it lacked the mandate to form a government. Again, it forced politicians from the different camps to enter a process of coalition building for the second time. The Yulia Tymoshenko Bloc was more successful in the negotiation process, masterfully using public pressure against a resurrection of the 'wider coalition' between Yushchenko and Yanukovych. She helped the President to elect his protégé Arseniy Yatsenyuk as Chairman of the Verkhovna Rada, which gave him informal opportunities to intervene in the legislative process.

The end of political crisis proved that in Ukrainian politics after the Orange Revolution attempts of one force (at that time, the Party of Regions) to monopolize power created the opposite effect; therefore, no political force in Ukraine is able to monopolize power and dictate its will. Another important precedent was the recognition of elections as *ultima ratio* and a tool for avoiding dangerous civil confrontation. The early parliamentary elections emerged as a kind of nationwide plebiscite of confidence to the political parties and institutions.

The failure of Orange cohabitation and a constitutional crisis

The new edition of the 'Orange' coalition re-established a pattern of cohabitation when the country was run by the two centres of power. However, this time, the position of the President had to be strengthened by the loyal Speaker of the parliament and a quota of 50 per cent in the Cabinet of Ministers. The law on the Cabinet of Ministers was amended to grant the President the right to make orders to the government by issuing decisions of the National Security and Defence Council. Moreover, the Head of the Secretariat of the President organized an informal group within the coalition, consisting of 10 deputies, who could abstain from voting if there was no agreement between the President and Prime Minister.

But, again, the Prime Minister started to conduct her own policies and initiatives which did not conformed to the President's plans. They clashed over the new gas deal with Russia during the January 2009 gas crisis, over methods of privatization, microeconomic and budgetary policies. The confrontation between Yushchenko and Tymoshenko grew by the middle of 2008 into a constant conflict over power. It was also overburdened by the issue of who would be the main presidential candidate in 2009–2010 campaign. . The practice of cohabitation proved to be irrelevant for Ukrainian real politics.

Unable to find a compromise, both Tymoshenko and Yushchenko, again, just as in the crisis of 2005, started to seek separate agreements with the oppositional forces. Cooperation between the President and Viktor Yanukovych was undermined by their polarized approaches to the Russia–Georgia war in August 2008. On the other hand, the Prime Minister persuaded Yanukovych that they could unite their efforts to amend the Constitution and establish a parliamentary republic with the presidential election in the Verkhovna Rada, which would make it possible for Yanukovych to return to power. However, the situational alliance between BYuT and the Party of Regions lasted for just a month. It was broken by disagreements over the necessity to start an impeachment procedure against Yushchenko and growing demands from Yanukovych to extend the powers of the President.

The crisis was sharpened by the President's attempt to dissolve the parliament and prevent an amendment of the Constitution. Nevertheless, his decision was blocked in the courts and the escalating financial crisis made him drop it. However, the danger of the early elections antagonized a majority of the deputies in the President's faction 'Our Ukraine – People's Self-Defence' and forced them and the other faction of the Lytvyn Bloc to reconstitute the coalition with BYuT, which had no majority of votes anyway as long as the minority of the 'Our Ukraine – People's Self-Defence' refused to vote for governmental initiatives. The coalition's disability became very visible when it adopted the state budget only with the help of Communists in December 2008 and could not appoint a new Minister of Finance after the previous one had resigned. On the other hand, oppositional forces in January 2009 did not manage to arrange a vote of non-confidence for the government, which made it impossible to resolve the standoff in accordance with constitutional provisions until the end of presidential elections in 2010.

The Russian factor

It is commonplace to acknowledge the importance of Russian influence in Ukraine. However, as we see from the foregoing, Moscow could not prevent the Orange revolution and since that time the two countries, in their institutional designs, have moved in different directions. Nevertheless, Ukraine could not make 'a great leap forward' to the EU and NATO and it seems that, in many cases, its leaders, despite official Euro–Atlantic slogans, are continuing the 'multi-vector policy' of the Kuchma era taking into account vast Russian interests in Ukraine.

Kuchma's 'responsive strategy': Causes and implications of its failure

Through 1991–2004 the Ukraine–Russia relations had been characterized by asymmetric interdependence between a former metropolis and a colony, which was also true for other former Soviet republics. However, it is also true that the Russian ruling elites during the first decade of new relations recognized the right of sovereign existence of Ukraine as a nation-state equal to the right of Russia to develop its own project of a nation-state. As Roman Szporluk points out,

> In 1991, the Russian Federation had played a crucial role in the peaceful dissolution of the USSR and in Ukraine's gain of independence, and it seemed then that its leaders and its people had abandoned the goal of imperial restoration and an authoritarian form of government, in short – had agreed to become a "normal" nation, similar to other "post-imperial" nations.
>
> (Szporluk 2007)

This assumption was the main reason why, in spite of the various problems in bilateral relations, Ukrainian and Russian leaders managed to keep domestic radical forces down and continue to seek compromises after occasional political and economic disputes.

However, Ukraine was regarded by Moscow as a strategic territory that in numerous aspects is vital for keeping Russian national interests safe. Even after 18 years of independent coexistence, there are public demands on Ukraine, such as staying out of 'US-sponsored alliances and organizations aimed against Russia' (which included not only NATO but also GUAM and the Community for Democratic Choice) (Primakov 2009), keeping favourable conditions for the Black See Fleet in Sevastopol before and even after 2017, 'respecting' the Russian language, the religious and other 'cultural rights' of the Russian minority, and allowing Russian businesses to secure control over key Ukrainian industrial assets. As far as Kyiv demonstrated its willingness to take these into account, Moscow showed its good will in considering Ukrainian concerns. If they had not been met, Moscow would have punished Ukraine by harsh political and economic means. This can be well illustrated by the history of trade wars and gas disputes between Russia and Ukraine since 1991.

Thus, the problem was that both sides constantly debated about limits and the scope of reciprocity in their relations. From the Ukrainian side, the core of the so-called 'multi-vectored' foreign policy of Kuchma in 1995–2000 was to be closer to the party, which demanded smaller concessions.[5] An example of such bargaining is the 1997 Treaty on Friendship, Partnership and Cooperation signed in the package with a General Agreement on the Stationing of the Black Sea Fleet in Sevastopol, which was also linked to the agreement of restructuring Ukrainian debts for natural gas consumed in 1992–94. It was obvious that due to the institutional and international weakness of Russia, Ukraine was able to counterbalance its pressure by strengthening ties with the European Union and the

United States. For instance, in 1997, after signing a treaty with Russia, Kuchma blessed the Charter on Distinctive Partnership with NATO.

After 2002, the concept of 'multi-vectored' foreign policy was to a great extent exhausted because of intentions of the Western governments to review relations with Ukraine as a result of a series of corruption scandals and steps by President Kuchma to intimidate political opposition and the press. Another important factor was growing Russian strength and Putin's intention to restore an exclusive sphere of influence in the boundaries of the former USSR. If Kuchma had chosen to confront or ignore new integrationist initiatives from Moscow, he would have lost support from political parties and financial–industrial groups dependent on close ties with Russia. Weakened by the tapegate scandal and growing isolation from the West, President Kuchma decided to play his own game with Russia, demanding economic benefits, such as the liberalization of bilateral trade and low energy prices, for the formal support of Kremlin integrationist efforts. Growing national economies with their mutually interconnected demands and favourable oil market conditions created a 'window of opportunities' both for Kuchma and Putin to arrange a set of intergovernmental agreements concerning gas and oil supply and transits through Ukraine as well as a demarcation of the Ukraine–Russia land border, etc.

Therefore, in 2001–4, the Ukrainian President introduced a new 'responsive' strategy towards the big neighbour, which contemplated meeting evident wishes and interests of Moscow before they had grown to the point of becoming hazardous for the interests of current Ukrainian elites and before they would be dictated without possible compensation. The best examples were the negotiation processes of creating an International Gaz Consortium, and the Single Economic Space, a kind of economic union to substitute the incapacious CIS, initiated in June 2002 and January 2003, respectively.[6]

In the next two years, the Ukrainian President was able to control the agenda and speed of negotiations on these issues. Kuchma also succeeded in resolving border disputes with Russia over the Kerch Strait and Tuzla Island without heating up bilateral tensions. In that critical situation, when an occasional parliamentary majority supported appeals to western powers to protect Ukraine from Russian threats, Kuchma again proved to be a player who avoided open conflict and alienation from Kremlin's interests. That is why, as Arkady Moshes pointed out, between 2000 and 2004 'Russia has unambiguously chosen to deal with the regime. Although this choice entails a risk of strategic loss if the opposition comes to power, it gives Russia tactical gains both in the economic and political spheres' (Moshes 2002).

Rapprochement on the sensitive bilateral issues was followed by a fading transparency of negotiations and increased involvement of Russian top officials in Ukrainian internal politics. 'The unique arbiter role that Russia plays in Ukrainian politics' (Moshes 2002) was built on the vested interests of the Ukrainian political parties and informal pressure groups (known as financial–industrial groups). Among the most active, one must mention the left-wing opposition parties, such as Communists, Socialists, as well as 'oligarchic' parties – the Donetsk-based

Party of Regions (leader Viktor Yanukovych), and the Social–Democratic Party (United) (leader Viktor Medvedchuk) and the political group of Volodymyr Lytvyn (speaker in 2002–6). All of them endeavoured to attract the attention of confidants of Putin, or of Putin himself, to their political and economic interests and achieve an alliance and/or win Kremlin support in the struggle with their rivals. In return, Ukrainian politicians and business groups offered their assistance and protection in their lobbying for privatization and acquisition bids in the parliament, the cabinet, the presidential administration and even in the law enforcement and security services.

Consequently, the possibility to make 'shadow deals' with Ukrainian elites at the expense of Ukraine's interests was something the Russian authorities interpreted as an open 'window of opportunities' for the lobbying of long-desired Russian projects in Ukraine. Some of the main evidence for this was the appointment of Viktor Medvedchuk to the position as head of the presidential administration. Medvedchuk was accused of a number of initiatives that hurt and sacrificed Ukraine's interests in relation to Russia,[7] such as the removal of 'NATO membership' as an official goal from the military doctrine of Ukraine, holding a so-called 'Year of Russia in Ukraine', which resulted in a strong Russian media influence on Ukraine and the promotion of pro-Russian views of 'common, fraternal co-existence' of two nations in one state, simultaneously with condemnation of the 'nationalistic' vision of Ukrainian history. He emulated the Kremlin system of media censorship and established it in key Ukrainian TV channels and for this purpose hiring political advisors from the Putin's Administration (Gleb Pavlovskiy, Igor Shuvalov and Marat Gelman) (Gelman's and Shuvalov's interview 2004). In 2004, they legalized their presence in Ukraine as members of the NGO 'Russian Club' and participated in the presidential campaign against Viktor Yushchenko.

These kinds of new relations between political actors in both countries ended up in a number abuse cases which could not but hurt public opinion. Moreover, the opposition used evidence of obscurity of 'responsive strategy' to shape the estrangement of a great share of the Ukrainian population to a 'Russian turn' of Kuchma and the oligarchs.

The climactic tensions over the consequences of the domestic dimension of the 'responsive strategy' was achieved during the presidential election campaign of 2004, when the Ukrainian prime minister and appointed successor, Viktor Yanukovych, introduced issues of dual citizenship and Russian as a second state language together with a demonstration of the more than ever closest economic relations with Russia as his bid for presidency. This program was actively supported by the Kremlin and by the media which were controlled by or loyal to pro-Russian oligarchs and to Putin himself. However, these issues polarized public opinion. The arrogant and manipulative manner of the promotion of such ideas added to an aversion of Kuchma's foreign policy among the population and politicians.

During the Orange Revolution and the discussions about a third round of presidential elections, Russian interference took an open form when Putin twice

(!) congratulated Yanukovych on his victory and when, on 1 December, Putin supported Kuchma's intention to restart whole election process (which would mean a prolongation of Kuchma's incumbency for another five to six months). In the Yanukovych camp pro-Russian radicals started an ill-thought attempt to threaten with the secession of the eastern and southern regions of Ukraine if 'nationalist' Yushchenko were to win.

The defeat of Yanukovych (Kuchma's designated successor) by the orange camp depreciated a 'responsive strategy' as an option for dealing with Russia. In spite of the fact that it helped to normalize bilateral relations, it was also poisoned by efforts of the Kremlin to thrust upon the Ukrainians its interests regardless of opinion and the choice of the majority of the population. To sum up, Ukraine–Russian relations in 2001–4 appeared to be neither a 'strategic partnership' nor a 'surrender of national interests', but, rather, an egoistic policy of Kuchma's administration that hoped to receive political support and exclusive economic benefits from ties with Russia.

Ukraine–Russia relations after the orange revolution

The new Ukrainian authorities tried to prevent Russia's future attempts of a direct political intervention. This view determined Yushchenko's policy of achieving goals of integration in NATO as a guarantee of sovereignty. At the same time, due to the complex network of relations inside the new 'orange' establishment and controversial goals of the respective political groupings surrounding Yushchenko, there were numerous attempts at using the same mechanism of 'responsive diplomacy' to derive political and economic benefits from the Russians. Such an evident contradiction accompanied with a poor performance of the executive branch of power could not but weaken the position of Ukraine and lead to the gradual freezing of contacts on the highest level and repeated standoffs in trade and energy relations.

2005–2006: Lost opportunities

The events of 2004 forced the new Ukrainian authorities to believe that Russia would look for political compensation for the loss of its stakes in the presidential elections. Such a perception was formed not only by Putin's open interference in the Ukrainian election process. A year after the elections, members of the Russian government (whether the presidential administration or the Council of Ministers) continued to call the Orange revolution 'an unconstitutional *coup d'etat*'. This definition was used by the secretary of the Russian Security and Defence Council, Igor Ivanov (Bykov 2005) and the senior deputy head of the presidential administration, Vladislav Surkov (Melikova 2005).

Another sphere of Ukrainian concern was Russian attempts to influence key European states and persuade them that Ukraine was an unstable, divided country with an unpredictable future. During his visit to Germany in April 2005, Putin said that 'if Ukraine enters the Schengen zone one certain problem will emerge. There, in Ukraine, as far as I know, live no less than 17 per cent of Russian people. That will mean a division of a nation! That will resemble the partition of Germany in the Eastern and the Western parts!' (Putin 11 April 2005). In October, the Russian Minister of Foreign Affairs, Sergey Lavrov, delivering a speech at the French Foundation of Political Studies expressed an official opinion even in a more undisguised manner: 'Estrangement to Ukraine would have been unbelievable for the Russian population, considering the long-standing ties between the two nations. It is fair enough to bear in mind that half of its (Ukrainian) population are ethnic Russians' (Lavrov 2005) which in reality is only 17 per cent.

Such an attitude made the new Ukrainian authorities very anxious about the Kremlin's plans. In 1997, Sherman Garnett predicted:

> The relative gap between Russian and Ukrainian political, economic, and military power at present may be as narrow as it has ever been, despite Ukrainian weakness, simply because of Russian own internal preoccupations and challenges. As the two states stabilize, this gap will grow again. The most dangerous period in Ukraine–Russia relations may well lie ahead.
>
> (Garnett 1997: 48–49)

Therefore, some people in Yushchenko's entourage expected that in 2005 Russia, affected by the defeat of the pro-Russian candidate in the presidential elections and outraged by an intensified Western-oriented, especially Euro–Atlantic, policy of the new President of Ukraine, would decide to use 'sticks' instead of 'carrots'. Two centres which influenced presidential views on foreign policy existed around the Secretary of the Security and Defence Council, Petro Poroshenko, and the Minister of Foreign Affairs, Boris Tarasiuk.[8] Under the influence of the Poroshenko and Tarasiuk groups, Yushchenko took several steps to prevent possible overpressure from Moscow and, at the same time, ensured the Russian government that its interests in the sensitive energy sector (such as transits of gas and oil to Europe via Ukraine) were protected. The subsequent events proved that the Ukrainian new ruling team overestimated Western support and underestimated the vast Russian opportunities and resources, as well as overestimated the coherence and efficiency of its own policies.

The 'restrictive strategy' of the President, formulated by the Ministry of Foreign Affairs, presumed that Ukraine could protect its interests and prevent Russian interference by posing high demands in spheres where the Kremlin had not been troubled since the late 1990s. In such a situation, it was argued, Russia would be preoccupied with responding to these demands, while Ukraine would abandon some of them in return for concessions on selected security and economic issues. The list of demands included conditions of the stationing of the Black Sea Fleet in Sevastopol and its withdrawal from Ukrainian territory in 2017 (according to the 1997 treaty), the division of USSR assets and property abroad, the strengthening of GUUAM (Georgia, Ukraine, Uzbekistan, Azerbaijan and Moldova) and a project of Caspian oil transits via Ukraine, the future status of Transdniestria as part of Moldova, the demarcation of the state borders and the signing of a readmission agreement.

Integration in NATO was regarded, especially by the Foreign Ministry team, as a strategic goal that would be pursued irrespective of the well-known Russian negative attitude to NATO enlargement and the Ukrainian decision to become a member in the future. In April 2005, cooperation between NATO and Ukraine was spurred by the adoption of the Intensified Dialogue, a preparation for the acceptance of the Membership Action Plan the following year. Tarasiuk spent all his tenure visiting capitals of key NATO countries to promote the idea of the Ukrainian MAP. The gas dispute between Russia and Ukraine in January 2006 sent a message from the Ukrainian Ministry of Foreign Affairs to western governments about the realism of Russian threats to Ukrainian independence and security. This impression was supported by evidence of Russian interference in unrest in Crimea against Ukraine–US military exercises in June 2006. Even after the emergence of the 'anti-crisis' coalition in August, Tarasiuk used the Universal of National Unity to demonstrate to his western colleagues that Ukraine's new government would continue its movement towards NATO. At that time, the MAP still seemed not only desirable but also a realistic next stage of cooperation between Ukraine and the Alliance.

Another approach (formulated by Poroshenko) pushed the idea of creating the Interstate High Commission chaired by Yushchenko and Putin themselves. He believed that only presidents (with the assistance of their closest men) on the personal level were able to solve existing bilateral problems. He was also at pains to establish permanent political contacts between Yushchenko's newly founded party 'People's Union Our Ukraine' and Putin's 'United Russia'. He even managed to sign a statement of intention with Boris Gryzlov, chairman of 'United Russia'. However, his initiative did not get support from the Cabinet of Ministers, the Ministry of Foreign Affairs and the Secretariat of the President because of his permanent conflicts with the chief executives of these bodies. In September 2005, Poroshenko was accused by Yulia Tymoshenko of lobbying for the interests of two Russian strongmen, Viktor Vekselberg and Sergey Abramov.

As a result, Poroshenko lost his position and his access to Yushchenko, but his approach to the construction of interstate relations did not cease but seized a strategic dimension of Ukraine–Russia coexistence – the natural gas trade. For four years, 2005–8, gas deals have been the most shady business of the 'orange' team. Relations in the gas sphere can be described as a 'black box', where one can only see 'inputs' in the form of statements or interviews of Gazprom and Naftogaz officials and 'outputs' as being the decisions and contracts between two parties and two Presidents.

Moreover, the lack of cohesion and strategic thinking in the Yushchenko team also disrupted plans to introduce transparency in the bilateral gas relations. In November–December 2005, Russia and Ukraine started very tough negotiations on gas supply and transits to EU countries. To demonstrate its capability to go from threats to actions on 1 January 2006, 'Gazprom' reduced the pressure in the pipelines. After European countries started to complain about the undersupply of Russian gas, on 4 January 2006, representatives of the two governments reached a sudden agreement to end the dispute. The negotiations were closed and resulted in the creation of a non-transparent Russia–Ukraine scheme detrimental to Ukrainian security. Former senior vice-minister of foreign affairs, Oleksandr Chalyi, who took part in very tough gas negotiations in 2000–2002, even called it the 'Pearl Harbor of Ukrainian energy diplomacy' (Zerkalo nedeli 2006). What the majority of experts and political observers knew was only the title of an intermediary company 'RosUkrEnergo' (RUE).[9]

RUE's influence on Russia–Ukraine relations grew even more after the resolution of the gas dispute. It became the main importer of Russian and Central Asian gas for Ukraine, although Gazprom had the right and opportunity to do it on its own. Putin publicly admitted that RUE existed 'due to the wish of the Ukrainian side' (Putin 7 February 2006). This made RosUkrEnergo the only but very striking example of a readiness of the Kremlin to do business with the 'orange', 'unconstitutional' and 'pro-Western' government in Ukraine.

The same is true for Yushchenko. The protection of RUE and the shady deals with Russia was incompatible with his proclaimed 'new standards of politics' and a 'European choice'. Again, Ukrainian elites and top officials, facing a choice between strategic gain (to reduce dependence on Russian energy resources) and tactical victory (to avoid a shocking increase of prices) preferred the latter because of its importance for the internal struggle. In such a situation, the Russian side could use permanent political conflicts in Ukraine aggravated by the institutional weakness and division of power among several groups to compel Ukrainian authorities to sacrifice strategic state goals for temporary advantages.

Although a paradox, during 2005–7, the Ukrainian President had no officially developed doctrine or concept of foreign policy toward Russia. In his official address to Verkhovna Rada in 2006, Yushchenko included two controversial messages which coexisted as a result of political competition within his team described earlier. On the one hand, he confirmed that Ukraine and Russia would continue to develop their 'strategic partnership' and use existing opportunities for economic cooperation (NISD 2006: 23). On the other hand, Yushchenko subordinated relations with Russia to the Ukrainian policy of European and Euro–Atlantic cooperation (NISD 2006: 139). The Ukrainian President acknowledged that the two countries had contradictory interests but expressed his hope that Ukraine would find arguments to persuade Russia that its integration with the West would not have anti-Russian effects. But Kyiv's steps were viewed by Moscow as anti-Russian. Since January 2005, Yushchenko intensified cooperation with Georgia and its President Saakashvili, who was seen by Moscow as an American-guided leader. The Ukrainian President limited the idea of creating a Single Economic Space with Russia, Belarus and Kazakhstan to a free trade zone. In April, Yushchenko blamed the Black Sea Fleet for illegally possessing a thousand acres of Crimean soil. During February–August 2005, Yushchenko tried to approach Turkmenistan President Niyazov to establish an alliance in the gas sphere. He supported Poroshenko's plan on reintegration of Transdniestria into

Moldova and ordered the enforcement of customs control on the border with the unrecognized republic.

Inconsistent steps of the authorities helped Ukrainian political opposition to develop its own alternative approach to Russia. Considering the Kremlin's hostile attitude towards the new President, it was comfortable for Yanukovych to predict a worsening of relations between the two countries. One of the main arguments in 2005 was a gas dispute and an almost twofold increase in gas prices. The other was Russia's decision to restrict imports of Ukrainian products, which started a new period of the 'trade wars'. The Party of Regions used these as proof of 'disastrous outcomes of orange anti-Russian policies'. It also continued campaigning for the introduction of the Russian language as a second state language and supported the idea of an all-Ukrainian referendum on NATO membership. However, after the 2006 elections, in coalition negotiations with Yushchenko, it had to give up some of its demands and even agreed tacitly with the President's intention of continuing movement towards membership in NATO and signed the respective provisions in the Universal of National Unity. This again showed Moscow that Ukrainian politicians continued to use the 'Russian factor' in a bid for power but were not going to fully yield to Russian demands.

2006–2007: Cohabitation in Ukraine and the 'Russian factor'

The inability of the President and Prime Minister to maintain a system of political cohabitation and to agree on key domestic policies completely blocked the development of foreign policy of Ukraine in two important directions: the European and the Russian. In his struggle with Yushchenko, Yanukovych looked for support in Russia, while the President positioned himself in the West as a more democratic and civilized leader compared with Yanukovych. The personal political standoff multiplied at the institutional level. The President's plans of spurring NATO integration and his attempt to revise treaties on the Black Sea Fleet were blocked by the Cabinet of Ministers. On the other hand, many coalition laws and governmental decisions were vetoed by the President.

It appeared that Yanukovych was able to prevent Kyiv from joining the MAP in the fall of 2006, but he developed cooperation with NATO in many spheres,[10] and he could not push for further integration with Russia. Moreover, in the sphere of trade policy the Yanukovych government and 'anti-crisis coalition' followed Yushchenko's policy of accession to the WTO without the 'coordination' demanded by Russia. In some sensitive issues, such as the President's intention to achieve international recognition of the famine of 1932–33 as an act of genocide (which had been evaluated in Russia as preparation for bringing charges against it as descendant of the Soviet Union), the prime minister and coalition failed to break Yushchenko's plans.[11] During the 14 months of his premiership, Yanukovych had to deal with pro-Western and loyal pro-presidential ministers of Foreign Affairs and Defence.

At the same time, the President became an observer of the prime minister's active and numerous visits to Moscow and his private consultations with Putin.

Yanukovych tried to position himself as 'pragmatic' in contrast to Yushchenko, who 'ruined' bilateral relations. From August 2006 to October 2007, Yanukovych had six 'eye-to-eye' meetings with Putin, while Yushchenko had talks with the Russian president only on five occasions during the whole period of 2005–8. Thus, the Kremlin made a very clear hint as to what Ukrainian leadership it preferred. After the appointment of Yanukovych, Russia lifted restrictions on trade with Ukraine, although gas prices for 2007 were again increased by forty five per cent.

Yet, the prime minister and the anti-crisis coalition did not manage to develop their own approach in Ukrainian foreign policy towards Russia. They had no idea of how to offer the Ukrainian population an attractive and coherent project of a common future with the Russian Federation. That is why, in the 2007 pre-term parliamentary elections, the Party of Regions decided to exploit the old idea of a referendum on 'Ukraine's neutrality' and for an introduction of the Russian state language. Evidently, this left the Party of Regions without representation in Western and Central Ukraine and again confirmed the dependency of the party on the neighbouring state, which is no strong point for any political force in Ukraine.

Even in spite of the fact that Yanukovych lost the power struggle to Yushchenko after the dissolution of the parliament and after the pre-term elections, the Kremlin demonstrated its support of the Party of Regions. On 9 October, a day after Yushchenko had demanded from the largest parliamentary factions to 'unite in the name of the national interests of Ukraine', Putin met Yanukovych and endorsed his bid for reappointment as Prime Minister by saying that 'Russia is very interested in the development of very close relations, especially in the economic realm, with such a sister nation as Ukraine' while Yanukovych stressed that, '/w/e (the Party of Regions) will do everything to provide stability of power and stability of relations with our strategic partner – our sister nation – Russia (Yanukovych 2007). Putin also expressed satisfaction in the rapid resolution of the emerged gas debt problem by the Yanukovych government.

After the early parliamentary elections, the 'gas lobby' (RosUkrEnergo) again played a major role in bridging the differences between the President and the Prime Minister. RUE owners sought to form this alliance because a coalition between Yushchenko and would be Prime Minister Tymoshenko meant new trouble for their business. Yushchenko and Yanukovych worked together during October–November 2007 to solve the problem of debts to Gazprom and to fix the gas price for the next year. The President approved a debt payment deal with Gazprom. Prime Minister Yanukovych acknowledged his subordinate position to the President and confirmed his willingness to stay in office in the event of forming a coalition between the Party of Regions and the presidential bloc 'Our Ukraine'. Yanukovych and Yushchenko coordinated their efforts during the ecological catastrophe in the Kerch Strait: both of them chose a firm position in demanding compensation from Russian ship owners for the ecologic damage. Strongmen in the Party of Regions suggested that such a coalition would be optimal for Ukraine if the President decided to delay his policy of integration into NATO as long as this issue was not supported by the majority of the population. However, the period of normal cohabitation and cooperation in foreign policy was too short and limited for the issues of energy policy to make general assumptions about the possible creation of a new approach to Russia.

2008–2009: Back to 'responsiveness'?

After two months of serious hesitations, Yushchenko finally adopted, in December 2007, a new edition of the 'orange' coalition with premier Tymoshenko. In its turn, the Tymoshenko Bloc agreed to back Yushchenko's plan of preparing Ukraine for getting the MAP in April 2008. The respective provision was included in the coalition agreement. Later in January, Tymoshenko put her signature, together with the President and the new speaker of the Verkhovna Rada, under an official letter of request to the Secretary-General of NATO confirming Ukraine's wish to join the MAP. Furthermore, Tymoshenko publicly admitted that she supported all priorities of presidential policies. Thus, Yushchenko again, since 2005, had the opportunity to represent the whole of Ukraine in its contacts with Russia and had a free hand to review any concessions made by the previous government. However, unlike the situation in 2005, he had to share power with Prime Minister Tymoshenko and could not give direct orders to ministers. Tymoshenko used the new constitutional rules to conduct her own policy in energy relations with Russia, which she failed to do in 2005.

Tymoshenko did everything in her power to get rid of RosUkrEnergo in gas supplies schemes to Ukraine. RUE struck back with a harsh PR-campaign in Russia and Ukraine against Tymoshenko. Representatives of RUE took part in a special sitting of the National Security and Defense Council (NSDC) chaired by the President, where Tymoshenko was charged with lobbying in the interests of the gas company close to her. When she ignored a decision of the NSDC, the President did not permit her to visit Moscow before his own negotiations with Putin. In February 2008, Yushchenko and Putin signed a political memorandum of resolution of the gas dispute. At first, Tymoshenko backed it, but, later in February and March, she conducted negotiations with Gazprom without reference to this document. Finally, in spite of criticism from the President and threats from Gazprom, the prime minister persuaded the Russian side to establish gas relations on principles different from those mentioned in the memorandum. Although the new agreement with Gazprom was signed only by the end of 2008, it was more transparent than the memorandum of the two presidents and allowed Ukraine to regain control of its own internal gas market from RUE and its subsidiary company.

The clear diplomatic defeat of the President against his own prime minister must have influenced the Russian view of the internal situation in Ukraine. It must be admitted that Tymoshenko during the gas negotiations took a less enthusiastic approach to NATO and adopted a position that to a certain extent resembled the position of the Party of Regions (that is, joining NATO only after a referendum, which was actually her electoral platform in 2007). In the next few months, she twice met with Putin as Russian prime minister. In August, after the beginning of the war in Georgia, she did not endorse the President's condemnation of Russian aggression but copied a balanced EU response on the events and supported the plan of French President Sarkozy and Russian President Medvedev (Tymoshenko 2008). In September, her faction voted for the law on special investigation commissions that enforced the initiative of the Party of Regions to investigate arms sales to Georgia (a step provoked by allegations of the Ukrainian president in selling arms to the Georgian Army that had been voiced by Putin and Medvedev). These steps became a pretext for presidential forces to break up the governmental coalition, and the new Ukrainian political crisis started from September 2008.

Independently from the President, the foreign policy of Tymoshenko added to a growing misunderstanding and a political struggle between the President and the Cabinet of Ministers. In general, her behaviour reminded one of Kuchma's 'responsive strategy'. Tymoshenko's 'partnership' bid in relations with Russia after the resolution of the gas dispute and after the war in Georgia looked very attractive in contrast to Yushchenko's rigidity about Russian security concerns as regards NATO enlargement and the status of the Black Sea Fleet. Many Ukrainian analysts suggested that she had no precise foreign strategy and acted from the point of view of solving problem after problem without looking too far ahead. Her stance on Russia in 2008 changed dramatically compared with her views in 2007, which was revealed in the famous 'Foreign Affairs' article 'Containing Russia' (Tymoshenko 2007). At that time, she used publications to shape the minds of American, not Russian, decision-makers about her own style in politics. She needed the United States's positive neutrality to show Yushchenko that the legally questionable decision to dissolve parliament and prevent Yanukovych from forming a constitutional majority would be understood in the West. She was also in a great need of US backing to put pressure on Yushchenko to stop protecting the 'gas lobby', which was described as the main tool of Russian political influence in Ukraine.

As soon as Tymoshenko returned to the position of Prime Minister, she softened her attitude to Russia, and, at the same time, demanded the removal of RosUkrEnergo from the Ukrainian gas market. She met Gazprom's wishes and opened direct access of its subsidiary company to the Ukrainian market, although limiting its quota to 25 per cent of the RUE quota (25 per cent of the industrial consumers' market). Tymoshenko's aim was twofold: to demonstrate that she can solve the gas issue in relations with Russia and to increase her electoral support in Eastern and Southern Ukraine.

The collapse of the orange coalition in September 2008 and the dissolution of the parliament in October (although suspended by the President in November due to the necessity of passing anti-crisis legislation in order to fight the negative impact of the global financial crisis), again interrupted the process of developing a coherent Ukrainian policy towards Russia, which could at least be coordinated between the President and the Prime Minister.

Again, the political turmoil in Ukraine prevented the West from giving a positive answer to President Yushchenko about the prospects of Ukraine joining the MAP, this time in December 2008. The Kremlin made a pause and returned to its usual policy of destroying the international image of Ukraine as being an 'unstable', 'unpredictable' country. The 'Russian card' would be played again in electoral debates in Ukraine. However, whoever wins presidential elections it seems that the basic trend would remain the same: the gradual Ukrainian movement towards Europe under Kyiv's official discourse of European integration while continuing its 'delicate balancing act' with Russia.

Notes

1 Although most Russian analysts considered Viktor Yanukovych as a possible 'pro-Russian' candidate for the Ukrainian presidency, this was a simplification. As an example, Ukraine was ahead of Russia in the process of joining the WTO, and the Yanukovych government continued negotiations with the WTO in this direction. The Russian side demanded 'coordination' (disclosure of the Ukrainian documents signed with the WTO countries, which contradicts WTO practice), but the Ukrainian side refused. When Russia laid territorial claims to the tiny but strategically important island of Tuzla in October 2003, this caused stormy protests in Ukraine, including Ukrainian officials. Even more important: it was clear that the Donetsk group would first of all defend their own interests which often contradicted the interests of more powerful Russian oligarchs and state-controlled monopolies. For more on Russia–Ukraine relations under Kuchma, see Hayoz and Lushnycky 2005: 189.
2 On real, imagined, and cleavages artificially deepened by politicians, see the special issue of the journal *National Security and Defence* published by the Ukrainian Center for Economic and Political Studies named after Olexander Razumkov (National Security and Defence 2006).
3 For more information on the topic, see policy paper prepared by the Stefan Batory Foundation (Boratynsky and Szymborska 2006).
4 Polls by the Kiev International Institute of Sociology and 'Democratic Initiatives' Foundation (27 April–4 May) showed that only 38 per cent of the population supported the 'grand coalition' between former rivals, Yanukovych and Yushchenko, and 47 per cent preferred a restoration of the 'Orange coalition'.
5 For more on Kuchma's foreign policy, see Wolczuk (2003).
6 For more on this topic, see Haran, Tolstov (2002).
7 Symbolically, Putin and Medvedev's wife became godparents of Medvedchuk's daughter.
8 Prime Minister Yulia Tymoshenko was, in 2005, not permitted to interfere in the foreign policy of Ukraine even when it touched upon economic and energy issues, and her cabinet was used as a technical apparatus to implement presidential decisions.
9 RUE was founded in December 2004 – January 2005, when Dmitriy Firtash, a co-owner and partner of Gazprom, according to his own revelations to the 'Financial Times' approached Viktor Yushchenko and talked about the 'importance of stability of Russian gas supplies to Europe through the territory of Ukraine'. His trusted person, Yuriy Boyko, continued to serve as chairman of the state energy monopolist 'Nafrogaz Ukrainy' until late March 2005, whereas other former 'kuchmists' elsewhere were already fired by the new team. Even after he was replaced by one of Yushchenko's appointees, 'RosUkrEnergo' was still represented in the Ukrainian government by Ihor Voronin, senior deputy chairman of 'Naftogaz' and a member of the steering board of RUE. The interests of RUE were lobbied and protected in the secretariat of the President by the senior aide to President Oleksandr Tretyakov and the former senior aide to Kuchma and chief advisor to the speaker of the Verkhovna Rada Serhiy Liovochkin. The omnipotence of the secret 'gas lobby' was demonstrated in August 2005, when the President allegedly ordered the security service of Ukraine to halt investigations and abstain from the arrest of Voronin.

10 On 15 June 2007 during the meeting of Minister of Defence, Anatoliy Hrytsenko admitted that the Yanukovych government implemented the Action Plan Ukraine–NATO better than the two previous 'orange' governments (cited from Ukrainian: Hrytsenko 2007).

11 Parliamentary opposition with the help of the Socialists passed a resolution of recognizing the 'Great famine' as genocide, which enabled the promotion of this idea in the international community.

12 Riding three horses

Moldova's enduring identity as a strategy for survival

Patricia Fogarty

Introduction

Moldova is one of a handful of countries that occupies geopolitical and ideological space between the European Union and Russia. Moldovan citizens are fully aware of their position as a small, poor, former Soviet state at the crossroads of Eurasia. They have felt economic and political pressure from Russia in the form of a ban on wine and agricultural imports in 2006–7 and an increase in natural gas prices from Russia, both of which were cited as causes for Moldova's 3.5 per cent decline in economic growth in the same period (DFID 2009). From Western Europe, the United States, and other donor countries and organizations, Moldova receives billions in foreign aid and loans (World Bank 2009, MSIF 2009a). That aid comes with different pressures on the government to implement Western standards of democracy, transparency and market economy principles.

Moldova's national economy could not survive without its neighbours to the figurative East and West. The country's largest exports are its agricultural products and its people, but wine is Moldova's signature export. Russia imposed a ban on Moldovan wine and other agricultural products between 2006 to 2008, which drastically impacted Moldova's trade balance, and Moldova has always had difficulties penetrating European markets. Moldova's wine and beverage exports accounted for 27.4 per cent of all exports in 2005, but declined to 12.3 per cent by 2008 (Moldovan Statistical Bureau 2009a). During the same period, Moldova's total in exports dropped by nearly one-third, while its trade deficit deepened by nearly US$600 million (Moldovan Statistical Bureau 2009b). Russia, Ukraine, Belarus and other Commonwealth of Independent States (CIS) members account for the majority of Moldova's exports of wine and other alcoholic beverages (Moldova.org 2009). Eastern EU countries, such as Poland and Romania import many Moldovan wine products, but Western European countries still prefer their own or other countries' merchandise (Korshak 2009; Moldova.org 2009).

In addition, approximately 25 per cent of Moldovan citizens work abroad in states spanning Europe and Eurasia, from Ireland, Portugal and Italy to Russia, Turkey and the Levant (CIA 2009; Moldova Azi 2007). Many work illegally or have been victims of human trafficking. However, remittances from foreign sources have fuelled Moldova's economic growth, reaching nearly 23 per cent of

GDP in 2003 (IMF 2005, Polinin 2004). The economic downturn of 2008–9 has hit Moldovans especially hard. Workers returning from joblessness abroad have come home, lowering remittances; at the same time, a regional drought lowered food supplies in 2007–8 (UNDP 2008). This combination of economic factors has left Moldova more vulnerable to political pressures from Russia and the EU.

Nevertheless, both the EU and Russia have something to gain by courting Moldovan favour. Russia seems to want to expand its influence over former territories to its southwest in order to secure natural gas transit routes through Ukraine and Moldova to other Eastern European and Black Sea states (Eke 2009), and in order to bring Russian minorities under its protection (Kolstø 2000). Russia also benefits politically by maintaining its role as a key negotiator in the unresolved conflict over Transdniestria, a separatist region in the eastern part of Moldova, where the Russian 14th Army is still garrisoned (Popescu 2005).[1] Since the European Union's expansion in 2007 to include Bulgaria and Romania, Russia has strengthened its influence in peace negotiations concerning Transdniester, effectively using the conflict as a bargaining chip with the EU and OSCE (Socor 2009b). However, Russia woos Moldova's government with a hammer rather than a flower. Rising gas prices and the agricultural product ban, as already noted, seemed to be penalties that could easily have been revoked if Moldova had overtly aligned itself with Russia and accepted Transdniestria as an equal partner in peace negotiations.

The European Union also needs a friendly neighbour, a buffer zone between its newest eastern border in Romania and the rest of the former Soviet Union. Political stability in Moldova would mean a resolution to the Transdniestrian conflict, the removal of Russia's 14th Army from Europe's eastern front, and a national government with Western governance standards. While Moldova and other states, such as Ukraine and Georgia, have been considered for NATO and possible EU membership for security reasons, the EU would also accrue economic benefits by inviting Moldovan favour. Economic stability would lead to a better market for European products, a site for further capitalist expansion and a reduction of illegal immigration to EU states.

With such influences and their livelihoods at stake, it is easy to see why Moldovans tend to vacillate between the strong powers to their East and West. Scholars have described how Moldova's location, history and multiple identities have shaped its ethnic troubles, internal governance and foreign policy (Kennedy 2007; King 2000). National opinion polls also show Moldovans' reticence to choose one 'side' over the other.[2] Another way to delve into the question of Moldova's status is to ask how daily life in Moldova reflects Moldovans' attitudes towards West and East. For example, what cultural references to Russia and the EU can one find in specific situations? What values are attributed to Western and Russian ideologies and actions? And to what extent have Moldovans integrated foreign ideologies into their own cultural identities?

The data presented here arise from ethnographic fieldwork conducted over eighteen months between 2003 and 2006.[3] During that time, I explored how everyday life in Moldova, as well as the general workings of a multilateral development agency, are shaped by the complex tensions between Western, post-Soviet, and local discourses on development, citizenship, and national identity. The development agency in question is the Moldova Social Investment Fund (MSIF or SIF). I chose the development arena to examine the heavy influence of Western states on Moldova since independence, but I found that former Soviet, and now local, ideas about development equally pervade daily life. The tensions present in everyday discourse represent Moldova's in-between status as much as geopolitical analyses and economic statistics, and further illustrate how Moldovans navigate relationships with their neighbours and see their future.

SIF's goals are to reduce poverty by 'empowering communities and their institutions to administer to their most important needs' (MSIF 2008). Namely, SIF subprojects improve social services by renovating or building items of infrastructure, such as schools, kindergartens, roads, roofs for public buildings, sewage or garbage management systems and water and natural gas systems. SIF is officially a Moldovan government program, but 80 per cent of its funds come from external sources, primarily loans and some grants from the World Bank and several donor countries (not including Russia). For its part, the Moldovan government provides 5 per cent of the local financing, while 15 per cent of the cost of each subproject comes directly from the community that implements it (MSIF 2009b). The multilateral donor organizations therefore have the power to influence the goals of the project and SIF's operating procedures. It is through a belief in these goals and procedures that the SIF team operates to instil in subproject participants the value of their communities and, by extension, their state. SIF employees interact regularly with representatives from the World Bank, the United Nations Development Program (UNDP), UNICEF, the Swedish International Defense Agency (SIDA), the United States Agency for International Development (USAID), and other international and foreign national organizations. They have become comfortable using development discourse in their work and as an ideology to guide their vision for Moldova's future.

The term 'discourse' is used here to mean social and linguistic practices that may also be part of wider systems of thought and action (following Fairclough 2001; Foucault 1972; Foucault and Gordon 1980). In Moldova, development as a discourse includes not only the use of vocabulary terms such as focus group (*focus grupuri*), sustainability (*durăbilitatea*) and transparency (*transparentă*), but more importantly, also involves myriad ways of acting and speaking that have become synonymous with Western-centred development agencies and projects. In their communication with representatives of the World Bank and donor organizations, SIF staff members show they are modern and motivated to help Moldovan people improve their lives through World Bank-approved development projects. In addition, all SIF staff members are Moldovan citizens and they interact regularly with Moldovan citizens from rural villages and towns of every region of the country. In their everyday work and personal lives, SIF employees draw upon their shared experiences with other Moldovans to signal belonging to the state and to specific groups within the state, and to maintain their identities in the changing cultural 'marketplace'. SIF employees are among a privileged group of NGO and development agency workers, who, in their everyday professional interactions, merge Western (and global) discourses of development, democracy, and citizenship with Soviet- and post-Soviet-influenced Moldovan discourses of culture, community and the state.

Social scientists have studied the concepts, practices, and discourses of development and how development discourses have shaped Western understandings of people and institutions in 'developed' and 'developing' countries (Cooke and Kothari 2001; Edelman and Haugerud 2005a; Escobar 1995; Ferguson 1990). Discourses of development have been constructed over time from the colonial period of Europe to the post-World War II period and the Bretton Woods system to the present day and are described in great detail elsewhere (World Bank 2008; Birdsall and de la Torre 2001; Edelman and Haugerud 2005b; Escobar 1995; Evans 2005; Leys 2005; Preston 1996). However, Moldovans' local practices have also arisen from Soviet discourses of development used in conjunction with plans to modernize and collectivize its people and economy (Fitzpatrick 1999; Humphrey 1983; Verdery 1996). Here, the issue is the influence of such historical and current discourses on Moldovans' daily life, and their construction of identities.

At the time of this writing, Moldova has undergone a few turbulent months since the April 2009 Parliamentary elections. People opposed to the Communist Party's electoral victory publicly protested the elections' outcome. They accused the Party of rigging the elections and using voter lists that included deceased citizens. Protests became riots and the Parliament building was partially burned. At least two of the protesters were killed and there were many accounts of police brutality (BBC 20 April 09). Former President Voronin has accused Romania of instigating the protests, an accusation supported by the Russian foreign ministry (BBC News 8 April 2009). Because of these recent developments, one may argue that people's feelings towards Russia and/or the EU have changed and may challenge the validity of the data presented here. However, the effects and influences of the EU and Russia on Moldovan daily life have been building over years and decades and will resist rapid change. Current affairs add a layer of complexity to the question of Moldova's place in Europe and Eurasia and will become a part of the answer to that question in years to come.

To illustrate how Moldovans experience their in-between-ness and how foreign discourses permeate their lives and their own talk, this paper attempts to trace connections between powerful discourses of development (both Soviet and Western) and the ways Moldovans utilize those discourses to situate themselves between dominating powers. First, I elaborate on the theoretical and methodological framework in which this study is situated. Second, I further explore the influences of Soviet and Western discourses of development. I argue that the effect of both is to construct Moldova as a problematic and malleable hinterland. In the third part, I present examples of communication and patterns of interaction between SIF employees and rural subproject participants. The examples show that the 'messiness' of Moldovan identity is not an epiphenomenon, since it is inextricably linked to Moldova's history, location and political future.

A growing Moldovan national identity can be described in terms of an expanding sense of citizenship alongside the village, regional and ethnolinguistic identities that have governed people's lives in the post-independence period. In the state that soon will celebrate its twentieth year of independence, a citizen identity provides a small point of stability, I argue, but a growing one. This stability seems to be a recognition of people's relationship to each other as fellow citizens of the state, and a recognition of what the state can do for them, with their input. Because SIF is an institution of the state, because it endeavours to maintain equality in its interactions with all citizens, and because SIF has helped people implement subprojects in hundreds of villages, towns and municipalities across the country, citizens who work together with SIF share a common set of relationships and experiences with at least one state institution and its workers. The work presented here shows that Moldova is on the road to legitimating the state and building a common identity for its citizens primarily through cooperation with international development organizations heavily influenced by Western ideologies. The construction of this kind of citizen identity may hold promise for strengthening Moldova's position as a state in-between the powerhouses of the EU and Russia.[4]

Theoretical framework

The present work builds on anthropological studies of people and their cultures, with a more specific focus on people's identities as part of contemporary states. Ethnographic methods in anthropology and other social sciences are guided by a focus on small groups of people within larger societies. Since an ethnography takes place among people over an extended period, the researcher has the chance to test hypotheses about people's behaviour through a variety of methods, including observation and recording of actual, reported and ideal behaviours. An ethnographer also participates in a culture in order to gain an insider's perspective. The interpretation of the insider's perspective illuminates cultural beliefs and how daily ways of acting, being and speaking (sometimes termed *habitus* (Bourdieu 1977)) contribute to a culture at large and to people's sense of belonging to a culture or their cultural identity.

In addition, anthropologists regularly study the practices of states and government actors as part of the study of human culture (Ferguson and Gupta 2002; Gal and Kligman 2000; Geertz 1994; Gupta 1995; Nugent 1997; Scott 1998; Verdery 1991). Citizens come into contact with the state through the many institutions, organizations, offices and community centres where state employees act on behalf of the government. The study of such interactions allows one to examine how the state and its representatives enact state and national ideologies on a small scale, and the reactions of citizens to the practices of the state. In such interactions, citizens are able to imagine the entity known as the state or the government and their place in relation to it in the conduct of their everyday lives.

However, powerful institutions are not only located within individual states. All people, even those in powerful countries, experience some impact of the decisions taken by leaders of foreign states and international organizations. Ferguson and

Gupta draw attention to the influence of transnational organizations and their neo-liberal governmental ideology on states and their sovereignty (2002). The authors note that the definition of 'civil society' often excludes internationally funded and transnational organizations. Ironically, in poorer countries, such transnational organizations usually sustain the grassroots and voluntary organizations that make up traditional definitions of civil society. Therefore, most studies also omit the influence of those foreign states, institutions and organizations on the domestic politics and institutions of a 'weak' or 'failed' state, such as many African states and, in this case, Moldova.[5]

Powerful state and transnational organizations have led the global trend (Western and Soviet regimes inclusive) since the 1940s to support the 'development' and 'modernization' of states, economies and peoples.[6] For example, modernization and development projects across the expanse of Eurasia were directed by the Soviet government until 1991. Currently, development projects are led globally by international organizations such as the World Bank and the UN; state-run agencies, such as SIDA, USAID and the Japan Social Development Fund; and non-governmental organizations, such as Transparency International (Edelman and Haugerud 2005a; Ferguson 1990; Grillo 1997). The majority of donor states and leaders of international organizations are Western European and North American, which makes for strong ties between contemporary development discourse and Western ideals of citizenship and government. Other organizations within poorer states may be hybrids of state and international institutions and NGOs: government organized NGOs (GONGOs), donor organized NGOs (DONGOs) and quasi-autonomous NGOs (QUANGOs). SIF fits into this hybrid group of institutions and practices. The Moldovan government officially directs SIF, yet SIF's activities and funding are tied much more closely to conditions of operation set out by foreign entities. This research takes SIF as a case study of the intersection of international, state and local discourses and practices of development.

More specific to this study's focus on language are linguistic anthropologists' examination of 'the place of powerful institutions ... in the construction and maintenance of ... linguistic hegemonies' and the ways that '*specific practices* within institutions give value to different languages and to different ways of using language' (Spitulnik 1998: 164–65, emphasis original). Therefore, an ethnographic and linguistic approach to studying a state's institutions and its people may analyse '*everyday practices* of local bureaucracies' and may problematize 'the *discursive construction* of the state' (Gupta 1995: 375, emphasis original). This approach follows a model of discourse and discursive relationships that acknowledges the power inherent in all social relationships and allows for a dialectical influence between language and society (Foucault 1972; Foucault and Gordon 1980). Focusing on everyday practices allows us to understand how citizens experience the state, the historical and ideological context that guides people's understandings of the state and the many social groupings through which people contest both the ideologies and practices of the state. Emphasis on the *habitus* of daily life allows us to investigate the ways that discursive relations are embodied in practices that build over time into common ways of being in the world and even become traditions, identities and ideologies (Bourdieu 1977). I use the set of theories and methods described here to look at the negotiation of a 'citizen' identity in Moldova, and how citizens are influenced by both Western and Russian ideas about who Moldovans are and where their allegiances lie.

Situating Moldova in space and time

In the early 1990s, Western advisors and donors arrived in the newly independent countries of the former Soviet Union to meet people who were primed to receive them enthusiastically (Wedel 1998). Most of the Moldovans who encountered people from the West had never before met someone from a non-Soviet state. Moldova had been a relatively productive Soviet state (important to food and alcohol production, space industries and televisions), yet citizens were accustomed to obeying authority from afar. The new state government hastened to distance itself from Soviet ideology, yet was in disarray. People had lost nearly all of their savings with the changeover from the Russian rouble to the Moldovan leu. Trade and markets had been disrupted, the government needed cash and advice to continue operating, and there was general uncertainty about Moldova's future. This confluence of circumstances allowed Westerners a great deal of leverage to influence the direction of economic and political changes in the target states. With representatives in Moldova from international bodies such as the World Bank, the IMF and the United Nations; with embassies and consulates in Moldova representing at least fourteen countries; and with neighbouring Romania now a member of the European Union, Moldovans' lives have changed dramatically since independence.

Moldovans also feel Russia's influence both economically and culturally, which affects Moldovans' view of themselves as much as similar influence from Europe and the United States. Economically, Russia has raised natural gas prices for Moldova almost to the same price that EU countries pay; and Russia's 16-month ban on wine and agricultural imports shocked Moldova's economy. The fact that Russia also banned wine from Georgia led Moldovans to think that there were political reasons for the ban, not the health and safety issues cited by the Russian government. Meanwhile, Russia's cultural influence is still strong in Moldova, in the form of Russian media outlets with Russophile news desks and popular television programming, Moldovan radio and television broadcasts in Russian and local cultural and state traditions that still resonate with the Soviet era. Many Moldovans recognize the positive cultural influence that Russia had and continues to have in Moldova; nearly everyone I met in the country, even fervent nationalists, acknowledged the positive changes that the Soviet Union brought to the region, such as the infrastructural development that is now crumbling, and the identity of being associated with the great arts and artists of Russia and the Soviet Union.

Moldova is often described as the 'crossroads' of Europe and Eurasia, and of Latin and Slavic cultures (Republic of Moldova 2007; High Anthropological School 2009; UN 2009). Indeed, the history of the region, having been conquered or annexed a number of times, and the ethnolinguistic groups coexisting in Moldova, representing all neighbouring states and even some farther away, makes it hard to deny Moldova's hybrid heritage (Kaneff and Heintz 2006; King 2000). However, Moldova-as-a-crossroads is not always a positive metaphor. The country's political fluctuations since 1991, especially the election of the Communist Party to power in 2001 and debates over national language and history education, usually produce much discussion and confusion in foreign observers (Chinn 1997; Ciscel 2005; Crowther 1998; King 2000). Outsiders would like to categorize Moldova as holding stronger allegiances to either East or West, but Moldova has consistently 'pursued a dual policy promoting good relations with the West while maintaining close ties with Moscow' (Chira-Pascanut and Schmidtke 2009: 2). The historical case of Macedonia may be instructive to understanding the source of confusion over Moldova's loyalties.

During the 1800s, the area surrounding the region of Macedonia was a borderland between Europe and the Ottoman lands to the East, much as Moldova is a borderland today between the EU and the CIS. Western Europeans used the term 'Balkan' to describe the region that divided Europe from the Ottoman Empire; by the twentieth century, the word 'had become a general pejorative meaning backward, and, especially, subject to political disorder and disintegration' (Irvine and Gal 2000: 63). Considering that Moldovans have often been described to me as 'simple', 'criminal', 'dangerous', 'backward', 'uncivilized' and 'peasants' by people from Russia, Romania, France, the United States and even by Moldovans,[7] this case bears further comparison.

The people living in the Macedonian region were a heterogeneous mixture of ethnic groups speaking a number of languages and dialects of Slavic, Balkan, Turkish and other origins. Western Europeans were not able to easily categorize the population either by language or by territory. They perceived the people as disordered and uncivilized, in implicit contrast to their 'orderly' world in Western Europe (which they failed to recognize was equally heterogeneous). The disorder of languages and territory was projected onto the people of the region, who Western Europeans came to consider as having loyalties to multiple political and ethnic groups, and therefore, untrustworthy. The outsiders 'missed – and their representations erased – the local logic by which the inhabitants of Macedonia understood categories of language and identity' (Irvine and Gal 2000: 65). As noted, many have described Moldovans as having mixed allegiances and being unable to decide on the future of their country. Some Moldovans have even come to believe this about themselves, rather than recognizing the strengths of their position.

Riding three horses

To give an example of how the Moldovan state's indecision between East and West is portrayed in Moldova, I can describe the visit to Moldova of an American foreign policy advisor and lobbyist in 2006. This person had travelled to Moldova for meetings with several Moldovan government officials, including President

Voronin, and had agreed to stop by the State University to speak to about fifteen students and professors from the American Studies Program. The topic arranged beforehand was Moldova's place in relation to the European Union and the possibility of Moldova ever joining the EU. During his lecture, the speaker elaborated several foreign policy strategies that Moldova had been following lately: leaders had met with Balkan states concerning joining the EU, with the Black Sea states concerning energy policy and with Russia concerning the conflict in Transdniestria. Moldova is 'riding three horses' as he put it, which is 'a hard way to get into Europe'.

The prospect of Moldova joining the EU is a common topic at policy discussions such as these, as well as at policy and NGO/donor conferences, in my experience. In fact, the speaker did not raise the possibility that Moldova would *not* want to join the EU, which may have arisen from his position as an American working in the foreign policy arena, or from his actual discussions with Moldovan government leaders. Nevertheless, he noted several actions that Moldova would have to take before the EU would consider admitting the former Soviet state: a) stop electing the Communist Party (or elect them under a different name, he suggested); b) raise defence and security spending ('Moldova has a conflict over an interior border and still has the lowest defense spending [in the region]'); c) become a member of NATO (no other country has recently joined the EU without first joining NATO); and d) improve Moldova's 'economic conversion' in order to raise GDP significantly to match that of EU states. First, by referring to Moldovans' election of the Communist Party into power, the speaker suggested that they may not really *want* to shed the Soviet past and/or move away from Russia; their allegiances are unclear. Second, he suggests that the Moldovan government does not share the same ideas about solving and preventing conflict that Western governments do, since Moldova's defence and security spending is lower than Westerners expect. Then, he directly recommends that Moldova follow a prescribed economic path to join the EU. He did not raise the possibility of other foreign policy options and considered the 'three horse' strategy as too scattered to be taken seriously.

After commenting on some areas in which Moldova is performing well, such as working towards the UN's Millennium Challenge goals, the policy advisor turned to the cultural meaning of joining Europe:

'And finally, because we have the opposition ... of essentially very scared Western Europeans, we do need to make the case that this is a European country. And it's very hard, um, I don't know how to explain this in an academic setting. [inaudible word] A lot of people think that Lithuania got into NATO and the EU because they play such good basketball, and you'd be surprised how ... important these shared cultural things are, you know, the voices of poets, playwrights, students, how much more important they are than hearing [inaudible word] from the foreign ministry.... We often say this, memberships in Europe and NATO and all these things are not really treaties between foreign ministries. They are marriages between cultures and civilizations. They're permanent, so they want to know people, and I would not underestimate how important the social contexts that sustain this are. It will inevitably come up in the next five to seven years.'[8]

The speaker called these the 'intangibles of Europe': shared cultures and civilizations, poets, playwrights, students and basketball teams. It would be hard to argue that Moldova does *not* share these very tangible items with other European countries. In fact, Moldovan citizens believe they share a 'cluster' of cultural markers with Europeans, including their 'language, descent, history … and religion' (Blommaert and Verschueren 1998: 194). If not for the approximately fifty years of Sovietization, the region at least on the right bank of the Nistru River could already be a part of the EU. Therein lies the problem. Despite the invited speaker's declaration that cultural factors are often more important than treaties, the speaker referred several times in other parts of the lecture to Moldova's economy, the problem of corruption and the re-election of the Communist Party to power. It was as if political and economic issues are inherently connected to the culture of the people who inhabit Moldova. In spite of sharing cultural features such as language, ethnicity and religion in common with many people considered 'European' and with countries that are already part of the EU, Moldovan citizens are currently more associated with the economic and political difficulties of the post-socialist transition.

Attendees of the small discussion responded with questions for the speaker about Moldova's best approach to foreign policy (i.e. which of the three horses to choose), how to deal with the wine embargo, the speaker's opinion of the President and government and his confidence in their abilities and about the possible dangers of joining the EU. The questions show us the foremost problem on Moldovans' minds: survival. They were wary of Russia, interested in regional alliances and yet also cautious of the EU and what being a member might imply for the survival of their state and people. There was not *one* clear path, and the discussion in the room reflected that fact.[9]

Kaneff and Heintz offer reasons why Europeans, Americans, Russians and many others have difficulties understanding the people of the region:

> If borders and boundaries are a resource to be used strategically, then in Bessarabia,[10] the power of this resource resides in its ambiguity. … Competing loyalties and multiple identities, the historically based connections tying the region to various powers at various periods in the past mean that there is no one centre of economic and political domination.
>
> (2006: 15)

Most European states feel they are making progress in integrating national and ethnic identities into a common identity of 'citizen' within their borders. The fact that Moldova's government and its people are still vacillating on this issue (at least from a Western point of view that disregards the messiness within one's own borders) may lead outsiders to think that Moldova still needs to improve its internal cohesion in order to fit in with Western ideals of acceptable governance. Thus, foreigners see Moldova as disordered and malleable, much like they perceived

240   *Patricia Fogarty*

Macedonia in the 1800s. The trend shows that more established states do not trust areas and peoples they cannot easily identify. It makes sense that superpower states, donors and international organizations consider these qualities weaknesses, and therefore, try to mould, shape and influence the Moldovan state in the way they see fit. As Kaneff and Heintz point out, however, the ambiguity of Moldovan identities are a source of power to the people in the region, able to be drawn upon in different situations, with different interlocutors (2006).

The remainder of this paper illustrates how discourses about Moldovan daily life, citizenship, the Moldovan state and its relationships with Russia and Western entities circulate in Moldova between people of different social, economic and political status. To do so, I describe one of the many meetings that I observed at the SIF offices in 2006: a visit by a rural community's subproject implementation committee to the SIF Executive Committee meeting, where the former make their case to receive funding for a gas line subproject in their village. Certain patterns of communication and bureaucratic practices in this example show the influences of Western ideologies and governance practices on SIF staff and villagers' debate, as well as the impact of Russian foreign policy and local–historical discursive ways of speaking and being.

The SIF Executive Committee meeting[11]

Each week, the SIF Executive Committee meets between one and six subproject proposal teams, for projects concerning infrastructure and community rebuilding. In this section of the paper, I show that the work done through SIF constitutes a dialogue about Moldova and being a Moldovan citizen between two very different sets of actors in Moldovan society, with the heavy influence of multilateral and international development organizations. The example of the SIF Executive Committee meeting shows that the state has begun to legitimate itself to citizens, in a quiet way, through the institutional apparatus of the Social Investment Fund. At the same time, both state and citizens are working hand-in-hand with the World Bank and thus with a Western-centred international development institution.

In order to understand the claims made here about citizenship and identity, it is necessary to understand the long process of applying for and implementing a SIF-funded subproject.[12] The process is heavily influenced by Western ideals of development, economic accountability and citizenship, since SIF is run according to World Bank operational standards. I see this progression of contact between villagers and SIF representatives as a dialogic process in which villagers are brought into the world of SIF and SIF staff learns more about this village. This process also engenders a debate about Moldova's and Moldovans' place in the world and each group puts forth its vision of what Moldova can be and should be. There are several steps in the process; I will only discuss those leading up to the approval of the subproject. First, village political and social leaders are invited to regionally held seminars where SIF staff members explain what SIF is, who funds it, what kinds of subprojects it funds and the first steps in the application process. Second, village leaders must hold at least three focus groups (*'focus grupuri'* in

Romanian, a foreign term and experience) to discuss the principal problems of the community, at the end of which they vote on the most critical issue that should be addressed in a subproject grant proposal.

Third, a SIF worker, usually a sociologist by training, comes to the village to observe a 'General Meeting' of the village, at which there are four orders of business: a) the mayor of the village and the SIF representative introduce the process of obtaining a SIF grant to villagers, especially those who were not invited to take part in a focus group; b) the results of the focus groups are announced; c) those present vote on the subproject that the village will undertake to implement; and d) they vote to elect a local steering committee, called an 'implementation agency', that will be responsible for the subproject's implementation. At this meeting, the SIF representative verifies that the village has maintained certain procedures required by the World Bank. For example, 40 per cent of the population of the village must be present; 30 per cent of those present must be women; and the mayor's office must certify that the object of renovation will not be privatized for 15 years, ensuring that it stay community property for the near future.

Fourth, after the General Meeting, the implementation agency members have two weeks to put together the paperwork for their subproject proposal and submit it to the SIF office. The official form for the proposal asks the local committee to: a) 'argue the community's need to realize this subproject'; b) 'indicate the problems that will be solved if this project is implemented'; c) tell 'the objectives of the subproject'; and d) 'the strategies and methods of realizing these objectives'. Instead of having the justification for the project explained to them by the state government, villagers must affirm the points for themselves, in their own words. The community estimates the cost of the subproject and enters their estimated contribution in cash from the population, from the local mayoralty's budget, and/or in building materials or labour. The World Bank requires a contribution of 15 per cent, but the proportion of cash they need to amass varies for different types of subprojects. Next, the villagers must explain how they will sustain the subproject, in both qualitative (service-related) and quantitative (monetary) ways. Thus, the community must show its preparedness in being able to justify the need for the subproject, to put together an estimated budget, and to think forward to how they will maintain the World Bank's, the Moldovan government's, and the community's investment, once the subproject is completed.

The fifth step is the 'evaluation stage' in the life of the subproject, at which a SIF team comes again to visit the town or village. The team usually consists of a sociologist, an engineer and sometimes a knowledgeable contractor. While the engineer and the contractor go with certain members of the implementation agency to survey the project's territory (school building, road and terrain for water lines), the sociologist speaks with the remaining members. Usually, the sociologist asks questions of the group to gauge their preparedness, such as: Have you registered your beneficiaries' association? Did you hold focus groups? Who made up the focus groups? How did people decide what to do? The sociologist's goal is not only to make sure that SIF requirements have been met, but also to remind the villagers of what steps they have taken, in a kind of verbal reinforcement of the value of these requirements. The sociologist also wants to see that the community is 'active' ('*activ*') and that they understand the reasons why they are completing these tasks. The sociologist later fills out a form with his or her perceptions of the group, commenting on whether the implementation agency is ready to move forward to the Executive Committee meeting. The sociologist gauges this by the readiness with which the group answers questions, the knowledge the villagers show of SIF and its procedures and whether more than one person is answering the questions.

Later, the engineer returns to the meeting room[13] to give his or her opinion about the viability of their proposed project and its budget. The engineer begins by describing the project as the villagers proposed it. Then, he[14] tells the group quite blatantly, what the realities of their situation are. He couches the talk in phrases such as, 'you have to decide' and 'it's your project', but in essence, he is giving them his best advice on how to spend their money wisely and have a good, sustainable result (one that does not fall apart in a short period, with good quality workmanship and parts).

Then, if there is time, the SIF staff will prepare the villagers for the next major hurdle to implementing their subproject: defending their subproject proposal before the SIF Executive Committee (EC). This usually takes place within the next two weeks. The SIF team prepares them by suggesting what information they should present, such as the building/project specifications, the budget, and the sustainability plan. On the day that the villagers come before the EC, they arrive in Chisinau before 10 am, which means they must travel for 1 to 6 hours to arrive on time, depending on how far they are from the capital and on their mode of transportation.[15] When they arrive at the SIF offices, the villagers report to the engineer or sociologist assigned to them, who checks over their documents and the items they have prepared to elaborate on before the Committee. The SIF representative again asks who will speak, what facts and numbers they will present and tells them what to expect when they enter the Director's office.[16]

The mood of the morning pre-Committee meetings can best be described as tense, since the participants are usually nervous about going before the EC. On the other hand, participants are sometimes joking and jovial, a common way to alleviate tensions in this type of situation. Still, when participants enter the Director's office, which doubles as the Committee meeting room, with its formal table arrangement and committee of senior SIF staff, the mood changes again. Most subproject participants begin the meetings with shaking hands and timid voices, at best. Only those who have been to the Committee on a prior occasion are more confident. Participants' timidity in the face of authority figures is a behaviour and belief common during the Soviet era (Simirenko 1982). By contrast, the Director claims a sense of partnership with the villagers.

In the transcript, we find excerpts from one Executive Committee meeting (see endnote 11). For the two implementation agencies from the neighbouring villages of Nucareni and Alexeevca,[17] which are subsumed under one mayor's office, this day was much awaited. They had worked for more than six months to arrive at this stage of the process, hoping to implement a subproject to run gas lines through their villages. The transcript of the Executive Committee meeting shows the common procedure for such interactions. In this case, each village made a separate presentation, although I address here only the subproject presentation of the first group. The Executive Director directs the flow of the meeting from beginning to end. Subproject participants are asked to introduce their teams, Committee members make a formal opening statement, and then the teams are invited to present their subproject proposals. Most groups accomplish this by listing the items in their proposal budget and arriving at the total cost of their proposal. Then, the Director asks questions of the implementation agency, and solicits questions from his staff members for the villagers. The questions pertain to how well the participants have thought their proposal through, how they have planned for the project's implementation and its durability. At the end, the Director makes the final decision whether or not to approve the proposal.[18]

This particular meeting took place in June 2006, as tensions were escalating between Russia and Moldova: the wine embargo had begun in March, and natural gas prices were on the rise. Gas prices went up from US$80 per 1,000 cubic metres to $110 by the time of this meeting and were set to go to $160 by the end of 2006 (Socor 2007). During the summer months, I observed the Executive Director and Committee becoming more and more stringent with the approval of gas subproject proposals. He questioned the villagers for longer periods concerning their village's commitment and ability to paying higher gas prices in the future. He wanted to make sure that SIF's and the villagers' investment would not be lost. In addition, the Director used these meetings as platforms to educate villagers on alternative types of fuel and heating systems. For example, he often informed those present that SIF had helped implement one heating-system subproject in the southern part of the country that used new technology from Germany to extract long-lasting heat from straw. He asked the committee members if they had researched any other technologies or types of heating systems. Most had not looked beyond the three common types of heating materials: coal, wood and natural gas, a fact that dismayed the Director.

The central negotiation over the subproject is between the Director, who speaks more than any other SIF staff in this meeting, and the village leaders. The negotiation concerns the feasibility of the gas subproject and whether or not the village will be able to pay for gas in the long run. The Director's argument revolves around the fact that gas prices have gone up and may rise astronomically in the future, whereas the prices for resources inside Moldova (such as wood and coal) are more stable. His argument introduces knowledge of the market economy in Moldova and geopolitical circumstances of the region (lines 23–35).

The Director mentions Gazprom by name twice (lines 160, 221), in a disparaging way: once concerning gas prices and a second time along with a reference to corrupt relationships between Gazprom, local Moldovan gas companies and their staff. What is left unsaid in the exchanges is that Moldova's relationship with Russia is also tenuous and that the gas prices are a symptom of the shaky ground between the two countries. For the Director, therefore, gas subprojects represent participation in the international market economy beyond Moldova's borders, Moldova's wavering relationship with Russia and Moldova's resulting unstable future. In other words, he brings the state into the discussion; he wants the villagers to think beyond the boundaries of their village and recognize the notion of Moldova as a political, economic and cultural entity that exists in a complex international arena.

On a more practical level, instability and risk are not sound bases on which to invest money. Twice the Director asks for the villagers' calculations and estimates for other types of heating systems, claiming that their appeals for gas are based on emotions, not on reason (lines 80–81, 124). He shows concern, however, in the Moldovans before him (lines 68–74, 92–93, 103–9) – how will they be able to pay for gas that will become more and more expensive? His concern may be with these villagers, but in a way, his concern is for the nation of people just like them.

For the villagers, the gas subproject has less to do with the national economy and their compatriots in other regions and more to do with improved living conditions. Gas is easier to access, easier and cleaner to work with (they simply turn it on and off versus obtaining wood or coal, and building, stoking and maintaining fires in stoves), and better for their health. As one woman states, 'with gas ... at least one can rest' (lines 156–57). The Director provokes them repeatedly as in:

> 'Ok, that's emotions, emotions, what you're saying ... but your calculations/ estimates, where are they?' (lines 80–82);
> 'Now you, you don't see any risks at all?' (lines 162–63);
> 'We will invest, we will spend money ... you ... for you to begin to have gas in your houses, and it could happen that you lose, that you won't use the gas' (lines 165–69); and
> 'So, furthermore – ... How will people make the connection [of the gas line] to each house? Have you studied, do you have a strategy, a concept measured/gauged – [for] those who will connect?' (lines 178–83).

At each point, the villagers attempt to show that they have, in fact, estimated the costs and weighed the risks versus the benefits of bringing gas to their community. One of their counterarguments is that other fuel prices have gone up, too, and there is much discussion of these factors (lines 137–64). The two competing arguments revolve around expertise of prices and local living conditions, but also have to do with the influence of Gazprom and Russia on local prices. To come to an agreement, the villagers must show that they are aware of the expense and are ready for it. They have planned for their future by growing crops that sell for higher prices (lines 117–20) and by setting the gas subproject in motion two years before (lines 125–31). They would like to convince the Director that they are not simply asking for a better life based on their emotional irrationality, but that they have also taken the market economy into consideration. For his part, the Director shows that he is willing to accept their expertise on the local population and its abilities as more than just their opinion. Perhaps he accepts their local version of participation in the market economy, and their recognition of the geopolitical situation outside their village.

From the transcript, we can also see the dialogic process of identifying oneself with different parts of society, and with world powers outside Moldova. First, and most simply, we see people relating to those around them by using 'we/our' vs. people not considered part of their group ('you/your'), thereby creating relational in-groups and out-groups (lines 8 and 33, for ex.). Most of the time, the Director puts himself in an out-group to the villagers. Three times, however, he puts himself and the villagers in similar groups: when he brings up the median salary of all Moldovans and their likely expenses (lines 98–109), when he speaks of Auntie Iliana (lines 70–74), and when he honestly wonders at the villagers' ability to pay (line 239). In bringing up Auntie Iliana, the Director perhaps is calling upon all Moldovans' knowledge of elderly people in their country, of the small government pensions the elderly receive, and of the knowledge people share of pensions not being paid on time throughout the 1990s. These three instances call upon common knowledge of recent economic conditions in their country.

Second, larger geopolitical comparisons and development ideals of participation in the market economy are at work. To begin, there is an implied comparison between Moldova and Russia, and Moldovan and Russian companies serve as proxies for their respective states (lines 24–28, 160, 221). Moldovans as citizens of Moldova are not part of Russia any longer, a point made clear by Gazprom's gradual price increase for Moldova, which will be equal to the price paid by the rest of Europe by 2011. This allows the people present (who are all native speakers of Moldovan/Romanian, by the way) to see themselves as one group, with the 'others' being Russia.[19] The price comparison even allows the Moldovans to envision themselves as part of Europe.

In addition, the SIF Director repeatedly asks the villagers to show their familiarity with SIF procedures (bidding, hiring of contractors and transparency – lines 178–87, 212–38), as well as their knowledge of international and local economic market conditions (lines 55–58, 86–94). Together, this body of knowledge, as well as their preparation for this moment through the SIF and World Bank procedures, connects the villagers to ideals of Western and international development, notably the participation of people in market economies, driven by competitive capitalism. Their knowledge draws them in to a shared world of focus groups, tendered contracts, and project ownership, even if they do so unwillingly or only in order to receive their grant. Moreover, the Director's call to rational calculations versus irrational emotions recalls the rational choice paradigm of economics, despite the World Bank's focus on teamwork and community development.

Third, by participating in SIF (and thus Moldovan government) development projects, the villagers are taking part in daily activities of citizenship and becoming more familiar with the state. All of the SIF procedures reinforce a democratic process, values of community building, the participation of less powerful community members and obeisance to the state: from meeting and voting as groups, to choosing items of social infrastructure as objects of renewal, to following the bureaucratic procedures of a government agency. The village implementation agencies include local public administration and must register their agencies with the government, reinforcing a hierarchy of state administrative power. The villagers must interact with local companies in the marketplace according to legal ways of hiring contractors and paying workers. Moreover, they are able argue their case before a body of people who represent the state (and multilateral donors), which gives the villagers a chance to see the state as people interested in their welfare, not merely a heavy bureaucracy that asks them to fill out forms and hands down decisions from above. According to the principle of *habitus*, following such daily practices and procedures over time may reinforce and generate notions of citizenship even in people who do not feel especially tied to their state or its government (Bourdieu 1977; 1984).

As for the SIF employees, they are also acting as Moldovan citizens. I saw in their daily work that their main concerns are for Moldovans and the betterment of Moldova. SIF workers as citizens are subject to the same influences of the political and economic situation of Moldova, but they are more entrenched in the requirements and ideals of the World Bank and the mindsets of international development than are villagers. Because they live in the capital city and their salaries are higher, they have wider awareness of foreign ideals and influences, and they can interact in the market in different ways than other Moldovans. Still, they are also enmeshed in local networks of family and social connections that give them something in common with the villagers beyond residing within the same state boundaries.

However, SIF employees symbolically represent the Moldovan state. They use their positions as intermediaries between international, national and local worlds (which are not bounded, by any means) to educate other Moldovans on community development, on participation in the market economy and on ways to conduct themselves transparently in their interactions with each other and with local businesses. The SIF employees model citizenship behaviours for other Moldovans, and the ways that they question the villagers' knowledge of the business and development environment reinforces that modelling.

Therefore, the two groups in this exchange are talking about the same subproject proposal from different perspectives that are couched in their past experiences, their socioeconomic status, and their current daily life. The villagers focus on their obligations to their locality and its residents, while the SIF Director favours the state as a whole. By invoking Auntie Iliana, family obligations, Moldova and Moldova Gas, the Director seems to see the village and its inhabitants as representative of *and* responsible to the state. *Who* is represented (the village) becomes *what* is represented (the state). If the villagers cannot pay for their gas, the state will suffer along with these particular residents, since Moldova will be further indebted to Gazprom and further subject to Russian displeasure. Both Russia and the EU are silent interlocutors in the conversation. Russia is not only represented by Gazprom, but also by the interactions between the more and less powerful participants of the meeting, which reflect a genre of communication common during the Soviet era (Fogarty forthcoming). The EU is represented by the host of practices, vocabulary items and expectations of proper economic dealings that the SIF procedures embody.

With SIF as the authority operating with the Moldovan government and the World Bank behind it, SIF is in a position of great coercive and economic power. SIF employees use that power to promote the democratic process and promote an identity of common citizenship that articulates with democratic and liberal economic principles. The villagers and the SIF staff presented in this dialogue know they are a part of the state of Moldova; they act within the state and its boundaries, whether with patriotism or without. The principle of *habitus* allows that their self-identification exists partly through participation, through their subjection to/by the market economy that acts within the state boundaries, and through their subjection to business entities such as Gazprom and Moldova Gas and their representatives (who are citizens/registered entities of states and therefore subject to state laws). As we have seen from the villagers' perspective, the ties to localities and the people within them are still strong and influential in Moldova. Nevertheless, SIF is sponsoring long-term interactions that encourage linking local identities to legitimate national government processes at the same time that community ties are strengthened through people's interactions as citizens and community members.

The concepts and discourses illustrated through SIF's daily interactions with villagers connect Moldovan development to Russian power, Westernized living conditions, international development ideologies and the regional and global market economy. Common talk about Moldova's prospects of joining the European Union, and the influence of Russia on the Moldovan economy and peace talks with Transdniestria seem to direct Moldovans' focus on the negative: what they need to do to meet others' standards or what they should do to stay alive next winter. However, the work done daily by institutions such as SIF, its staff and the citizens with whom they interact is a more positive way of building awareness of what citizens can do (and are doing) about Moldova's in-between-ness. Thus, Moldova's foreign policy of 'riding three horses' represents Moldovans' geopolitical stance *and* the identity of the Moldovan people. Though it may not quickly gain Moldova access to the EU or build favour with Russia, the three-horse strategy may be akin to a three-horse-drawn *troika*, carrying Moldova and its citizens together toward their future.

Notes

1 This issue deserves more space than given here; see King (1995) and King (2000) for more details.
2 Since 2000, the Barometers of Public Opinion have shown that Moldovans vacillate between favouring NATO, the CIS and the EU. The highest numbers of people have recently leaned toward maintaining a neutral stance and not joining any specific coalition (IPP 2009).
3 I also lived and worked in Moldova from 1994–96.
4 I am deeply indebted to the staff of the Moldova Social Investment Fund for allowing me to conduct research on the work they do in Moldova. The opinions presented in this

paper are my own, as are any errors in representation of SIF's work. Emory University's Institutional Review Board for protection of human research participants approved the research, and funding was graciously provided by a Fulbright-Hays Doctoral Dissertation Research Grant and Emory University.

5 Ferguson and Gupta (2002) use African states as their examples; Parmentier (2003) describes Moldova as a failed state. Foreign Policy journal ranked Moldova 49th out of the 60 most 'Failed States' of the world in 2008; other former Soviet states that made the index are Uzbekistan (26th), Tajikistan (38th), Kyrgyzstan (39th), Turkmenistan (46th), Belarus (53rd) and Georgia (56th). Moldova improved from 48th to 49th place over 2007 (Foreign Policy 2008).

6 Eighteenth-century to early-twentieth-century modernization projects associated with colonialism are excluded in this discussion.

7 Of course, there are different contextual reasons for using such descriptions. Moldovans describing people from the countryside, from a different region of Moldova, or from different ethnic groups in a negative light are likely to compare themselves positively to 'Others' within their own country. My goal in this section is to focus on foreign views of Moldovans, and how international linguistic ideologies place Moldovans with respect to Europe and Russia, which many consider more 'developed' and 'cultured'. These descriptions and beliefs also circulate among Moldovans, who adapt them to contexts and situations closer to home.

8 Transcription conventions are discussed in Footnote 20.

9 Other authors who give relevant examples of similar behaviour are Anderson (2005) and Cash (2007).

10 Bessarabia, or Basarabia, is the term for the region bounded by the Nistru and Prut Rivers. The Republic of Moldova contains most of Bessarabia within its political borders.

11 A transcript of the meeting in the original Moldovan/Romanian, as well as a translation into English, is available at: http://patriciafogarty.wordpress.com. Please email the author at fogarty_t@yahoo.com for access.

12 Here, I mean only Component 1.1 of the SIF project, that is, the component that funds projects in villages that have never before received SIF funding.

13 These are sometimes rooms in the mayor's building (*primaria*, similar to a town hall) or rooms in the building that is to be renovated.

14 I simplify the gender of the engineer to 'he' here because all but one of the engineers are male.

15 People travel by public bus, private minibus or private car. Most seemed to have arranged one or two private cars for their transportation; people in buses often have to stand for their entire journey.

16 One SIF sociologist noted that she always encouraged subproject participants to present the social value of their projects first, which not all subproject participants did. She felt that the community's need for the subproject should be highlighted before the committee, not merely the facts and figures of their proposal.

17 Both village names are changed. I have never been to the actual villages of Nucareni and Alexeevca.

18 Most proposals are approved; if not approved, they are usually sent for revisions, to be brought again before the Committee. Complete denials for funding are rare and usually indicate that a village has egregiously violated SIF's required operational procedures.

19 I should note, however, that the Executive Director did not vary his stance on gas subprojects if the villagers present belonged to other ethnolinguistic groups and the Committee meeting was conducted in Russian.

13 Conclusions and outlook

Kjell Engelbrekt and Bertil Nygren

This final chapter will briefly assess the results of the analysis with regard to the three parameters around which the volume is organized, namely 'ideational friction', great power relations between Russia and the EU 'Big Three', as well as the role and significance of countries and major issues located somewhere in between these two vast entities. It will then move on to discuss the prospects for a constructive but engaged relationship between Europe and Russia over the next ten to fifteen years on the basis of those results.

First of all, in selecting values, norms and institutions as the first parameter and analytical focus we inevitably highlight patterns of continuity, as the former cannot be altered overnight. However, this makes sense to the extent that it is precisely the continuity of certain aspects of Russian political attitudes and behaviour, at a particular historical juncture, that we are trying to explain. Rieber (2007) has cautioned us not to underestimate the 'persistent factors' in Russian foreign policy, several of which are relevant for relations with Europe. In this volume, Russell Bova lists personalized authority, statism, the concept of *sobornost* and the unity of Church and state as important attributes of Russian political culture that impede the emergence of democracy and robust institutions of civil society. Accepting that these attributes are only gradually and partially malleable, they cannot help influence the broader relationship to Europe, both in terms of bilateral contacts and by way of the EU and its key bodies. Bova concludes that 'a large gap between the political cultures of Russia and Western Europe remains', and cautions us to measure progress 'in generations' rather than in years or electoral mandates of political leaders.

Bertil Nygren's chapter explores the 'ideational friction' at a more operational level, as reflected in the formal and informal rules that shape the electoral process in the Russian Federation. Nygren emphasizes the negative developments in the areas of mass media, the judiciary and the electoral process. While there is some level of a real choice in Russian elections, that choice is circumscribed by limited press freedom, the lack of independent courts and a widespread sense that the electoral process is strongly tilted in favour of the incumbents. It is not that European political leaders and independent observers are not ready to acknowledge the tremendous challenges facing Russian society in these various areas, and consequently adopt a relatively constructive, long-term stance. Nevertheless, as Nygren suggests, the frustration on the part of Europeans stems from recent attempts by the Russian establishment at justifying and solidifying illiberal political practices. Coining the term 'sovereign democracy' as emblematic of a Russian form of democratic rule, with more than a nod at the political culture of authoritarianism and statism, is a case in point.

A dialogue founded on conventional intergovernmental diplomacy, narrowly stressing the interests of both parties and whatever common denominator that can be identified, ought to work despite such differences in outlook. The EU–Russia relationship, however, includes several components that render the situation more complex. This is illustrated in Tatiana Romanova's juxtaposition of the contrasting interpretations of the notion of reciprocity that are promoted by Brussels and Moscow, respectively. In her reading of the cumbersome process of even agreeing on the basic terms of that dialogue, a wide array of institutional divergences creates serious obstacles. For instance, whereas Moscow is striving to strike limited deals on issue after issue, the internal structure and historical mission of the EU predisposes it towards trying to bring Russia into its institutional orbit. It seems that both sides will need to accommodate the other in order for the dialogue to be reinvigorated, and Romanova recommends 'positive reciprocity' that allows bottom-up initiatives and gradually could relax contradictions between the two parties.

At a politically symbolic (and philosophical) level, Sergei Prozorov makes a somewhat analogous argument. His own reading of the problem, based in part on Alain Badiou's social theory, is that a Europe–Russia discourse hitherto was built on a binary logic of inclusion/exclusion narratives. For a variety of reasons, neither inclusion nor exclusion is likely to take place through the integration or, respectively, non-integration of Russia into the EU institutional mould. It is within the logic of this very discourse, Prozorov explains, that Russia has developed an option of self-exclusion, an alternative narrative that similar to the previous two is unrealistic and unfruitful. The fundamental premise of the Europe–Russia discourse is one of political 'singularity', which in turn has generated a Russian narrative of counter-legitimacy and reinforced a politics of resentment towards the West more broadly. The only way to escape this binary logic and its corollary narratives, Prozorov concludes, is to try to establish an alternative discourse that would recognize other, intermediary forms of cooperation and exchange.

It is by now a well-supported observation that Moscow prefers not to deal with the EU as an aggregate entity but as individual countries, and above all with governments that it considers fall into a category of 'peers'. While the first section of this volume helps to illustrate the origins and reasons behind Russia's diplomatic 'unease', the second explores some of the substantive issues and shared interests that are being pursued in a more restricted circle of major powers in forums like the United Nations Security Council and NATO's Partnership Council, or bilaterally with the United Kingdom, Germany, France and, of course, the United States. Another illustration of Moscow's reluctance to deal with small players with leverage through memberships in key organizations, such as the EU and NATO, emerges in the way it handles international treaties and infrastructure projects of considerable regional importance.

Yuri Fedorov details developments within the area of 'hard security' issues, analysing Russia's 2007 suspension of the CFE Treaty, the missile defence controversy with Poland and the Czech Republic, the Russia–Georgia war of 2008, and the looming conflict over Russia's military presence in Crimea and Ukraine's sovereignty more broadly. According to Fedorov the Russian leaders have created a security dilemma for the major Western states in that they either need to confront the underlying challenge in these disputes or largely accept the military and political advances made by the former. The motives behind this challenge are supposedly a combination of apprehension on the part of the Russian military and the arms industry that they are becoming irrelevant, and a sense of humiliation of the demise of the USSR as a superpower, which they share with part of the electorate. In this situation, Fedorov argues, Europe might be well advised to abandon simplistic notions of constructively 'engaging with Russia', while at the same time avoid fanning the flames lit by leaders who believe greater confrontation will serve their interests. In formulating an assertive strategic doctrine in connection with the Russia–Georgia war of 2008, he concludes, Medvedev has sided – at least for the time being – with the latter.

Boris Frumkin demonstrates how Russia's fortunes of today are closely entangled with those of its European partners and neighbours, so that a sharp decline in the demand of its products precipitates an equally sharp drop in corporate and government revenues. Indeed, due to the overwhelmingly great size of energy resources in Russia's trade with EU countries, the effect is magnified in the national context, and in late 2008 and early 2009 such a decline wiped out much of the accumulated foreign currency reserves and caused a liquidity problem for Russian banks (and a corollary credit squeeze for companies). Frumkin points to the Russia–Baltic cooperation as an example of the considerable potential that exists when it comes to adjusting to the economic crisis and its repercussions through the renegotiation of transit and logistical arrangements, and underlines that this is something that Russian companies operating from the inside typically are prepared to do. However, unfortunately that type of pragmatism and flexibility rarely extend to governments or to large, Russia-based corporations. All actors tend to stick with a national, myopic view of business interests and thus fail to recognize the opportunities associated with creating a joint Russia–Baltic platform from which to draw mutual benefits in the European marketplace.

By contrast, Russia–Germany business relations are thriving and receive consistent political support from the two governments. Since the interaction between Russia and Germany, as Angela Stent notes, in many ways defined the history of the twentieth century, this is a promising and probably indispensable component of a productive, long-term Russia–Europe relationship. As Stent is able to show, the political leaderships of both countries have been cultivating the bilateral connection since the dismantling of the German Democratic Republic (GDR) and the Soviet Union in the early 1990s, quite deliberately and at times against the scepticism of domestic and foreign commentators. While Chancellor Angela Merkel never used the rhetoric of her predecessor, Gerhard Schroeder, in praising the Russian leadership's democratic credentials, she has supported the notion of tying the two countries more closely together in terms of energy supply, and at the same time rejected fears that Moscow at some point might be tempted to exploit this relative position of strength. In conclusion, Stent suggests that Germany's commitment to Russia has important political limitations, in that Berlin would never place Moscow's interests before those of the EU at large.

France seems destined to fall behind Germany in developing economic ties with Russia regardless of the will of its leadership or the activities of business circles. Isabelle Facon examines the ways in which Paris from time to time, including when France in July–December 2008 held the rotating presidency of the EU, has appeared to be more receptive to Russian desires and concerns than some other member states. Not too much should be made of common references to 'multipolarity' and the need for intermittent state intervention in the economy, though, Facon tells us. Instead, there is a sense of 'fatigue' on both sides given that neither in recent years was able to live up to the expectations of the other. In the case of France, the growing centrality of EU affairs in all foreign relations, along with the wish to mend fences with the United States after the diplomatic dispute concerning the Iraq war, indicates that Paris will not allow Russia to influence its overall approach on issues related to these two priorities.

The results of the analyses of the second parameter examined in this volume do therefore not affirm the worst fears of CEE states – member countries of the EU as well as non-members – that the tendency of the 'Big Three' European powers to occasionally circumvent EU structures to deal directly with Moscow will necessarily be reinforced. Although this tendency exists and will continue in the future, it is not a pattern that is likely to overshadow or even weaken the clout of EU aggregate institutions when a policy is in place or when vital interests of European industries or individual countries are at stake. A more likely prospect, however, is that non-vital interests of smaller or medium-sized businesses or countries within the community of EU member states could suffer. By the same token, moreover, there are few guarantees that the non-vital interests of non-member states will be respected by its more resourceful neighbours in the west of the continent.

Affecting both member and non-member states in Southeastern Europe, the Russian energy offensive during 2007 and 2008 proved successful using carrots rather than sticks, drawing on the goodwill of left-leaning political parties with a habit of accommodating Moscow. Kjell Engelbrekt and Ilian Vassilev show that the Kremlin in close cooperation with Gazprom, Rosneft and Transneft skillfully carried out a comprehensive plan to consolidate Russian delivery systems in Southeastern Europe. This was accomplished by way of offering Bulgaria, Serbia, Hungary and others package deals of infrastructure investments and a stake in downstream assets in exchange for transit charges. Until early 2009, this approach worked extraordinary well for Moscow, but the combination of political fallout of the Russia–Georgia war in August 2008 and the suspension of gas deliveries through Ukraine in January 2009 convinced EU governments to assume a more proactive stance on the supply of energy. Whether this will amount to cohesive Russia policy on the part of the EU is too early to say, though some elements towards one came into place during the first half of 2009. In addition, cash-strapped Russian energy companies will have serious problems delivering on their promises in the coming years.

Further to the east, Ukraine straddles the two 'universes' but has recently developed more westward ties. From the inception of Ukrainian independence in 1991 until 2004, the Russia factor was significant in domestic politics and economic relations. That asymmetric interdependence, Burkovsky and Haran point out, was qualified by Moscow's formal and explicit recognition that its southern neighbour was entitled to nominal sovereignty. Surprisingly for many experts, once a political movement away from the power orbit of Russia arose in the 2004, Moscow was unable to prevent the 'orange revolution' from taking place. Despite attempts by Russian leaders to directly influence electoral and legislative processes, Ukrainian politicians tried to balance Moscow's desires with the interests of the domestic electorate and the business community. Whereas the prospects for full EU or NATO membership are unfavourable at present, Burkovsky and Haran conclude that the basic westward direction of Ukrainian society over the past five years or so is likely to continue despite the potential turmoil that may ensue from the 2010 Ukrainian elections.

By many accounts, Moldova finds itself in an existentially more precarious situation, virtually a 'failed state' torn by the discrepancies that arise between two opposing alternatives with respect to its economic and political future. Patricia Fogarty describes how Moldovans are well aware that they occupy an uneasy boundary position and adjust accordingly in a variety of ways. Due to the adverse economic conditions of Moldova and the breakaway Transdniestria region, the dependence on large Russian corporations – especially Gazprom as the main energy provider – and Western donor agencies is particularly high. The economy consists of agricultural production, the related food and wine industry and minor manufacturing businesses. As a result, labour emigration, human trafficking, arms and drugs trade loom large in Moldovan society. In the ethnographic case study reported by Fogarty, the dominant non-government organization charged with distributing Western aid becomes the symbolic bearer of a vision of (re)consolidated Moldovan statehood.

Overall, the implications of changes related to the third parameter have received surprisingly little attention in Moscow, Brussels and other European capitals. Given the potential of 'in-between-Europe' for shaping the premises of strategic relations and substantive negotiations through political clout within the EU, or through its geographic location on Russia's doorstep, key political actors in Central and Eastern Europe (including Ukraine) could play a pivotal role in shaping the wider relationship. Russian leaders in particular still seem reluctant to accept this change of scenery. Moscow continues to act outside EU institutions and tries to deal directly with British, German and French governments, sometimes to the (at least apparent) detriment of smaller states in its vicinity. To the extent that this approach prompts disenchanted decision-makers in CEE countries to begin consistently blocking deals that benefit Russia, this is a shortsighted and ineffective type of diplomacy that will sooner or later have to be abandoned.

In Southeastern Europe, though, Russian assertiveness may have a better chance to pay off. One reason for this is the less historically charged relationship with countries like Bulgaria, Serbia, Slovakia and Hungary. Another factor is the reasonably attractive incentives offered to domestic businesses and political elites. A third, and arguably decisive, factor is that the EU so far has been unable to develop an effective counterstrategy to Moscow's initiatives, as the Union needs much more time to adjust its policies than does its counterpart.

However, as the role of the United States in Europe has diminished in recent years, it is conceivable that Russia would become the primary catalyst of growing 'actorness' on the part of EU institutions located in Brussels. Precisely because European great powers continue to pursue divergent approaches towards its eastern neighbour and Moscow remains astute at exploiting those differences, the debate on EU's ambitions to develop strategic policies on a global scale is quite likely, at least in part, to be about the coherence and robustness of Brussels' relationship to Russia. Until 2008–9, the EU may have lacked concrete policy challenges to tackle, but the Russia–Georgia war and the Ukrainian gas crisis of early 2009 presented two such issues on which elements of a more cohesive stance may be emerging.

One such element of a more cohesive approach to Russia could be inherent to the 'Eastern Partnership' launched in May 2009, with the aim of promoting political and socioeconomic reforms in six partner countries and facilitating long-term legal and institutional approximation with EU member states. Not to be mistaken for a substantive policy, the 'Eastern Partnership' initially earmarked only 600 million Euro, yet, might be a sufficiently solid platform for collaborative projects driven by the six countries invited, all trying to rebalance their position in the 'in-between' category of European nations. Together with the previously launched 'European Neighbourhood Policy' projects, the concrete partnerships forged between EU and partner countries might help underwrite the sovereignty and territorial integrity of EU-friendly CIS countries that cannot aspire to full EU membership in the immediate future.

In order to be effective, the second element of a more coherent Russia policy on part of the EU should nevertheless be specific and ambitious, preferably oriented towards economic projects that could help stabilize 'in-between-Europe' in the medium to long term. The single most interesting idea in this context is no doubt that of an EU–Russia consortium to refit and modernize Ukraine's massive gas and oil pipeline system. Apart from solving the financial problems an undertaking so massive that Ukraine has little prospects to pull it off, a Russia–EU energy consortium would be in a position to reconcile Russian interests for security of demand with European concerns about security of supply. Even if the exact formulas for renovation costs, transit fees and delivery prices will be difficult to negotiate equitably, the benefits of a stable arrangement would be shared by all parties.

In that sense, the two main crises in Europe–Russia relations 2008–9 might be understood to harbour the seeds of a new era of closer engagement. By opening up the possibility for concrete collaboration between EU and partner countries in the CIS region, and perhaps with an energy pact involving Ukraine as well, Brussels and Moscow would set some key priorities for Europe–Russia relations for the next ten to fifteen years, concentrating on critical issues of mutual interest. At the same time, it would be an engagement that lives up to that of 'positive reciprocity' (Romanova) and evades the inclusion/exclusion narrative of European integration discourse (Prozorov) of the past. It would be a relationship focused on reaching agreements and building bridges, rather than digging trenches.

Primary sources

APSNYPRESS 2005, 23 March, *General-Lieutenant Anatoliy Zaytsev Appointed* [Home-page of APSNYPRESS], [Online]. Available: http://abhazia.com/news/comments. php?id=8539; http://daccessdds.un.org/doc/UNDOC/GEN/N08/454/58/PDF/N0845458. pdf?Open Element [2009, 06/28].

Athens News Agency 2009, 15 July, *Greece, Romania Promote Energy Security, Greece, Bulgaria Sign LNG Supply Agreement* [Homepage of ANA Daily News Bulletin in English], [Online]. Available: http://www.hri.org/news/greek/ana/2009/09-07-15.ana. html#12 [2009, 07/25].

BBC News 2009, 8 April, *Romania Blamed over Moldova Riots*. Available: http://news. bbc.co.uk/2/hi/europe/7989360.stm [2009, 06/25].

—— 2009, 20 April, *Moldova Police Face Brutality Allegations*. Available: http://news. bbc.co.uk/2/hi/europe/8007428.stm [2009, 06/25].

Biden, J. 2009, 7 February, *Speech at the 45th Munich Security Conference* [Homepage of Wall Street Journal], [Online]. Available: http://blogs.wsj.com/washwire/2009/02/07/ 8375 [2009, 06/22].

BIKI 2009, no. 26.

—— 2009, no. 30.

—— 2009, no. 32.

—— 2009, no. 38.

BP 2008, *BP Statistical Review of World Energy*, British Petroleum, London.

Britannica Concise 2009, n.a. [Homepage of Britannica Concise Encyclopedia], [Online]. Available: http://www.answers.com/library/Britannica%20Concise%20Encyclopedia-cid-67656 [2009, 06/29].

CFE 1990, *Treaty on Conventional Armed Forces in Europe*, international treaty, http:// www.osce.org/documents/doclib/1990/11/13752_en.pdf.

China Daily 2008, 8 December, *China-Russia Pipeline Boosts Energy Links*, Beijing.

CIA 2004, 11 May, *CIA World Fact Book: Ukraine* [Homepage of Central Intelligence Agency], [Online]. Available: http://www.umsl.edu/services/govdocs/wofact2004/geos/ up.html [2008, 10/20].

—— 2009, 30 April, *CIA World Factbook* [Homepage of Central Intelligence Agency], [Online]. Available: https://www.cia.gov/library/publications/the-world-factbook/geos/ MD.html [2009, 06/23].

—— 2009, 30 April, *CIA World Factbook: Moldova*. Available: https://www.cia.gov/ library/publications/the-world-factbook/geos/MD.html [2009, 06/23].

Civil Georgia 2005, 24 March, *Our Officers are Trained in Russia: Interview with Abkhaz Defense Minister* [Homepage of UN Association of Georgia], [Online]. Available: http://www.civil.ge/eng/article.php?id=9423 [2009, 06/29].

—— 2008, 27 October, *State Minister Testifies Before War Commission* [Homepage of UN Association of Georgia], [Online]. Available: http://www.civil.ge/eng/article.php?id=19842 [2009, 06/20].

Council for the National Interest 2006, 22 June, *Virtual Politics: Faking Democracy in the Post-Soviet World* [Homepage of Gale Group], [Online]. Available: http://www.thefreelibrary.com/Virtual+Politics%3a+Faking+Democracy+In+The+Post-Soviet+World.-a0143722964 [2009, 06/29].

DFID 2009, 30 April, *Where We Work: Moldova, Overview* [Homepage of Department for International Development (United Kingdom)], [Online]. Available: http://www.dfid.gov.uk/where-we-work/europe/moldova/ [2009, 06/23].

ENS 2007, 3 April, *Pan-European Oil Pipeline Agreement Signed* [Homepage of Environment News Service], [Online]. Available: http://www.ens-newswire.com/ens/apr2007/2007-04-03-03.asp [2008, 12/08].

EU Commission 2006a, *EU–Russia Energy Dialogue Seventh Progress Report*, European Commission, Moscow/Brussels.

—— 2006b, 4 July, *European Commission Approves Terms for Negotiating New EU–Russia Agreement* [Homepage of The European Commission's Delegation], [Online]. Available: http://www.delrus.ec.europa.eu/en/news_810.htm [2009, 06/29].

—— 2006c, *A European Strategy for Sustainable, Competitive and Secure Energy, COM(2006) 105 final*, European Commission, Brussels.

—— 2007a, *Proposal for a Directive of the European Parliament and of the Council Amending Directive 2003/54/EC Concerning Common Rules for the Internal Market in Electricity, COM(2007) 528 final*, European Commission, Brussels.

—— 2007b, *The European Interest: Succeeding in the Age of Globalisation. Contribution of the Commission to the October Meeting of Heads of State and Government, COM(2007) 581 final*, European Commission, Brussels.

—— 2008a, *A Common European Approach to Sovereign Wealth Funds. Communication from the Commission to the European Parliament, the Council, the European Economic and Social Committee and the Committee of the Regions, COM(2008) 115 provisional*, European Commission, Brussels.

—— 2008b, 5 November, *Review of EU–Russia Relations, MEMO/08/678* [Homepage of European Commission], [Online]. Available: http://europa.eu/rapid/pressReleasesAction.do?reference=MEMO/08/678 [2009, 06/29].

—— 2008c, *EU–Russia: First Round of Negotiations for the New Agreement*, Brussels, 3 July, IP/08/1099.

—— 2008d, *Second Strategic Energy Review: An EU Energy Security and Solidarity Action Plan, SEC(2008) 2794*, European Commission, Brussels.

EU/RF 1994, *Agreement on Partnership and Cooperation Establishing a Partnership between the European Communities and their Member States, of the One Part, and the Russian Federation, of the Other Part*, European Union and Russian Federation, Corfu.

EurActiv 2008a, 8 May, *EU Launches Eastern Plan in Russia's Backyard*, www.euractiv.com.

—— 2008b, 22 September, *France-Russia Business Booming Despite Georgia Crisis*, www.euractiv.com.

—— 2009, 14 July, *EU Countries Sign Geopolitical Nabucco Agreement*, www.euractiv. com.

European Council 2007, 14 December, *EU Declaration on Globalisation, Presidency Conclusions, 16616/1/07 REV*, Council of the European Union, Brussels.

Eurostat 2008, 'The Tourist Accommodation Sector Employes 2,3 Million in the European Union', *Eurostat News Release*, [Online], no. 90. Available from: http://epp.eurostat. ec.europa.eu/cache/ITY_OFFPUB/KS-SF-08-090/EN/KS-SF-08-090-EN.PDF. [06/27/ 2009].

—— 2009, *EU-27 Trade with China and Russia in 2007. Statistics in Focus 9/2009*, Eurostat/European Commission, Luxembourg.

Federal Law 1999, *On the Election of Deputies of the State Duma of the Federal Assembly of the Russian Federation (unofficial translation), approved on 2 and 9 June by the State Duma and the Federation Council, respectively. Published 24 June. No. 121-F3*, http:// www.democracy.ru/english/library/laws/dumaelect_eng/index.html.

—— 2001, 11 July, *Law on Political Parties, Russian Federal Law No. 95-FZ, 2001*, Russia, Moscow.

—— 2005, 18 May, *Federalny Zakon nr.51-FZ 'O vyborakh deputatov Gosudarsvtennoi Dumy federalnogo Sobraniia Rossiiskoi Federatsii 18 May 2005 (Federal Law on Elections of Deputies to the State Duma, Russian Federal Law No. 51-F3, 2005)*, http:// www.cikrf.ru/law/2/zakon_51.jsp.

—— 2008, Enacted 7 May, *Federalniy zakon osushestvleniya inostrannyh investitsiy v hoziaistvennye obshestva, imeyusheie strategicheskoe znachenie dlya obespecheniya oboroniy straniy i bezopasnosti gosudarstva (Federal Law on the Regulation of Foreign Investments in Economic Organizations with Strategic Significance on Safeguarding the Defence and Security of the State), adopted 29 April 2008*, N 57-F3, Moscow.

Finansovye Izvestiia 2003, 17 October, *Do sinkhronizatsii energosistem Rossiia i EC neobkhodimo reshit 'problemy vzaimnosti' (Before the synchronization of the energy system of Russia and the European Union it is necessary to solve 'the problem of commonality')*.

Foreign Policy 2008, *The Failed States Index 2008: The Rankings* [Homepage of ForeignPolicy.com (United States)], [Online]. Available: http://www.foreignpolicy. com/story/cms.php?story_id=4350&page=1 [2009, 06/25].

France/Russia 2008, 15 June, *Mutual Facilitation Agreement on Entry, Travel and Exit Requirements for the Nationals of the Two Countries, 15 June 2004* [Homepage of Foreign Ministry of France], [Online]. Available: http://www.diplomatie.gouv.fr/en/ country-files_156/russia_399/france-and-russia_3057/index.html [2009, 07/25].

Freedom House 1998, *Freedom in the World 1996–1997*, Transaction Publishers & Freedom House, Edison, NJ.

—— 2006, *Freedom in the World 2006*, Freedom House, New York.

—— 2008, 'Russia' in *Freedom in the World 2007*, Rowman & Littlefield Publishers, Washington, D.C.

French Foreign Ministry 2009, 26 May, *Note on Political Relations between France and Russia*. Available: http://www.diplomatie.gouv.fr/fr/pays-zones-geo_833/russie_ 463/france-russie_1214/presentation_3040/index.html [2009, 07/06].

Gazprom 2009, *Yamal Megaproject* [Homepage of Gazprom], [Online]. Available: http:// www.gazprom.com/eng/articles/article32739.shtml [2009, 12/09].

Handelsblatt 2009, 27 April, *Russische Gasexport halbiert sich (Russia Has Export Cut in Half)*, Düsseldorf.

High Anthropological School 2009, *Moldova on the Crossroads of Millennia Project.* Available: http://www.ant.md/en/projects/moldova.htm [2009, 07/04].

IDEA 2002, *International Electoral Standards. Guidelines for reviewing the legal framework of elections*, International Insititute for Democracy and Electoral Assistance, Stockholm.

IEA 2000, *Black Sea Energy Survey*, International Energy Agency, Paris.

—— 2007, *Member Countries and Countries Beyond the OECD* [Homepage of International Energy Agency], [Online]. Available: http://www.iea.org/Textbase/country/index.asp [2008, 11/28].

—— 2008, *Key World Energy Statistics 2008*, International Energy Agency, Paris.

IISS 2008, *The Military Balance 2008*, International Institute for Strategic Studies & Routledge, London.

Illarionov, A. 2008a, 24 October, *Situatsiya v Yuzhnoi Osetii i Gruzii (The situation in South Ossetia and Georgia)* [Homepage of Echo Moskvy], [Online]. Available: http://echo.msk.ru/programs/razvorot/548457-echo.phtml [2009, 20 June].

—— 2008b, 13 August, *The Second Georgian War. The Preliminary Results.* [Homepage of RFE/RL], [Online]. Available: http://www.rferl.org/content/Preliminary_Conclusions_From_The_War_In_Georgia/1190743.html [2009, 20 June].

IMF 2005, *Republic of Moldova: Financial System Stability Assessment, including Reports on the Observance of Standards and Codes on the following topics: Monetary and Financial Policy Transparency, Banking Supervision*, International Monetary Fund, Washington, D.C.

Interfax 2008, 29 October, *Transneft, CNPC Sign Agreement to Build Pipeline to China.* Available: www.interfax.cn/news/6477/ [2008, 12/08].

IPP 2009, *Barometer of Public Opinion, March 2009. Dynamics, Part 2*, Institute for Public Policy and Centre for Sociological Surveys and Marketing CBS-AXA, Chisinau. Available: http://www.ipp.md/barometru1.php?l=en&id=35.

Iraq Mission Activities 2003, 5 March, *Joint Declaration by France, Germany and Russia (on Iraq Mission Activities)* [Homepage of United Nations, French Permanent Mission], [Online]. Available: http://www.un.int/france/documents_anglais/030305_mae_france_irak.htm [2009, 07/04].

ISPP 2004, *Poll by Instytut Sotsialnoi ta Politychnoi Psyhologii (Institute of Social and Political Psychology)*, Institute of Social and Political Psychology, Kyiv.

ITAR-TASS 2008, 15 July, *France-Russie: Il Importe D'accroître les Investissements (France-Russia: It Is Important to Increase Investments)*, http://www.itar-tass.com.

Kamynine, M.L. 2008, 7 March, *Interview with Kamynine M. L. (official representative of the Russian Foreign Ministry)*, RIA Novosti.

Kommersant 2007a, 13 July, *Total Wins Share in Shtokman*, Moskva.

—— 2007b, 13 July (122), *Prizvaniye Rossii – Povernutsiya Litsom k Evrope (Russia's Calling – to Turn its Face Toward Europe)*, Moskva.

L'Expansion 2009, *Nucléaire – Les Coulisses de la Guerre Franco-Russe (Nuclear – Behind the Scenes of the Franco-Russian War)*, 1 April, L'Expansion.com/L'Express, http://energie.lexpansion.com/articles/enquetes/2009/04/Nucleaire – Les-coulisses-de-la-guerre-franco-russe/.

Lavrov, S. 2005, 26 October, Reprint of Sergei Lavrov's speech at the Foundation for Political Studies, *Paris, 11 October 2005, Ukrainska Pravda*.

—— 2008a, 9 August, *Interview of Minister of Foreign Affairs of the Russian Federation to the BBC*, BBC.

—— 2008b, 8 April, *On Russian–American relations. Interview with Foreign Minister Lavrov.* [Homepage of Ekho Moskvy], [Online]. Available: http://www.echo.msk.ru/programs/beseda/506017-echo.phtml.

Law Dictionary 2003, *edited by Stephen H. Gifis*, 5th edn, Baron's Legal Guides, New York.

Levada Center 2007, 14 February, *Voices from Russia: Society, Democracy, and Europe*, http://www.levada.ru/eng/.

Medvedev, D. 2006, 24 July, *Dlya Protsvetaniia Vsekh, Nado Uchitivat Interesy Kazhdogo (For Everyone to Thrive, the Interest of Everyone Should be Taken into Account). An Interview with Vice-Premier Dimitry Medvedev.* Available: http://www.expert.ru/printissues/expert/2006/28/interview_medvedev/ [2009, 07/29].

—— 2008, 5 June, *Speech at Meeting with German Political, Parliamentary and Civic Leaders; Press Statement and Answers to Questions following Russian–German Talks* [Homepage of Kremlin], [Online]. Available: http://www.kremlin.ru/eng/text/speeches/2008/06/05/2203_type82912type82914type84779_202153.shtml; http://www.kremlin.ru/eng/speeches/2008/06/05/2135_type82914type82915_202127.shtml [2008, 07/11].

—— 2008, 15 July, *Speech at the Meeting with Russian Ambassadors and Permanent Representatives to International Organisations* [Homepage of Russian Foreign Ministry], [Online]. Available: http://www.president.kremlin.ru/eng/text/speeches/2008/07/15/1121_type82912type84779_204155.shtml [2008, 11/15].

—— 2008, 26 August, *Interview with Russia TV Channel Today.* Available: http://www.president.kremlin.ru/eng/speeches/2008/08/26/2003_type82915type82916_205778.shtm [2008, 15 November].

—— 2008, 31 August, *Interview with Television Channels One, Rossia, NTV* [Homepage of President of Russia], [Online]. Available: http://www.president.kremlin.ru/eng/text/speeches/2008/08/31/1850_type82912type82916_206003.shtml [2009, 20 June].

—— 2008, 25 September, *Beginning of Meeting with Crew Members of the Strategic Nuclear Cuiser Submarine St. George the Victorious* [Homepage of President of Russia], [Online]. Available: http://www.president.kremlin.ru/eng/speeches/2008/09/25/1000_type84779_206915.shtml [2009, 20 June].

—— 2008, 30 September, *Speech at the Ceremony for Officers Who Have Been Newly Appointed to Senior Command Positions and Received High (Special) Ranks* [Homepage of President of Russia], [Online]. Available: http://www.president.kremlin.ru/eng/speeches/2008/09/30/1359_type82912type82913 _207068.shtm [2009, 20 June].

—— 2008, 11 October, *Conversation with the Crew of the Aircraft Carrier Admiral Kuznetsov* [Homepage of President of Russia], [Online]. Available: http://www.president.kremlin.ru/eng/speeches/2008/10/11/2144_type84779_207843.shtml [2009, 20 June].

—— 2008, 5 November, *Address to the Federal Assembly of the Russian Federation.* Available: http://www.president.kremlin.ru/appears/2008/11/05/1349_type63372type63374type63381type82634_208749.shtml [2009, 06/20].

Merkel, A. and N. Sarkozy 2009, 3 February, 'La sécurité, notre mission commune (Security, Our Common Mission)', *Le Monde*.

Mezhdunarodnaia Zhizn 2006, 'Rossiisko-Germanskie Otnosheniia v Kontekste Mirovoi Politike (Russian–German Relations in the Context of World Politics)', no. 14 July, pp. 120–60.

Moldova Azi 2007, 30 March, *Remittances from Moldovans Working Abroad Increased by about a Quarter* [Homepage of Moldova Azi, citing http://info-prim.md/], [Online]. Available: http://old.azi.md/news?ID=43822 [2009, 06/23].

Moldova.org 2009, 2 February, *Wine exports rose 44.5% in 2008*. Available: http://economie.moldova.org/news/wine-exports-rose-445-in-2008-180582-eng.html [2009, 06/23].

Moldovan Statistical Bureau 2009b, 3 June, *Activitatea de Comerţ Exterior a Republicii Moldova în Januarie–Aprilie 2009* (*Foreign Trade of the Republic of Moldova January–April 2009*) [Homepage of National Statistical Bureau of the Republic of Moldova], [Online]. Available: http://www.statistica.md/newsview.php?l=ro&idc= 168&id=2626 [2009, 07/04].

—— 2009a, 3 June, *Exportul Republicii Moldova Stucturat pe Trimestre şi Grupe de Mărfuri* (*2005–2009*)(*Exports of the Republic of Moldova in Trimesters, Groups of Commodities* (*2005–2009*)). In *3_Export_Sec_Cap_trim_2005–2009.xls*. Available: http://www.statistica.md/category.php?l=ro&idc=336& [2009, 07/04].

Moscow News 2007, 13 July, *Russia's New Black Sea Base Complete by 2012*, Moscow News, http://mnweekly.ru/news/20070713/55261987.html.

Moscow Times 2004a, 10 March, *OSCE Observers*, Moscow.

—— 2004b, 2 March, *Top Channels Reject Rybkin Commercials*, Moscow.

MSIF 2008, not available, *Objectives* (*Obiective*) [Homepage of Fondul de Investitii Sociale din Moldova (Moldova Social Investment Fund)], [Online]. Available: http://www.msif.md/user/despre_fism/obiective.shtml [2009, 06/25].

—— 2009a, not available, *Moldova Social Investment Fund: Donor Countries* [Homepage of Fondul de Investitii Sociale din Moldova (Moldova Social Investment Fund)], [Online]. Available: http://www.msif.md/user/parteneri_fism/donors.shtml [2009, 06/23].

—— 2009b, not available, *Sursele de Finantare ale Proiectului FISM* (*MSIF Project Sources of Funding*) [Homepage of Fondul de Investitii Sociale din Moldova (Moldova Social Investment Fund)], [Online]. Available: http://www.msif.md/user/despre_fism/sursele_de_finantare.shtml [2009, 06/23].

Nabucco n. a., *Project Description/Pipeline Route* [Homepage of Nabucco Gas Pipeline International], [Online]. Available: http://www.nabucco-pipeline.com/project/project-description-pipeline-route/project-description.html [2008, 12/09].

National Security and Defence 2006, 'Common Identity of Ukrainian Citizens: Specificity and Problems of Formation', *National Security and Defence*, [Online], vol. 29, no. 7. Available from: http://www.uceps.org.ua/eng/). [2009/06/29].

News.Rin.ru 2008, 27 February, *InoSMI Otsenivaiot Vooruzhennye sily Gruzii ki Abkhazii* (*InoSMI evaluate the armed forces of Georgia and Abkhazia*) [Homepage of News Rin ru], [Online]. Available: http://news.rin.ru/news/155951 [2009, 06/20].

NISD 2006, *Poslannia Presidenta Ukraini Verkhovny Radi Ukrainii pro Zovnizhne ta Vnutrenne Stanovishche Ukraini u 2005 Godi* (*The address of the President of Ukraine to the Verkhovny Rada of Ukraine on the External and Internal Situation in Ukraine in 2005*), NISD, Kyiv.

North Atlantic Council 2008, 3 April, *Bucharest Summit Declaration. Issued by the Heads of State and Government Participating in the Meeting of the North Atlantic Council in Bucharest*, Bucharest.

NTI 2008, 8 May, *NTI Country Profiles: Iran* [Homepage of The Nuclear Threat Initiative], [Online]. Available: http://www.nti.org/e_research/profiles/Iran/index.html [2009, 06/28].

Oil and Energy Trends 2008, 'Russia Ties-Up More Oil and Gas', *Oil and Energy Trends: A Monthly Publication of International Energy Statistics and Analysis*, [Online], no. 15 February. Available from: http://www.wiley.com/bw/journal.asp?ref=0950–1045.

OSCE 1999, n.a., *Agreement on Adaptation of the Treaty on Conventional Armed Forces in Europe*. In: *The Istanbul Documents 1999*. Available: http://www.osce.org/documents/mcs/1999/11/4050_en.pdf [2009, 07/05].

—— 2005, *OSCE Human Dimension Commitments. Volume 1. Thematic Compilations*, OSCE Office for Democratic Institutions and Human Rights, Warsaw.

—— 2009a, *Democracy and the Rule of Law* [Homepage of OSCE Office for Democracy Institutions and Human Rights], [Online]. Available: http://www.osce.org/odihr/13492.html [2009, 06/22].

—— 2009b, *Politically Binding Commitments* [Homepage of OSCE Office for Democratic Institutions and Human Rights], [Online]. Available: http://www.osce.org/odihr/13493.html [2009, 06/29].

—— 2009c, *About Election Observation; Methodology* [Homepage of OSCE Office for Democratic Institutions and Human Rights], [Online]. Available: http://www.osce.org/odihr-elections/17781.html; http://www.osce.org/odihr-elections/17783.html [2009, 06/22].

—— 2009d, *How An Election Missions Works* [Homepage of OSCE Office for Democratic Institutions and Human Rights], [Online]. Available: http://www.osce.org/odihr/17791.html [2009, 06/22].

Pew Research Center 2006, *Russia's Weakened Democratic Embrace. 2005 Pew Global Attitudes Survey*, Pew Research Center, Washington D.C.

Prime-TASS 2009, 3 February, *Statistics Agency Says Russia's Economy Up 5.6% in 2008*. Available: http://www.prime-tass.com/news/show.asp?id=451479&topicid=0 [2009, 07/25].

Putin, V. 2000a, *Ot pervogo litsa: Razgovory s Vladimirom Putinym (First Person: Conversations with Vladimir Putin)*, Vagrius, Moskva.

—— 2000b, *First Person*, Public Affairs, New York.

—— 2000, 8 July, *Poslanie Federalnomu Sobraniiu Rossiiskoi Federatsii (Address to the Federal Assembly of the Russian Federation)* [Homepage of President of the Russian Federation], [Online]. Available: http://www.kremlin.ru/text/appears/2000/07/28782.shtlm [2000, 07/08].

Putin, V. 2001, 25 September. *Rede im deutschen Bundestag (Speech in German Parliament)*. Available: www.bundestag.de/parlament/geschichte/gastredner/putin/putin/wor.hml

—— 2004, 16 May, *Poslanie Federalnomu Sobraniu Rossiyskoj Federatsii (Address to the Federal Assembly of the Russian Federation)* [Homepage of President of the Russian Federation], [Online]. Available: http://www.kremlin.ru/text/appears/2004/05/64879.shtml.

—— 2005, 12 April, *Press Conference of Russian President Vladimir Putin, 11 April 2005*, Kommersant, http://kommersant.ru/doc.aspx?DocsID=569269.

—— 2006, 7 February, *Putin nazval 'zhulynichestvom' situatsiyo v gazovih otnosheniyah s Ukrainoi (Putin Refers to Gas Relations with Ukraine as 'Deceitful')*. RIA Novosti, http://www.rian.ru/politics/20060207/43403429.html.

—— 2006, 9 September, *Meeting with Participants in the Third Meeting of the Valdai Discussion Club* [Homepage of President of Russia], [Online]. Available: http://www.kremlin.ru-eng/speeches.

—— 2006, 1 October, *Beginning of Meeting with the Permanent Members of the Russian Security Council* [Homepage of President of Russia], [Online]. Available: http://www.president.kremlin.ru/eng/speeches/2006/10/01/1059_type82913_111844.shtml [2009, 06/20].

—— 2007, 14 September, *Meeting with Members of the Valdai International Discussion Club 14 September 2007 in Sochi* [Homepage of President of the Russian Federation], [Online]. Available: http://www.kremlin.ru/eng/speeches/2007/09/14/1801_type82917 type84779_144106.shtml [2009, 07/25].

Putin V. Press conference, Presidents Sarkozy and Putin, Moscow, Kremlin, 10 October 2007 http://weng.kremlin.ru/speeches/2007/10/10/1431_type82914type82915_147984.shtml.

—— 2007, 26 October, *Vstupitelnoe Slovo na Dvadtsatom Sammite Rossiya – Evrosouz (Opening Remarks at the Twentieth Russia–European Union Summit)*, *Mafra*. Available: http://www.kremlin.ru/eng/speeches/2007/10/26/2054_type82914_149665.shtml.

—— 2008, 30 January, *Speech at an Enlarged Session of the Federal Security Service (FSB) Presidium, January 30 2008*. Available: http://www.kremlin.ru/eng/speeches/2008/01/30/1525_type82913_158762.shtml [2009, 07/29].

—— 2008, 29 August, *Ne Nado Nikogo Pugat. Ne Strashno Sovsem (Nobody Needs to be Frightened. It Is Not That Scary). CNN Interview, Sochi, recorded 28 August* [Homepage of President of Russia], [Online]. Available: http://kreml.org/interview/190498074 [2009, 07/23].

—— 2009, 6 April, *Report to the State Duma on the Russian Government's Performance in 2008* [Homepage of The Government of the Russian Federation], [Online]. Available: http://www.premier.gov.ru/events/2490.html [2009, 07/25].

Rambler.ru 2008, 5 September, *Glava SKP RF Bastrykin Prizval ne Spekulirovat na Temu Chislennosti Zherstv v Uy.O (Head of the Investigative Commission of the Procurement Office Batrykin Calls for End to Speculation of Number of South Ossetia Casualties)* [Homepage of Rambler], [Online]. Available: http://www.rambler.ru/news/events/russiageorgia/13379592.html [2009, 06/20].

Rehn, O. 2004, 10 November, 'The Common Economic Space with Russia: State of Play of the Negotiations and the Role of Industry. EU–Russian Round Table of Industrialists, SPEECH/04/481', European Commission, The Hague.

Republic of Moldova 2007, 25 July, *History* [Homepage of Official website of the Republic of Moldova], [Online]. Available: http://www.moldova.md/en/istorie/ [2009, 06/23].

Reuters 2007, 5 December, *EU Split on Vote as Sarkozy Congratulates Putin*.

RFE/RL 2004, *Newsline*, 14 January, Radio Free Europe/Radio Liberty, http://www.rferl.org/.

—— 2004, *Newsline*, 17 February, Radio Free Europe/Radio Liberty, http://www.rferl.org.

—— 2004, *Newsline*, 1 March, Radio Free Europe/Radio Liberty, http://www.rferl.org.

—— 2004, *Newsline*, 11 March, Radio Free Europe/Radio Liberty, http://www.rferl.org.

—— 2004, *Newsline*, 12 March, Radio Free Europe/Radio Liberty, http:www.rferl.org.

—— 2008, n.a., *Russia Votes*. Available: http://www.rferl.org/specials/russianelection [2008].

—— 2008, 12 August, *Russia, Georgia Agree to Peace Plan to End Fighting*, Radio Free Europe/Radio Liberty.

Rosstat 2008, *Russia in Figures 2008: Statistical Handbook*, Federal State Statistics Service, Moscow.

Russian Constitution 1993, *The Constitution of the Russian Federation, adopted 12 December/enacted 25 December*, http://www.constitution.ru/en/10003000–3001.htm.

Russian Defence Ministry 2008a, 6 August, *Podvedenie itogov antoterroristich-eskogo ucheniia 'Kavkaz-2008' (Drawing Conclusions from the Anti-Terrorist Exercise 'Caucasus-2008')*. Available: http://www.mil.ru/info/1069/details/index.shtml?id=49146 [2009, 06/29].

—— 2008b, 8 August, *Novosti* (*News*) [Homepage of Ministry of Defence of the Russian Federation], [Online]. Available: http://www.mil.ru/info/1069/details/index.shtml?id= 49292 [2009, 06/20].

Russian Foreign Ministry 2007, 11 January, *Information Note on Russian–French Relations* [Homepage of Ministry of Foreign Affairs of the Russian Federation], [Online]. Available: www.mid.ru.

Russian President 2007, *Decree on Suspending the Russian Federation's Participation in the Treaty on Conventional Armed Forces in Europe and Related International Agreements*, Decree, Federal, http://www.kremlin.ru/eng/text/docs/2007/07/137839. shtml.

Russian Foreign Ministry 2008a, 10 August, *Record of the Presentation and Answers to the Questions by the Deputy Minister of the Foreign Affairs of the Russian Federation G.B. Karasin at the Press-Conference in the RIA Novosti Information Agency* [Homepage of Ministry of Foreign Affairs of the Russian Federation], [Online]. Available: http://www.mid.ru/bl.nsf/78b919b523f2fa20c3256fa3003e9536/ 1da41391c654a206c32574a2004917bc/$FILE/11.08.2008.doc [2009, 07/25].

—— 2008b, 8 July, *Statement of the Russian Ministry of Foreign Affairs Concerning the Signing of the US–Czech Agreement on Deployment of Elements of the US Global Missile Defense System on the Territory of the Czech Republic* [Homepage of Ministry of Defence of the Russian Federation], [Online]. Available: http://www.mid.ru/brp_ 4.nsf/e78a48070f128a7b43256999005bcbb3/2fcca1548822c351c32574810036877c? OpenDocument [2009, 06/20].

—— 2008c, 12 July, *Foreign Policy Concept of the Russian Federation* [Homepage of President of Russia], [Online]. Available: http://www.kremlin.ru/eng/text/docs/2008/07/ 204750.shtml [2009, 07/05].

Russian Taxpayer's Union 2007, 'Evropa Mozhet Zablokirovat Ekspansiiu Rossiiskogo Buzinesa (Europe Can Bloc the Expansion of Russian Business)', *Narodnye Dengi*, [Online], no. 4 January. Available from: http://www.peoplesmoney.ru/news/?id=1408.

Sarkozy, N. 2007, 9 August, *Interview with the French President*, Rossiyskaya Gazeta, Moskva.

—— 2007, 27 August, *Press Conference* [Homepage of President of France], [Online]. Available: http://www.elysee.fr.

—— 2007, 9 October, *Press Conference* [Homepage of President of France], [Online]. Available: http://www.elysee.fr.

Sarkozy, N. and V. Putin 2007, 10 October, *Press Conference with President Putin, Moscow, Kremlin* [Homepage of President of France], [Online]. Available: http:// www.elysee.fr/search?q=Kremlin&site=elyseev2&proxystylesheet=v2&output= xml_no_dtd&client=v2&lr=lang_fr&ie=utf8&oe=utf8&mode=html&getfields= video-image.nxtitle.nxcategory.topic.media-image [2009, 07/23].

Sarkozy, N. 2008, 27–29 August, *Conference with the Ambassadors, Elyseé Palace* [Home-page of French Minstry of Foreign Affairs], [Online]. Available: http://www.diplomatie. gouv.fr/en/ministry_158/15th-ambassadors-conference-27-29.08.07_9749.html [2009, 07/05].

—— 2008, 8 October, *World Policy Conference, Evian*. Available: http://www. worldpolicyconference.com/ [2009, 07/05].

—— 2008, 14 November, *Sarkozy Urged Russian President to Ensure Energy Supply to Lithuania*, Lithuanian News Agency, Vilnius.

—— 2008, 21 November, *Sarkozy Réinvente L'Amitié Franco-Russe* (*Sarkozy Reinvents Franco-Russian Friendship*) [Homepage of La LettreA], [Online]. Available: http://

www.lalettrea.fr/Identification/login_article.asp?rub=login&lang=FRA&service= ARL&context=REC&doc_i_id=51768825&Rubrique=action_publique   [2009, 07/25].

—— 2009, 3 April, *President Sarkozy Receives U.S. President Obama in Strasbourg* [Homepage of President of France], [Online]. Available: http://www.elysee.fr/webtv/au-jour-le-jour/le-president-barack-obama-a-strasbourg-ceremonie-officielle-d-accueil-video-1-1098.html [2009, 07/23].

South Ossetian TV and Radio 2008, 6 June, *Mezhdunarodnaia Konferentsiia 'Geneologiia Narodov Kavkaza. Traditsiia i Sovbremennost' Otkrylas bo Vladokavkaze* (*The International Conference 'Geneology of the Peoples of Caucasus' opened in Vladikavkaz*). Available: http://osradio.ru/news/genocid//12841.html [2009, 06/28].

Steinmeier, F. 2008, 4 March, *Auf dem Weg zu Einer Europäischen Ostpolitik. Speech by German Foreign Minister Steinmeier.* Available: http://www.auswaertiges-amt.de/diplo/de/Infoservice/Presse/Reden/2008/080304-BM-Ostpolitik.html   [2009, 06/22].

Tymoshenko, Y. 2007, 'Containing Russia', *Foreign Affairs*, vol. 86, no. 3 (May/June), pp. 69–83.

—— 2008, 20 August, *Tymoshenko Nraviatsa Printsipy Medvedeva i Zarkozy* (*Tymoshenko Likes Medevedev's and Sarkozy's Principles*), RosBalt Ukrainii.

Ukranian Public Radio 2004, 19 May, *Interview with Marat Gelman and Igor Shuvalov*, Kyiv, http://www.pravda.com.ua/archive/2004/may/19/2.shtml

UN 2009, *Permanent Mission of the Republic of Moldova to the United Nations: Country Facts.* Available: http://www.un.int/wcm/content/lang/en/pid/5305 [2009, 06/25].

UN General Assembly 1974, *Definition of Aggression. United Nations General Assembly Resolution 3314 (XXIV), 29th Session, 14 December 1974*, http://www.un-documents. net/a29r3314.htm.

UNDP 2008, 27 February, *UN helps Moldova Recover from Drought* [Homepage of United Nations Development Program], [Online]. Available: http://content.undp.org/go/newsroom/2008/february/un-moldova-drought-20080227.en [2009, 07/04].

UNOSAT 2008, 19 August, *Village Damage Summary: Kekhvi to Tskhinvali, South Ossetia.* Available: http://unosat.web.cern.ch/unosat/freeproducts/Georgia/Russia_ ConflictAug08/update2/UNOSAT_GEO_Damage_Assessment_EREDEVI_19aug08_ Lowres.pdf [2009, 06/22].

US Foreign Policy Encyclopedia 2009, n.a. [Homepage of The Gale Group], [Online]. Available: http://www.answers.com/library/US+Foreign+Policy+Encyclopedia-letter [2009, 06/29].

Vecernje novosti 2008a, 16 October, *Russian Gazprom Executive Discusses South Stream Project Implementation*, Belgrade.

—— 2008b, 5 December, *Russia Gets NIS, Serbia Gets Gas*, Belgrade.

Vedomosti 2009, 23 March, with Wall Street Journal & Financial Times, http://www. vedomosti.ru/newspaper/index.shtml?2009/03/23.

—— 2009, 30 March, with Wall Street Journal & Financial Times, http://www. vedomosti.ru/newspaper/index.shtml?2009/03/30.

—— 2009, 13 April, with Wall Street Journal & Financial Times, http://www. vedomosti.ru/newspaper/index.shtml?2009/04/13.

—— 2009, 21 April, with Wall Street Journal & Financial Times, http://www. vedomosti.ru/newspaper/index.shtml?2009/04/21.

—— 2009, 24 April, with Wall Street Journal & Financial Times, http://www. vedomosti.ru/newspaper/index.shtml?2009/04/24.

—— 2009, 30 April, with Wall Street Journal & Financial Times, http://www.vedomosti.ru/newspaper/index.shtml?2009/04/30.

—— 2009, 4 May, with Wall Street Journal & Financial Times, http://www.vedomosti.ru/newspaper/index.shtml?2009/05/04.

—— 2009, 7 May, with Wall Street Journal & Financial Times, http://www.vedomosti.ru/newspaper/index.shtml?2009/05/07.

—— 2009, 14 May, with Wall Street Journal & Financial Times, http://www.vedomosti.ru/newspaper/index.shtml?2009/05/14.

—— 2009, 19 May, with Wall Street Journal & Financial Times, http://www.vedomosti.ru/newspaper/index.shtml?2009/05/19.

—— 2009, 20 May, with Wall Street Journal & Financial Times, http://www.vedomosti.ru/newspaper/index.shtml?2009/05/20.

VNIKI 2007, "Vneshneekonomicheskiy Kompleks Rossii: Sovremennyoe Sostoyanie i Perspektivy (Russia's Foreign Economic Complex: Today's Situation and Perspectives), Moscow, no. 2, pp. 19, 26.

Voenno-promyshlennii kurier 2007, 'Iskander Vstupil v Psikhologicheskuiu Voinu (Iskander Enters into Psychological Warfare)", *Voenno-promyshlennii kurier*, [Online], vol. 29, no. 215. Available from: http://www.vpk-news.ru/article.asp?pr_sign=archive.2007.215.articles.conception_02. [06/20/2009].

Voprosy ekonomiki 2008, no. 2.

—— 2008, no. 11.

—— 2008, no. 97.

World Bank 2009, *Country Lending Summaries – Moldova*. Available: http://go.worldbank.org/JZVJ4EM7L0 [2009, 06/23].

—— 2008, 16 May, *About Us*. Available: http://go.worldbank.org/3QT2P1GNH0 [2009, 07/01].

World Values Survey 2009, [Homepage of World Value Survey Association], [Online]. Available: http://www.worldvaluessurvey.org/ [2009, 06/29].

Yanukovich, V. 2007, 9 October, *Yanukovich Poobeshal Putinu Obespechit Stabilnost Otnoshenii s Rossiei* (*Yanukovich Promised Putin to Keep Stability in Relations with Russia*), RIA Novosti, http://www.rian.ru/politics/20071009/83192464.html.

Zerkalo nedeli 2006, 14–20 January, Kiev.

Secondary sources

Ackeret, M. 2009, 14–15 February, 'Chinesiche Milliarden für rrussisches Erdöl (Chinese Billions for Russian Oil)', *Neue Zürcher Zeitung*, Zurich.

Albright, D. 2007, 15 March, *Iran's Nuclear Program: Status and Uncertainties* (*Testimony before the Committee on Foreign Affairs, U.S. House of Representatives*) [Homepage of Institute for Science and International Security], [Online]. Available: http://www.isis-online.org/publications/iran/AlbrightTestimony15March2007.pdf [2009, 06/30].

Amsterdam, R. 2006, October 5, *The Imperial Swagger of Sovereign Democracy* [Homepage of Permalink], [Online]. Available: http://www.robertamsterdam.com/2006/10/the_imperial_swagger_of_sovere.htm [2009, 07/25].

Anderson, E.A. 2005, 'Backward, Forward, or Both?', *European Education*, vol. 37, no. 3, pp. 53–67.

Arbatov, A. 1998, 'Russian Foreign Policy: Myths and Realities', in *Inside the Russian Enigma*, ed. A.A. Nadia, Europaprogrammet, Oslo, pp. 101–35.

Aron, L. 2008, 10 September, 'Russia's Next Target Could Be Ukraine', *The Wall Street Journal*.

Åslund, A. 2007, *How Capitalism Was Built: The Transformation of Central and Eastern Europe, Russia, and Central Asia*, Cambridge University Press, Cambridge.

Asmus, R.D. 2002, *Opening NATO's Door: How the Alliance Remade Itself for a New Era*, Columbia University Press, New York.

Avioutskii, V. 2008, 10–12 April, 'The Gas War between Russia and Ukraine: A Geo-Strategic Opposition', Annual Convention, Association for the Study of Nationalities (ASN), New York.

Badiou, A. 2005, *Being and Event*, Continuum, London.

—— 2006, *Logiques des Mondes* (*Logics of Worlds*), Seuil, Paris.

Balibar, E. 2003, *We the People of Europe: Reflections on Transnational Citizenship*, Princeton University Press, Princeton.

Baran, Z. 2007, 'EU Energy Security: Time to End Russian Leverage', *The Washington Quarterly*, vol. 30, no. 4, pp. 131–44.

Barber, T. 2008, 27 October, 'Europe Split on How to Restore Ties with Moscow', *Financial Times*, London.

Barysch, K. 2007, 'Three Questions that Europe Must Ask about Russia', *CER Briefing Note*, no. 16 May.

Beck, U. and E. Grande 2005, *Das Kosmopolitische Europa* (Cosmopolitan Europe), Suhrkamp, Frankfurt am Main.

Bensmann, J. 2008, 17 November, 'Neue Wege für das Erdöl aus Kasachstan' (New Routes for Kazakh Oil) *Neue Zürcher Zeitung*, Zurich.

Bilefsky, D. 2006, 21 October, 'Putin Rejects EU Demands that Russia Ratify Energy Charter', *International Herald Tribune*, Paris.

Birdsall, N. and A. de la Torre 2001, *Washington Contentious: Economic Policies for Social Equity in Latin America*, Carnegie Endowment for International Peace and Inter-American Dialogue, Washington D.C.

Blau, P. 1964, *Exchange and Power in Social Life*, Wiley, New York.

Blommaert, J. and J. Verschueren 1998, 'The Role of Language in European Nationalist Ideologies', in *Language Ideologies: Practice and Theory*, ed. B.B. Schieffelin, K.A. Woolard and P.V. Kroskrity, Oxford University Press, New York, pp. 189–210.

Bonse, E. 2009, *Steinmeier warnt vor Eskalation in Ukraine (Steinmeier Warns of Escalation in Ukraine)*, 28 April, *Handelsblatt*, Düsseldorf.

Boratynsky, J. and A. Szymborska 2006, *Neighbours and Visas: Recommendations for a Friendly European Union Visa Policy*, Stefan Batory Foundation, Warsaw.

Bordachev, T. 2003, 'V Objatiah Civilian Power' (In the Embrace of Civilian Power), *Pro et Contra*, vol. 8, no. 1.

Bordachev, T. and A. Moshes 2004, 'Is the Europeanization of Russia Over?', *Russia in Global Affairs*, no. 2.

Bourdieu, P. 1977, *Outline of a Theory of Practice*, Cambridge University Press, Cambridge and New York.

—— 1984, *Distinction: A Social Critique of the Judgement of Taste*, Harvard University Press, Cambridge, MA.

Bova, R. (ed) 2003, *Russia and Western Civilization: Cultural and Historical Encounters*, M.E. Sharpe, Armonk, NY.

Braghiroli, S. and C. Carta 2008, *The EU Attitude Towards Russia: Condemned to be Divided?* [Homepage of CIRCAP University of Siena], [Online]. Available: www.jhubc. it/ecpr-riga [2008, November].

Brandenburg, U. 1993, 'The "Friends" Are Leaving: Soviet and Post-Soviet Troops in Germany', *Aussenpolitik (english edition)*, vol. 44, no. 1, pp. 76–88.

Bremner, C. 2008, 14 November, *Vladimir Putin Wanted to Hang Georgian President Saakashvili by the Balls*, The Times, http://www.timesonline.co.uk/tol/news/world/ europe/article5147422.ece.

Bretherton, C. and J. Vogler 1999, *The European Union as a Global Actor*, Routledge, London.

Browning, C. 2003, 'The Internal/External Security Paradox and the Reconstruction of Boundaries in the Baltic: The Case of Kaliningrad', *Alternatives*, no. 28, pp. 545–81.

Browning, C. and P. Joenniemi 2008, 'Geostrategies of the European Neighborhood Policy', *European Journal of International Relations*, vol. 14, no. 3, pp. 519–51.

Brunwasser, M. and J. Dempsey 2008, 18 January, *Russia Signs Deal to Build Gas Pipeline through Bulgaria*.

Byers, M. 'Reciprocity and the Making of International Environmental Law', [Online]. Available from: http://www.law.duke.edu/news/papers/strategyandpersuasion.pdf.

Bykov, M. 2005 (April), 'Interview with the Secretary of the Russian Security Council Igor Ivanov', *Strategia Rossii*, [Online], no. 4. Available from: http://www.fondedin. ru/sr/new/fullnews_arch_to.php?subaction=showfull&id=1114768703&archive= 1114768827&start_from=&ucat=14&.

Carothers, T. 2002, 'The End of the Transition Paradigm', *Journal of Democracy*, Vol. 13, no. 1, pp. 5–21.

Carter, C.L. 2002, Winter, *Theories of Media: Keywords Glossary. Reciprocity* [Homepage of The University of Chicago], [Online]. Available: http://humanities.uchicago.edu/faculty/mitchell/glossary/reciprocity.htm#_ftn1 [2009, 06/28].

Cash, J. 2007, 'Origins, Memory, and Identity: "Villages" and the Politics of Nationalism in the Republic of Moldova', *East European Politics and Societies*, vol. 21, pp. 588–610.

Chadaev, A. 2006, *Putin: Ego Ideologiya* (Putin: His Ideology), Evropa, Moskva.

Cherkasov, P. 1998, 'Russia: Europe or Asia?', in *Inside the Russian Enigma*, ed. A.A. Nadia, Europaprogrammet, Oslo, pp. 19–20, note 5.

Cherniaiev, A. 1993, *Shest' Let s Gorbachevym* (Six Years with Gorbachev), Progress, Moscow.

Chesnakov, A. 2008, 'O Evrope – Bez Ugryzenia Sovesti' (On Europe: Without Burdening the Conscience), *Russkiy Zhurnal*.

Chinn, J. 1997, 'Moldovans: Searching for Identity', *Problems of Post-Communism*, vol. 44, pp. 43–51.

Chirac, J. 1997, 26 September, *Speech at MGIMO, Moscow*.

Chira-Pascanut, C. and O. Schmidtke 2009, 'Contested Neighbourhood: Moldova, at the Crossroads between the EU and Russia', *Paper presented at the South East and Eastern European Countries EU Accession Quandary, International Workshop* http://web.uvic.ca/~polisci/verdun/conference/papers/Schmidtke.pdf.

Chirot, D. 1996, 'Herder's Multicultural Theory of Nationalism and Its Consequences', *Eastern European Politics and Societies*, vol. 10, no. 1, pp. 1–15.

Ciscel, M.H. 2005, 'Language and Identity in Post-Soviet Moldova', in *The Consequences of Mobility*, ed. B. Preisler et al., Roskilde University, Roskilde.

Cooke, B. and U. Kothari 2001, 'The Case for Participation as Tyranny',in *Participation: The New Tyranny?*, ed. Cooke, B. and U. Kothari, Zed Books, New York, pp. 1–15.

Cowan, G. 2008, 'Rival Partners: French and Russian Defence Co-Operation', *Jane's Defence Weekly*, no. 3 March.

Crawley, V. 2005, n.a., *U.S. Negotiating Use of Shared Military Bases in Romania, Bulgaria* [Homepage of US Info State Gov], [Online]. Available: http://www.globalsecurity.org/military/library/news/2005/11/mil-051104-usia01.htm [2005, November 4].

Crowther, W. 1998, 'Ethnic Politics and the Post-Communist Transition in Moldova', *Nationalities Papers*, vol. 26, no. 1, pp. 147–64.

Davies, N. 1996, *Europe: A History*, Oxford University Press, Oxford and New York.

—— 2000, 10 January, *Interview with John Tusa on BBC 3* [Homepage of British Broadcasting Corporation], [Online]. Available: http://www.bbc.co.uk/radio3/johntusainterview/davies_transcript.shtml [2009, 07/04].

Dawisha, K. and S. Deets 2006, 'Political Learning in Post-Communist Elections', *East European Politics and Societies*, vol. 20, no. 4, pp. 691–728.

DeBardeleben, J. (ed) 2005, *Soft or Hard Borders? Managing the Divide in an Enlarged Europe*, Ashgate, Aldershot.

Dempsey, J. 2008, *Russia Further Cuts its Oil Deliveries to Czech Republic*, 13 July.

Derrida, J. 1992, *The Other Heading: Reflections on Today's Europe*, Indiana University Press, Bloomington.

Diamond, L.J. 2002, 'Thinking about Hybrid Regimes', *Journal of Democracy*, vol. 13, no. 2, pp. 21–35.

Dugin, A. 2000, *Osnovy Geopolitiki* (*Basics of Geopolitics*), Arktogaia, Moskva.

Dvorkin, V., et al. 2007, 28 September, *Missile Defense in Europe: Dangers and Opportunities*.

Eckstein, H. 1998, 'Congruence Theory Explained', in *Can Democracy Take Root in Post-Soviet Russia?*, ed. H.e.a. Eckstein H. et al., Rowman and Littlefield, Lanham, MD, pp. 3–33.

Edelman, M. and A. Haugerud 2005b, *The Anthropology of Development and Globalization: From Classical Political Economy to Contemporary Neoliberalism*, Blackwell Publishers, Malden, MA.

—— 2005, 'Introduction: The Anthropology of Development and Globalization', in *The Anthropology of Development and Globalization*, ed. Edelman, M. and A. Haugerud, Blackwell Publishers, Inc, Malden, MA, pp. 1–74.

Eke, S. 2009, 15 May, *Russia Signs Gas Pipeline Deals* [Homepage of BBC News], [Online]. Available: http://news.bbc.co.uk/2/hi/europe/8051921.stm#map [2009, 06/23].

Ekiert, G., J. Kubik and M.A. Vachudova 2008, 'Democracy in the Post-Communist World: An Unending Quest?', *East European Politics and Societies*, vol. 21, no. 1, pp. 7–30.

Elbe, S. 2003, *Europe: A Nietzschean Perspective*, Routledge, London.

Elenski, O. 2006, 28 July–3 August, 'Novaya Voina po Staromy Stzenariyu (A New War Based on an Old Scenario)', *Nezavisimoye Voennoye Obozrenie*.

Engelbrekt, K. and J. Hallenberg (eds) 2008, *The European Union and Strategy: An Emerging Actor*, Routledge, London and New York.

Escobar, A. 1995, *Encountering Development: The Making and Unmaking of the Third World*, Princeton University Press, Princeton.

Evans, P. 2005, 'The Challenges of the Institutional Turn: New Interdisciplinary Opportunities in Development Theory', in *The Economic Sociology of Capitalism*, ed. Nee, V. and R. Swedberg, Princeton University Press, Princeton, pp. 90–106.

Facon, I. 2008a, 'The West and Post-Putin Russia: Does Russia "Leave the West"', *Notes de la Fondation Pour la Recherche Stratégique/Bertelsmann Stiftung*, no. 10.

—— 2008b, *Author's Interviews in Moscow*, December.

Fairclough, N. 2001, *Language and Power*, 2nd edn, Longman, London and New York.

Felgengauer, P. 2006, *Ob Agentah Zabili (Spies are Forgotten)*, 9 October, Novaya gazeta, http://www.novayagazeta.ru/data/2006/77/15.htm.

Ferguson, J. 1990, *The Anti-Politics Machine: 'Development,' Depoliticization, and Bureaucratic Power in Lesotho*, Cambridge University Press, Cambridge and New York.

Ferguson, J. and A. Gupta 2002, 'Spatializing States: Toward an Ethnography of Neoliberal Governmentality', *American Ethnologist*, vol. 29, no. 4, pp. 981–1002.

Ferrero-Waldner, B. 2009, 9 March, *After the Russia/Ukraine Gas Crisis: What Next? Speech in Chatham House, London* [Homepage of European Commission], [Online]. Available: http://europa.eu/rapid/pressRelease reference=SPEECH/09/100.

Fischer, Peter A. (2009) 'Chinesische Milliarden für russisches Erdöl' (Chinese Billions for Russian Oil), *Neue Zürcher Zeitung*, 19 February 2009.

Fitzpatrick, S. (1999) *Everyday Stalinism: Ordinary Life in Extraordinary Times: Soviet Life in the 1930s*. New York: Oxford University Press.

Fogarty, P. forthcoming, ' "It's Your Project": Examining the Intersections of Development Discourse, Local Culture, and an Emerging National Identity in Moldova', in *untitled*, ed. R. Mandel.

Fogarty, P. (forthcoming) *Building Moldova, Being Moldovan: Discourses and Practices of Identity, Development, and Citizenship*. Unpublished dissertation. Emory University.

Foucault, M. 1972, *Archaeology of Knowledge; and, The Discourse On Language*, Pantheon Books, New York.

Foucault, M. and C. Gordon 1980, *Power/Knowledge: Selected Interviews and Other Writings, 1972–1977*, Pantheon Books, New York.

Franzinetti, G. 2008, 'Mitteleuropa in East–Central Europe: From Helsinki to EU Accession', *European Journal of Social Theory*, vol. 11, no. 2, pp. 219–35.

Fredholm, M. 2005, *The Russian Energy Strategy & Energy Policy: Pipeline Diplomacy or Mutual Dependence?*, Conflict Studies Research Centre.

Fukuyama, F. 1996, *Trust: The Social Virtues and the Creation of Prosperity*, Free Press, New York.

Furman, D. 2008, 'Neupravlyaemii korabl (A Ship Out of Control)', 10 September, *Nezavisimaya gazeta*.

Fuster, T. 2008, 'Polen Hofft auf Stimmungswandel in Russland (Poland Hopes of Change of Attitude in Russia)', 7 February, *Neue Zürcher Zeitung*, Zurich.

Gal, S. and G. Kligman 2000, *The Politics of Gender after Socialism: A Comparative-Historical Essay*, Princeton University Press, Princeton.

Garnett, S.W. 1997, *Keystone in the Arch: Ukraine in the Emerging Security Environment of Central and Eastern Europe*, Carnegie Endwoement for International Peace, Washington D.C.

Gärtner, E. 2008, 21 October, *Kalter Krieg um Energie in Osteuropa (Cold War about Energy in Eastern Europe)*.

Geertz, C. 1994, 'Primordial and Civic Ties', in *Nationalism*, ed. Hutchinson, J. and A.D. Smith, Oxford University Press, New York and Oxford, pp. 29–34.

Goble, P. 2008, 1 November, *Putin, Popular Indifference Subverted Key Provisions of 1993 Constitution, Drafters Says* [Homepage of Window on Eurasia], [Online]. Available: http://windowoneurasia.blogspot.com/2008/11/window-on-eurasia-putin-popular.html [2009, 07/28].

Goldman, M.I. 2008, *Petrostate: Putin, Power and the New Russia*, Oxford University Press, Oxford.

Goldstein, J.S. 2008, 'Core Principles of International Relations Theory', in *International Relations*, ed. Goldstein, J.S. and J.C. Pevehouse, 8th, Longman, Harlow, Essex.

Gomart, T. 2007, 'Paris et le Dialogue UE-Russie: Nouvel élan avec Nicolas Sarkozy? (Paris and the EU–Russia Dialogue: A Fresh Start with Nicolas Sarkozy?)', *Russie.Nei. Visions*, no. 23.

Gorbachev, M. 1987, *Perestroika: New Thinking for Our Country and the World*, Harper and Row, New York.

—— 1995, *Zhizn i reformy* (Life and Reforms), Novosti, Moscow.

Gorst, I. 2004, *The Energy Dimension in Russian Global Strategy – Russian Pipeline Strategies: Business versus Politics*, The Baker Institute Energy Forum, Rice University, http://www.rice.edu/energy/publications/docs/PEC_Gorst_10_2004.pdf.

Gouldner, A.W. 1960, 'The Norm of Reciprocity: A Preliminary Statement', *American Sociological Review*, vol. 25, pp. 161–78.

Gress, D. 1998, *From Plato to NATO: the Idea of the West and Its Opponents*, Free Press, New York.

Grier, P.T. 2003, 'The Russian Idea and the West', in *Russia and Western Civilization: Cultural and Historical Encounters*, ed. R. Bova, M.E. Sharpe, Armonk, NY, pp. 23–77.

Grillo, R.D. 1997, 'Discourses of Development: The View from Anthropology', in *Discourses of Development: Anthropological Perspectives*, ed. Grillo D. and R.L. Stirrat, Oxford University Press, New York, pp. 1–34.

Groves, S. 2007, 28 November, *Advancing Freedom in Russia. Backgrounder #2088* [Homepage of Heritage Foundation], [Online]. Available: http://www.heritage.org/research/worldwidefreedom/bg2088.cfm [2009, 06/29].

Gupta, A. 1995, 'Blurred Boundaries: The Discourse of Corruption, the Culture of Politics, and the Imagined State', *American Ethnologist*, vol. 22, no. 2, pp. 375–402.

Hadfield, A. 2008, 'Energy and Foreign Policy: EU–Russia Energy Dynamics', in *Foreign Policy: Theories, Actors, Cases*, ed. Smith, S., A. Hadfield and T. Dunne, Oxford University Press, Oxford, pp. 321–38.

Hale, H.E. 2004, 'The Origins of United Russia and the Putin Presidency: The Role of Contingency in Party-System Development', *Demokratizatsiya*, vol. 12, no. 2.

Hanson, S.E. 2008, 'The Uncertain Future of Russia's Week State Authoritarianism', *East European Politics and Societies*, vol. 21, no. 1, pp. 67–81.

Haran, O. and R. Pavlenko 2003, 'Political Reform or a Game of Survival for President Kuchma?', *PONARS Policy Memo*, [Online], no. 294 (November). Available from: https://gushare.georgetown.edu/eurasianstrategy/Memos/2003/pm_0294.pdf.

Haran, O. and S. Tolstov 2002, 'Ukraine's Relations with Russia and Belarus: A Ukrainian View', in *The Slavic Triangle*, ed. Moshes A. and B. Nygren, National Defence College, Stockholm.

Harding, L. 2008, 'Russia: Any Country Could Be Next, Warns Ukrainian President', 28 August, *The Guardian*, London.

Hartlyn, J., J. McCoy and T.M. Mustillo 2008, 'Electoral Governance Matters. Explaining the Quality of Elections in Contemporary Latin America', *Comparative Politics Studies*, vol. 41, no. 1, pp. 73–98.

Hayoz, N. and A. Lushnycky (eds.). *Ukraine at a Crossroads*. Berlin: Peter Lang, 2005.

Hosp, G. 2008, 1 July, *Licht aus Für das Russische Strom-Monopol* (*Lights Out for the Russian Electricity Monopoly*), *Neue Zürcher Zeitung*, Zurich.

Hrytsenko, A. 2007, 15 June, *Ninysnii uriad Viktora Yanukovicha Ukhvalnoe Risheniia Shodo Realizatsii Planu din Ukraina – NATO Nabit Shvidshe za Poperedni 'Pomaranchevi' Uriadi* (*The Present Cabinet of Viktor Yanukovich Takes Decisions on Implementing Ukraine-NATO Plan Even Quicker Than the Previous Orange Cabinet*), Radio Svoboda, http://www.radiosvoboda.org/content/news/965318.html.

Humphrey, C. 1983, *Karl Marx Collective: Economy, Society and Religion in a Siberian Collective Farm*, Press Syndicate of the University of Cambridge, New York.

Huntington, S.P. 1996, *The Clash of Civilizations and the Remaking of World Order*, Simon and Schuster, New York.

Inglehart, R. and C. Welzel 2005, *Modernization, Cultural Change, and Democracy*, Cambridge University Press, New York.

Irvine, J.T. and S. Gal 2000, 'Language Ideology and Linguistic Differentiation', in *Regimes of Language: Ideologies, Polities, and Identities*, ed. P.V. Krosrity, School of American Research Press, Santa Fe, NM, pp. 35–84.

Kagan, R. 2008, 11 August, 'Putin Makes His Move', *The Washington Post*.

—— 2003, *Of Paradise and Power: America and Europe in the New World Order*, Knopf Publishers.

Kaneff, D. and M. Heintz 2006, 'Introduction: Bessarabian Borderlands, One Region, Two States, Multiple Ethnicities', *The Anthropology of East Europe review: Central Europe, Eastern Europe and Eurasia*, vol. 24, no. 1, pp. 6–16.

Karaganov, S. 2009, 'The Magic Numbers of 2009', *Russia in Global Affairs*, [Online], no. 2. Available from: http://eng.karaganov.ru/articles/204.html. [06/27/2009].

Kennedy, R. 2007, 'Trains, Trade and Transnistria: Russian Influence in Moldova', Annual Convention, 12–14 April, Association for the Study of Nationalities (ASN, http://polisci. osu.edu/grads/kennedy/Russia%20Pressurepoints.pdf.

Keohane, R.O. 1984, *After Hegemony: Cooperation and Discord in the World Political Economy*, Princeton University Press, Princeton.

—— 1986, 'Reciprocity in International Relations', *International Organization*, vol. 40, no. 1, Winter, pp. 1–27.

Keohane, R.O. and J.V. Nye 2001, *Power and Interdependence*, 3rd edn., Longman, New York.

Khristenko, V.B. 2004, 10 November, *Doklad na Shestom Obschem Sobranii Kruglogo stola Promyshlennikov Rossii i Evropeiskogo Souza (Speech at Sixth General Session of Round Table of Industrialists of Russia and the European Union)* The Hague .

Kimmage, D. 2005, *Endnote: Kyrgyztan Highlights Failure of Managed Democracy*, 13 April, Radio Free Europe/Radio Liberty, http://www.rferl.org.

King, C. 1995, *Post-Soviet Moldova: A Borderland in Transition*, Royal Institute of International Affairs/Russian and CIS Programme, London.

—— 2000, *The Moldovans: Romania, Russia, and the Politics of Culture*, Hoover Institution Press, Stanford, CA.

Kohn, H. 1953, *Pan-Slavism*, Vintage Books, New York.

Kolstø, P. 2000, 'Interstate Integration in the Post-Soviet Space: The Role of the Russian Diasporas', in *Building Security in the New States of Eurasia: Subregional Cooperation in the Former Soviet Space*, ed. Dwan, R. and O. Pavliuk, M.E. Sharpe, Armonk, NY.

Konnander, V. 2008, *Ryssland. En Suverän Demokrati? En Studie om Folkstyre och Statsmakt i Förfall (Russia – A Sovereign Democracy? A Study of Popular Rule and State Power in Decline*, FOI, Stockholm.

Korshak, S. 2009, 'Moldovan Wine Needs More Drinkers than Queen Elizabeth', *TopNews.in.*, [Online], no. 6 April, pp. 2009–06/23. Available from: http://www. topnews.in/moldovan-wine-needs-more-drinkers-queen-elizabeth-2147731.

Kozlov, V., et al 2009, 4 May, 'Zhyzn bez sprosa (Life without Demand), in part citing statistics from *Vedemosti*, ', *Expert*, no. 7, pp. 19–20.

Kramer, A.E. 2008, 12 July, *Czechs See Oil Flow Fall and Suspect Russian Ire on Missile System.*

—— 2009, 13 March, *Gazprom and Ukraine Work Out Details of Deal to Keep Russian Natural Gas Flowing.*

Krastev, I. 2006, 16 November, *Sovereign Democracy: Russian Style* [Homepage of open Democracy], [Online]. Available: http://www.opendemocracy.net/globalization-institutions_government/sovereign_democracy_4104.jsp [2009, 06/29].

—— 2007, 14 September, *Ideological Clashes between Russia and Europe* [Homepage of EurActiv/Open Democracy], [Online]. Available: http://www.euractiv.com/en/opinion/ideological-clashes-russia-europe/article-166723 [2009, 06/29].

Krutihin, M. 2009, 25 March, *Dareniy Kon: Evropeiskie Predlozheniya po Gazomu Tranzitu ne Utraivaiot Gazprom (Gift Horse: European Proposals Regarding Gas Transit Are Unacceptable to Gazprom)* [Homepage of RusEnergy], [Online]. Available: http://www.rusenergy.com/?page=articles&id=1022 [2009, 06/29].

Kulish, N. 2008, 'Germany Acts to Guide the West's Ties to Russia', 5 December, *New York Times*, www.nytimes.com/2008/11/06/world/europe/06pirate.html?partner=rssnyt&emc=rss.

Kundera, M. 1984, 'The Tragedy of Central Europe', *New York Review of Books*, vol. 31, no. 7, pp. 33–38.

Kuznetsov, A. 2008, 25 June, 'Propalo Sal'do' (The Balance Disappeared), *Rossiyskaya Gazeta*.

Lacan, J. 2001, *Ecrits: A Selection*, Routledge, London.

Laclau, E. 2005, *On Populist Reason*, Verso, London.

Lahille, E. 2007, 'Commerce Mondial des Armes: Bons Baisers de Russie' (World Arms Trade: Lots of Love from Russia)', *Le débat stratégique*, vol. 90, no. April.

Laqueur, W. 1990, *Russia and Germany*, Transaction Publishers, New Brunswick.

Larsson, R. 2007, *Nord Stream, Sweden and Baltic Sea Security*, Swedish National Defence Establishment.

Latynina, Y. 2008, 31 May [Homepage of Ekho Moskvy], [Online]. Available: http://www.echo.msk.ru/programs/code/517706-echo.html.

Le Coq, C. and E. Paltseva 2008, *Common Energy Policy in the EU: The Moral Hazard of the Security of External Supply. Report No. 1*, Swedish Institute for European Policy Studies, Stockholm.

Legvold, R. and C.A. Wallander (eds) 2004, *Swords and Sustenance: The Economics of Security in Belorus and Ukraine*, The MIT Press, Cambridge, MA.

Leonard, M. and N. Popescu 2007, *A Power Audit of EU–Russia Relations*, European Council on Foreign Relations, Brussels.

Leys, C. 2005, 'The Rise and Fall of Development Theory', in *The Anthropology of Development and Globalization*, ed. Edelman, M. and A. Haugerud, Blackwell Publishers, Malden, MA, pp. 105–25.

Linz, J. and A. Stepan 1996, *Problems of Democratic Transition and Consolidation*, Johns Hopkins, Baltimore, Md.

Little, A. 2008, March 4, *Russia Redefines Democracy* [Homepage of BBC News], [Online]. Available: http://news.bbc.co.uk/2/hi/europe/7274798.stm [2009, 06/29].

Lobjakas, A. 2004, 10 February, *EU: Brussels Prepares to Sharpen Policy in Russia*, Radio Free Europe, http://www.rferl.org/content/article/1051496.html.

——2007, 17 October, *Russia Weighs In On EU's Tough New Energy Policy* [Homepage of Radio Free Europe], [Online]. Available: http://www.rferl.org/content/article/1078971.html [2009, 07/29].

Lucas, E. 2008, *The New Cold War. Putin's Russia and the Threat to the West*, Houndmills: Palgrave Macmillan, New York.

Luhmann, N. 1995, *Social Systems*, Stanford University Press, Stanford.

Lukin, A. 2004a, 'Authoritarianism Deposing "Clan Democracy"', *Moscow Times*.

—— 2004b, *A Short History of Russian Elections' Short Life*, Moscow Times.

Lukyanov, F. 2008, 'Russia-EU: The Partnership that Went Astray', *Europe–Asia Studies*, vol. 60, no. 6, pp. 1107–19.

Lunghescu, O. 2009, 'Deal to Boost Key EU Gas Project', 8 May, *BBC News*, http://news.bbc.co.uk/2/hi/europe/8039587.stm.

Lunts, L.A. 1984, *Mezhdunarodnoe Chastnoe Pravo* (*International Private Law*), Yuridicheskaia literatura, Moscow.

Magun, A. 2006, 1 December, 'Suverennaya Demokratia ili Otchayanny Konservatizm (Sovereign Democracy or Conservatism of Despair?)', *Russkiy Zhurnal*, [Online], 2007. Available from: http://russ.ru/politics/docs/suverennaya_demokratiya_ili_otchayannyj_konservatizm.

Malfliet, K., L. Verpoest and E. Vinokurov (eds) 2007, *The CIS, The EU and Russia: Challenges of Integration*, Palgrave Macmillan, New York.

Mandelson, P. 2008, 'Russia and the EU: Building Trust on a Shared Continent. Conference paper', *Russia in the 21st Century*, Moscow.

Manners, I. 2002, 'Normative Power Europe: A Contradiction in Terms?', *Journal of Common Market Studies*, vol. 40, no. 2, pp. 235–58.

Margolina, S. 2008, 30 January, 'Gazprom und die Macht im Kreml (Gazprom and the Power of the Kremlin)', *Neue Zürcher Zeitung*, Zurich.

Mariton, H. 2009, 9 April, 'Quelle Coopération Entre la Russie et la France dans le Contexte de la Crise Financière et Économique Mondiale ? (What Cooperation between Russia and France in the Context of the World Financial and Economic Crisis?)', *News Press*.

McAllister, I. and S. White 2008, 'Voting 'Against All' in Postcommunist Russia', *Europe–Asia Studies*, vol. 60, no. 1, pp. 67–87.

McGregor, C. 2004, 13 February, 'Putin Kicks off Re-Election Drive', *Moscow Times*.

Meister, U. 2008, 1 September, 'Brown Fordert Gemeinsame EU-Energiepolitik (Brown Demands Common EU Energy Policy)', *Neue Zürcher Zeitung*, Zurich.

Melikova, N. 2005, 13 June, 'Suverenitet Vazhnee Demokrati (Sovereignty Is More Important Than Democracy)', *Nezavisimaya Gazeta*, Moskva.

Mezhuev, B. 2007, 'Razvod s Evropoi (Divorcing Europe)', *Russkiy Proekt*, [Online], no. 1 December. Available from: http://rus-proekt.ru/power/2505.print.

Mojsilovic, J. and B.K. Vukajlovic 2008, 28 January, *Politics Spurs Serbia-Russia Energy Deal* [Homepage of B92], [Online]. Available: www.b92.net/eng/insight/opinions.php? yyyy=2008&mm=01&nav_id=47204-.

Monaghan, A. and L. Montanaro-Jankovski 2006, *EU–Russia Energy Relations: The Need for Active Engagement. EPC Issue Paper No. 45*, European Policy Centre, http:// www.epc.eu/TEWN/pdf/89495137_EPC%20Issue%20Paper%2045%20EU-Russia%20energy%20relations.pdf.

Morozov, V. 2003, 'V Poiskah Evropy: Rossiysky Politichesky Diskurs i Okruzhajushy mir (Searching for Europe: Russian Political Discourse and and the Surrounding World)', *Neprikosnovenny Zapas*, vol. 4, no. 30.

—— 2008, 'Sovereignty and Democracy in the Contemporary Russia: A Modern Subject Faces the Post-Modern World', *Journal of International Relations and Development*, vol. 11, no. 2, pp. 152–80.

Moshes A. The Next President of Ukraine: Predicting the Unpredictable, *PONARS Policy Memo* No. 265, October 1, 2002. — Centre for Strategic and International Studies. Available: http://csis.org/files/media/csis/pubs/pm_0265.pdf

Moslakin, A. 2008, 6 June, 'Prezident Rossii Priekhal Ocharovyat' Nemtsev (Russian President Arrives to Charm Germans)', *Biznes i Baltiia*, http://dlib.eastview.com/sources/article.jsp?id=17572991.

Neumann, I. 1996, *Russia and the Idea of Europe*, Routledge, London.

—— 1998, *The Uses of the Other: The 'East' in European Identity Formation*, University of Minnesota Press, Minneapolis.

Nietzsche, F. 1998, *On the Genealogy of Morals*, Oxford Paperbacks, Oxford.

Nugent, D. 1997, *Modernity at the Edge of Empire: State, Individual, and Nation in the Northern Peruvian Andes, 1885–1935*, Stanford University Press, Stanford, CA.

Nygren, B. 2008c, *The Rebuilding of Greater Russia: Putin's Foreign Policy Towards the CIS Countries*, Routledge, London and New York.

—— 2008a, 'The EU's Democratic Norm Project for Eurasia: Will the Beauty Tame the Beast?', in *The European Union and Strategy: An Emerging Actor*, ed. Engelbrekt, K. and J. Hallenberg, Routledge, London and New York.

—— 2008b, 'Putin's Use of Natural Gas to Reintegrate the CIS Region', *Problems of Post-Communism*, vol. 55, no. 4, pp. 3–15.

Okara, A. 2007, 'Sovereign Democracy: A New Russian Idea or a PR Project?', *Russia in Global Affairs*, vol. July-September, no. 2.

Opalski, M. 2001, 'Can Will Kymlicka Be Exported to Russia?', in *Can Liberalism Be Exported? Western Political Theory and Ethnic Relations in Eastern Europe*, ed. Kymlicka, W. and M. Opalski, Oxford University Press, Oxford, pp. 298–319.

Ostrom, E. 1990, *Governing the Commons: The Evolution of Institutions for Collective Action*, Cambridge University Press, Cambridge.

Pain, E. and O. Volkogonova 2008, *Rossiiskaya Modernizatsia: razmyshlyaya o samobytnosti. (Russian Modeniation: Thinking about Distinctveness)* Moscow.

Parfitt, T. 2008, 13 August, 'Russia Exaggerating South Ossetian Death Toll, Says Human Rights Group', *The Guardian*, http://www.guardian.co.uk/world/2008/aug/13/georgia.

Parmentier, F. 2003, *La Moldavie à la croisée des chemins* (*Moldova at the Crossroads*), Editoo.com, Paris.

Petro, N.N. 1995, *The Rebirth of Russian Democracy*, Hardvard University Press, Cambridge, MA.

Polinin, A. 2004, 27 March, 'IMF Continues Not to Give Money to Moldova', *Moldavski Vedomosti*.

Popescu, N. 2005, *The EU and Transnistria: From Deadlock to Sustainable Settlement. IPF Policy Brief* (*final*), Central European University/Open Society Institute, http://www.policy.hu/npopescu/ipf%20info/IPF%201%20transnistria.pdf.

Preston, P.W. 1996, *Development Theory: An Introduction*, Blackwell Publishers, Cambridge, MA.

Primakov Ye. Mir bez Rosii? (The World Without Russia), *Rossijskaja gazeta* 21 January 2009. Available: http://www.rg.ru/2009/01/21/primakov-kniga.html.

Proedrou, F. 2007, 'The EU–Russia Energy Approach under the Prism of Interdependence', *European Security*, vol. 16, no. 3–4, pp. 329–55.

Proissl, W., C. Hecking and H. Wetzel 2009, 19 February, *Dossier Krampfader Ostseepipeline (The Varix File: The Status of the Baltic Sea Pipeline)*, Financial Times Deutschland.

Prozorov, S. 2004, *Political Pedagogy of Technical Assistance: A Study in Historical Ontology of Russian Postcommunism.* Studia Politica Tamperensis, Tampere.

—— 2006, *Understanding Conflict between Russia and the EU: The Limits of Integration*, Palgrave, Basingstoke.

—— 2007, 'The Narratives of Exclusion and Self-Exclusion in the Russian Conflict Discourse on EU–Russian Relations', *Political Geography*, vol. 26, no. 3, pp. 309–29.

—— 2008, 'Belonging and Inclusion in European-Russian Relations: Alain Badiou and the Truth of Europe', *Journal of International Relations and Development*, vol. 11, no. 2, pp. 181–207.

—— 2009, 'In and Out of Europe: Identity Politics in EU–Russian Relations', in *Identity and Foreign Policy: Baltic-Russian Relations and European Integration*, ed. Berg, E. and P. Ehin, Aldershot, Ashgate.

Przeworski, A. 1991, *Democracy and the Market: Political and Economic Reforms in Eastern Europe and Latin America*, Cambridge University Press, Cambridge.

Puddington, A. 2008, 27 March, *By the People: Democracy Isn't Everyone's Idea of Freedom*, Weekly Standard, Washington D.C.

Rahr, A. 2000, *Wladimir Putin: der "Deutsche" im Kreml* (*Vladimir Putin: "The German" in the Kremlin*), Universitäts Verlag, Munich.

Rajmaira, S. and M.D. Ward 1990, 'Evolving Foreign Policy Norms: Reciprocity in the Superpower Triad', *International Studies Quarterly*, vol. 34, no. 4, pp. 457–75.

Remington, T. 1999, *Politics in Russia*, Longman, New York.

Remizov, M. 2001, 1 December, 'Pohitim Evropu! (Let's Kidnap Europe!)', *Russkiy Zhurnal*.

—— 2002a, *Opyt Konservativnoi Kritiki (An Attempt at Conservative Criticism)*, Pragmatika Kultury, Moskva.

—— 2002b, 1 December, 'V Poiskah Yazyka Voiny (In Search of a Language of War)', *Russkiy Zhurnal*.

—— 2003, 4 April, 'Budushee odnoy sekty (The Future of A Sect)', *Russkiy Zhurnal*.

Rhodes, C. 1989, 'Reciprocity in Trade: The Utility of a Bargaining Strategy', *International Organization*, vol. 43, no. 2, pp. 273–99.

Riabchuk, M. 2002, 'Ukraine: One State, Two Countries?', *Tr@nsit online*, [Online], no. 23, pp. Available: http://www.eurozine.com/articles/2002-09-16-riabchuk-en.html.

Rieber, A.J. 2007, 'How Persistent Are Persistent Factors?', in *Russian Foreign Policy in the 21st Century and the Shadow of the Past*, ed. L. Robert, Columbia University Press, New York, pp. 205–78.

Röller, L.H., J. Delgado and H.W. Friederiszick 2007, *Energy: Choices for Europe*, Bruegel Blueprint Series, Brussels.

Romanova, T. 2007, 'Energy Partnership: A Dialogue in Different Languages', *Russia in Global Affairs*, vol. 5, no. 1.

—— 2009, 'The Political Economy of EU–Russian Energy Relations', in *Political Economy of Energy in Europe: Forces of Fragmentation and Integration*, ed. G. Fermann, Berliner Wissenschafts-Verlag, Berlin.

Romer, J. 2000, 'Les Relations Franco-Russes (1991–99), Entre Symboles et Réalités' (French–Russian Relations (1991–99), Between Symbols and Realities)', *Annuaire Français de Relations Internationales*, pp. 439–47.

Rose, R. 2009, *Understanding Post-Communist Transformation: A Bottom Up Approach*, Routledge, London and New York.

Rose-Ackerman, S. 2007, 'From Elections to Democracy in Central Europe: Public Participation and the Role of Civil Society', *East European Politics and Societies*, vol. 21, no. 1, pp. 31–47.

Rubchenko, M. and M. Talyskaya 2009, 'Vovremya Podderzhat (Timely Support)', *Expert*, no. 10, pp. 58.

Rubinskiy, Y. 2008, 'Moscow – Paris: The Dialogue Goes On', *International Affairs (Moscow)*, vol. 54, no. 1.

Rumer, E.B. 2007, 'Russian Foreign Policy beyond Putin', *Adelphi Paper*, no. 390.

Rutland, P. 2004, 'Transitions Online', *JRL*, vol. 8120, no. 9, March 16.

Ryabov, A. 2008, 'Tandemocracy in Today's Russia', *Russian Analytical Digest*, [Online], vol. 49, no. 5 November. Available from: http://www.res.ethz.ch/.

Sahlins, M. 1972, *Stone Age Economics*, Aldine-Atherton, Chicago.

Sakwa, R. 2005, 'The 2003–4 Elections and Prospects for Democracy', *Europe–Asia Studies*, vol. 57, no. 3, pp. 369–98.

—— 2008, *Putin: Russia's Choice*, Routledge, New York.

Schedler, A. 2006, *Electoral Authoritarianism: The Dynamics of Unfree Competition*, Lynner Rienner, Boulder, CO.

Schmid, U. 2008, *Gyurcsany und Orban Kreuzen die Klingen (Gyurcsany and Orban Cross Swords)*, 7 March, *Neue Zürcher Zeitung*.

Schmitt, C. 2003, *The Nomos of the Earth in the International Public Law of the Jus Publicum Europaeum*, Telos Press, New York.

Schroeder, G. 2001, 15 December, *Die Zeit*, Hamburg.

—— 2002, 'ARD TV Channel Interview with Gerhard Schröder', 23 November, *Hamburger Abendblatt*.

—— 2007, 'Message to Europe from Moscow', *Eurasia Daily Monitor*, [Online], vol. 4, no. 168, pp. 12 September.

Schroeder, H. (ed) 2010, *Länderbericht Russland* (*Country Report Russia*), Bundeszentrale für politische Bildung, Bonn.

Scott, J.C. 1998, *Seeing Like a State: How Certain Schemes to Improve the Human Condition Have Failed*, Yale University Press, New Haven, CN.

Seregin, A. et al. 2006, 'Russia Profile Experts Panel: The Debate on Sovereign Democracy,' *Johnson's Russia List*, September 22. Available: http://www.cdi.org/russia/johnson/2006-212-37.cfm.

Sergeyev, V. and N. Biryukov 1993, *Russia's Road to Democracy: Parliament, Communism, and Traditional Culture*, Elgar, Brookfield, VRT.

Sestanovich, S. 2008, 'What Has Moscow Done? Rebuilding U.S.-Russian Relations', *Foreign Affairs*, vol. 87, no. 6 (November/December), pp. 12–28.

Sherr, J. 2008 (October), 'Russia and Georgia: A Dangerous Game', *The World Today*, vol. 64, no. 10.

Shevstova, L. 2004, 'Russia's Electoral Time Bomb', 1 March, *Moscow Times*, Moscow.

—— 2007, *Russia – lost in transition. The Yeltsin and Putin legacies*, Carnegie Endowment of International Peace, Washington D.C.

—— 2008, 'Moscow's Domestic Policy: Russian Roulette', *The World Today*, vol. 64, no. 10.

Shlykov, V. 2005, 24 January, 'Russian Defense Economy and Structural Militarization', *Ezhednevny zhurnal*.

Shweder, R.A. 2001, 'Cultural Psychology', in *MIT Encyclopedia of the Cognitive Sciences*, Keil, F.C. and R.A. Wilson, MIT Press, Cambridge, MA.

Simirenko, A. 1982, *Professionalization of Soviet Society*, Transaction Publishers, New Brunswick, NJ.

Simons, T.W.J. 2008, *Eurasia's New Frontiers: Young States, Old Societies, Open Futures*, Cornell University Press, Ithaca.

Singhofen, S.C. 2007, *Deutschland und Russland Zwischen Strategische Partnerschaft und Neue Konkurrenz* (*Germany and Russia between Strategic Partnership and Renewed Competition*), Konrad Adenauer Stiftung, Berlin.

Sjursen, H. (ed) 2006, *Questioning EU Enlargement: Europe in Search of Identity*, Routledge, Oxon.

Skinner, R. 2006, June, *Strategies for Greater Energy Security and Resource Security* [Homepage of Oxford Institute for Energy Studies], [Online]. Available: http://www.oxfordenergy.org/presentations/BANFF_June_06-1.pdf#search=%22energy%20security%22 [2008, 11/15].

Socor, V. 2007, 'Gazprom Touts Agreement with Moldova as "Models" [Electronic version].' *Eurasia Daily Monitor*, 3. Retrieved March 1, 2008, from http://www.jamestown.org.

Socor, V. 2009, 'Russia Seeks "Guaranteed" Control over Moldova's Transnistria', *Eurasia Daily Monitor*, [Online], vol. 26 March, no. 2008, pp. 03/01. Available from: http://www.jamestown.org/edm/article.php?article_id=2371765.

—— 2009, 'Sochi Agreements and Aftermath Deflate South Stream Hype', *Eurasia Daily Monitor*, [Online], vol. 6, no. 101, pp. 20 June. Available from: http://www.jamestown. org/single/?no_cache=1&tx_ttnews%5Btt_news%5D=35043.

Sokolov, S. 2007, 'Rossiya-ES: sammit s osadkom', *Russkiy Zhurnal*, May http://russ.ru/ politics/docs/rossiya_es_sammit_s_osadkom [2009, 05/15].

Solovev, V. 2008, 14 April, *V stranakh SNG: obyavlena mobilizatsiia po vsem frontam* (*In the CIS Countries: Mobilization on All Fronts Has Been Declared*), Nezavisimaia Gazeta, http://www.ng.ru.

Spitulnik, D. 1998, 'Mediated Modernities: Encounters with the Electronic in Zambia', *Visual Anthropological Review*, vol. 14, no. 2, pp. 63–84.

Stent, A. 1981, *From Embargo to Ostpolitik: the Political Economy of West German–Soviet Relations, 1955–1980*, Cambridge University Press, Cambridge.

—— 1999, *Russia and Germany Reborn: Unification, the Soviet Collapse and the New Europe*, Princeton University Press, Princeton.

—— 2007, 'Reluctant Europeans: Three Centuries of Russian Ambivalence towards the West', in *Russian Foreign policy in the Twenty-first Century in the Shadow of the Past*, ed. R. Levgold, Columbia University Press, New York, pp. 393–442.

Stern, J. 2006, *The New Security Environment for European Gas: Worsening Geopolitics and Increasing Global Competition for LNG*, Oxford Institute for Energy Studies, Oxford.

Streltsova, Y. 2007, 'Nikolaï Sarkozy. Perspektivy Sotroudnitchestva Rossii-Frantsii-ES (Nicolas Sarkozy. Prospects for Russia-France-EU Cooperation)', *Mirovaya Ekonomika and Mezhdunarodnye Otnosheniya*, no. 11.

Stubb, A. 2008, *The First Post-080808 Diagnosis: Speech by the Minister of Foreign Affairs at the Annual Meeting of Heads of Missions, Helsinki 25 August 2008* [Homepage of Finnish Ministry of Foreign Affairs], [Online]. Available: http://formin.finland.fi [2008, 10/28].

Sutela, P. 2007, *Reciprocity in EU–Russia Relations. Evidence prepared for the Sub-Committee on Foreign Affairs, Defence and Development Policy of the House of Lords Select Committee on the European Union*.

Szporluk, R. 2007, 'The Western Dimension of the Making of Modern Ukraine.' Available: http://www.eurozine.com/articles/2005-07-22-szporluk-en.html.

Toynbee, A.J. 1948, *Civilization on Trial*, Oxford University Press, New York.

Traynor, I. and T. Harding 2007, 18 May, 'Eastern European Woes Ruin Merkel's Grand Plans for EU Alliance with Russia', *The Guardian*, London.

Trenin, D. 2005, 'Vneshnee Vmeshatelstvo v Sobytia na Ukraine i Rossiysko-Zapadnye Otnoshenia (Foreign Interference in Ukrainian Affairs and Russian–Western Relations)', *Moscow Carnegie Centre Briefings*, vol. 7, no. 2.

Trofimenko, H. 1999, *Russian National Interests and the Current Crisis in Russia*, Ashgate, Aldershot.

Tullock, G. 1993, *Rent Seeking. The Shaftesbury Papers, 2nd Volume*, Edward Elgar, Aldershot.

Tzermias, N. 2006, 19/20 January, 'Russland Stärkt Seinen Einfluss im Balkan (Russia Strengthens Its Influence in the Balkans)', *Neue Zürcher Zeitung*, Zurich.

Tziganok, A. 2007, 'The Eurasian Powder Keg, 8 February', *Nezavisimoye Voennoe Obozrenie*, http://nvo.ng.ru.

Valenta, J. 1991, *Soviet intervention in Czechoslovakia, 1968. Anatomy of a Decision*, 2nd, revised, The Johns Hopkins University Press, Baltimore and London.

van Bladel, J. 2008, *The Dual Structure and Mentality of Vladimir Putin's Power Coalition. User Report*, FOI-R-2519-SE.

Vassilev, I. 2008, 13–16 May, 'The Future of Energy Transit through South Eastern Europe', *Energy Security in the European Union: Green Linkages and the Black Sea Context*, Wilton Park Conferences, Bucharest.

Verdery, K. 1991, *National Ideology under Socialism: Identity and Cultural Politics in Ceausescu's Romania*, University of California Press, Berkeley, CA.

——1996, *What Was Socialism, and What Comes Next?* University of California Press, Berkeley, CA.

von Klaeden, E. 2008, *Kein Sonderzug nach Moskau: Deutsche Russlandpolitik muss Europaeisch Sein (No Special Train to Moscow: German Russia Policy Must be European)*, Edition Koerber-Stiftung, Berlin.

Waever, O. 1995, 'Securitization and Desecuritization', in *On Security*, ed. R.D. Lipschutz, Columbia University Press, New York, pp. 46–86.

Wagnsson, C. 2008, *Security in a Greater Europe: The Possibility of a Pan-European Approach*, Manchester University Press, Manchester.

Wedel, J.R. 1998, *Collision and Collusion: The Strange Case of Western Aid to Eastern Europe, 1989–1998*, St. Martin's Press, New York.

White, S. 1979, *Political Culture and Soviet Politics*, St Martin's Press, New York.

—— 2000, *Russia's New Politics: The Management of a Postcommunist Society*, Cambridge University Press, Cambridge.

White, S., R. Rose and I. McAllister 1997, *How Russia Votes*, Chatham House, London.

Wight, M. 1977, *System of States*, Leicester University Press, Leicester.

Woker, M. 2008, 2 September, 'Abgekühlte Freundschaft Serbiens zu Moskau (Serbia's Friendship with Moscow Cools)', *Neue Zürcher Zeitung*, Zurich.

Wolczuk, R. 2003, *Ukraine's Foreign and Security Policy 1991–2000*, Routledge, New York.

Yablokova, O. 2004a, 27 February, 'Putin's Words and Deeds Dominate Newscasts', *Moscow Times*.

—— 2004b, 5 March, *News Focus on Putin Obscures Campaign*, Moscow Times.

Yuriev, M. 2007, *Tretja imperia: Rossiya kotoraja dolzhna byt' (The Third Empire: The Russia that Must Be)*, Limbus Press, Moskva.

Zueva, K. 2008, 'Obraz Rossii: vzgliad iz Frantsii (Russia's Image: The View from France)', *Mirovaya ekonomika i mezhdunarodnye otnosheniya*, no. 2.

Zyuganov, G. 2004, *O Russkih i Rossiii (On Russians and Russia)*, Molodaya Gvardiya, Moskva.

Index